AF540619

ENCYCLOPAEDIA OF STEM CELLS - 5

SPECIFIC STEM CELLS

By

Dr. Amita Sarkar

Dept. of Zoology

Agra College

Agra (U.P.)

(India)

DISCOVERY PUBLISHING HOUSE PVT. LTD.

NEW DELHI-110 002

First Published-2009

ISBN 978-81-8356-401-4

Published by:
DISCOVERY PUBLISHING HOUSE PVT. LTD.
4831/24, Ansari Road, Prahlad Street,
Darya Ganj, New Delhi-110002 (India)
Phone: 23279245 • Fax: 91-11-23253475
E-mail: dphbooks@rediffmail.com
dphtemp@indiatimes.com
Website: www.discoverypublishinghouse.com

Printed at:
Sachin Printers, Delhi

Preface

The present title *Specific Stem Cells* is the amazing advancement of biotechnology. It provides the various fundamental aspects of stem cell technologies to be understood adequately. It has been compilated for graduate and undergraduate students, research scholars, teachers, practising biochemical engineers, biotechnologists, applied and industrial microbiologists, cell biologists and scientists involved in bioprocessing research and development. The basic concepts have been clearly explained and their functions are adequately highlighted. The presentation of the text is simple and systematic. The selection of chapters and topics is according to the specified syllabi of several Indian Universities and are so structured as to enable the student to move easily from the fundamental to the complex. It is our earnest hope that this title will be of great value to all our students.

To make the work more comprehensive and informative, the author has consulted many authoritative books, research journals, abstracts, monographs etc. He is grateful to all those great scholars whose work are cited or substantially reproduced.

There can be no claim to originality except in the manner of treatment and much of the information has been obtained from the books and scientific journals available in the different libraries.

The author expresses his thanks to his friends and colleagues whose continue inspirations have initiated him to bring out this book.

The author expresses his gratitude to Mr. Wasan and staff of M/s Discovery Publishing Pvt. Ltd. for their whole hearted co-operation in the publication of this book.

In the mean time, the author will remain sincerely responsible for any shortcomings of the book and the grateful to the readers for their suggestions and constructive criticism for the continuous betterment of the book. He takes this opportunity to appeal to the readers to send their suggestions straightway to his publisher.

Author

Contents

1

Introduction

This article describes the basic techniques required for successful cell culture. It also acts to introduce some of the other chapters in this volume. It is not intended to describe the establishment of a tissue culture laboratory, nor to provide a historical or theoretical survey of cell culture. There are several books that adequately cover these areas, including the now somewhat dated, but still valuable volume by Paul, that of Freshney, and the multi-authored volumes edited by Jakoby and Pastan, Davis, and Cells. Instead, this chapter focuses on the techniques for establishing primary rodent cell cultures from embryos and adult skin, maintaining and subculturing these fibroblasts and their transformed derivatives, and the isolation of genetically pure strains. The cells described are all derived from Chinese hamsters since to date, these cells have proved to be the most useful for somatic cell genetics. The techniques, however, are generally applicable to most fibroblast cell types.

Materials

1. Alpha minimum essential medium (α-MEM) containing penicillin and streptomycin: for economy, we buy prepared medium as powder in 20-L aliquots. A 44-g quantity of sodium bicarbonate is added, the powder is made up to 20 L in deionized distilled water, the pH adjusted to 7-4, and the media sterile-filtered through a 0.22-μM filter using a pressure vessel coupled to a filtration apparatus and driven by a pressurized 95.5% air CO_2 gas mix This gas mix maintains pH on preparation and storage The 500-mL bottles are stored at 4°C in the dark until use. Prepared media can also be purchased from many suppliers.

2. Growth medium: α-MEM plus 15 or 7 5% (v/v) fetal calf serum This is made up as required and stored at 4°C.
3. *Fetal calf serum* (FCS) should be pretested to ensure it supports optimal growth. It can also be heated at 56°C for 30 min to destroy complement if it is to be used for cultures where the presence of complement can cause complications. Sera should be aliquoted and stored at –20°C
4. Ca^{2+} Mg^{2+}-free phosphate-buffered saline (Dulbecco's PBS). 8 g/L NaCl, 0.2 g/L KCl, 0.2 g/L KH_2PO_4, 2.16 g/L Na_2HPO_4 $7H_2O$), pH 7 2.
5. PBS citrate. PBS + sodium citrate at 5.88 g/L.
6. Trypsin: One vial of lyophilized Difco (Detroit, MI) Bactotrypsin in 400 mL of PBS citrate (0.125% trypsin) or 10 times this concentration for the isolation of embryonic fibroblasts.
7. Counting fluid: PBS + 0 2% (v/v) FCS.
8. Formalin fixative: 10% (v/v) commercial formaldehyde (comes as a 40% [v/v] solution).
9. Methylene blue stain: 0.1% (w/v) methylene blue in distilled water filtered through a Whatman No. 1 filter.
10. Trypan blue: 0.5% (w/v) in PBS.
11. Colcemid: 10 μg/mL, store at 4°C.
12. Karyotype fix: Methanol acetic acid (3:1) made up on the day of use and kept on ice in a tightly stoppered bottle
13 Giemsa stain: Use commercial Giemsa concentrate diluted 3.47 parts in commercial Gurr's buffer (one tablet to 1 L distilled water). Alternatively, 10 mM potassium phosphate, pH 6.8, can be used as the buffer. The diluted stain is only stable for 2-3 mo.

Methods

Establishment of Primary Chinese Hamster Fibroblast Cultures

Embryo Culture

1. Kill a 12-d old pregnant Chinese hamster with ether.
2. Wash the animal in tap water and then with 70% ethanol.
3. Make a surgical incision on the dorsal side to expose the uterus using sterile instruments (these can be dipped in ethanol and flamed to maintain sterility during the operation).
4. Remove the uterus in *toto*, and transfer it to a sterile Petri dish. Dissect the embryos, and place them in a new sterile Petri dish.

5. Mince the embryos very finely, and while still in the Petri dish, wash the pieces with 5 mL of 0.125% Bactotrypsin at 37°C.
6. Tilt the Petri dish so that embryo pieces go to the side Remove the pieces into a 50-mL centrifuge tube using a wide-bore pipet.
7. Add 40 mL of fresh 1.25% Bactotrypsin, and incubate at 37°C for 5 min.
8. Regain the embryo pieces by centrifugation at 100g for 3.5 mm, and discard the supernatant.
9. Resuspend the pieces in 40 mL of fresh 0.125% Bactotrypsin, and incubate at 37°C for 25 min (this can be performed in a roller apparatus).
10. Neutralize the trypsin with 4 mL of FCS
11. Deposit the supernatant through a 100-μm sterile mesh into another centrifuge tube.
12. Centrifuge the supernatant for 5 min at 300g at room temperature.
13. Resuspend the pellet in 10 mL α-MEM plus 15% FCS, and count the cells in a hemocytometer at about 1/100 dilution.
14. Lay down 1.5×10^7 cells in 40 mL of α-MEM plus 15% FCS into a 75 cm^2 flask, and place it in a 37°C tissue-culture incubator.
15. The next day, replace the medium with an equal volume of α-MEM plus 15% FCS.
16. Forty-eight to 72 h later, the monolayer should be confluent, and at this point, the cells are ready for subculture. This is performed by incubating the monolayer with 4.5 mL of 0.125% Bactotrypsin at 37°C until the cells detach. Cell detachment can be visualized either by observing the cell monolayer in oblique light or directly under the microscope. When the cells have detached (~80%), add 0.5 mL FCS (10%), pipet up and down five times, and transfer contents to a 15 mL centrifuge tube.
17. Centrifuge the cells at 300g for 3.5 min at room temperature.
18. Remove the supernatant, resuspend the cell pellet m 5 mL α-MEM plus 15% FCS, and determine the cell concentration.
19. Resuspend the cells at 4×10^6/vial in α-MEM plus 15% FCS plus 10% (v/v) sterile dimethyl sulfoxide (DMSO), and freeze at –135 or –176°C. The cells will remain viable for several years.
20. The cells may also be subcultured at one-thud to one-tenth dilutions. They have doubling times of approx 36 h. At this point, start to calculate the number of mean population doublings by keeping careful records of subculture number and split ratio.

Skin fibroblasts

1. Kill and wash an animal as described for the isolation of embryonic fibroblasts. In fact, it is often convenient to prepare skin fibroblasts from the same animal as the one from which the embryos were obtained.
2. Cut small pieces (1-2 mm^2) of dermis from the exposed skin flaps using sterile instruments, avoiding any fur.
3. Place several (5-10) small pieces into a 25-cm^2 flask, and allow them to adhere for 30 min in a very thin film of medium (0.5 mL) at 37°C.
4. Once adhered, add 5 mL of growth medium to the opposite surface (i.e., top surface) of the flask to avoid washing off the skin pieces. Place the flask in the incubator in the upside-down position for 24 h (the surface tension holds a thin film of medium to the upper surface and stick the explants to the flask surface).
5. Once the explants are firmly stuck, gently invert the flask and return to the incubator.
6. The next day, it is often advisable to change the medium to remove any debris and unattached explants.
7. After several days, first "*epithelial*" type cells and then fibroblast will grow out of the explants. Let this process continue until most of the surface is covered with fibroblasts or until obvious necrosis is observed in the explant. It may be necessary to change the medium every week until substantial outgrowth is observed.
8. Remove the explanted material with a Pasteur pipet attached to a vacuum line leaving the adherent fibroblasts.
9. At this stage, depending on the density, the fibroblasts can either be trypsinized (>50% confluent) or allowed to continue to grow to form a monolayer before they are trypsininzed, subcultured, and frozen.

Maintenance and Subculture of Transformed Cell Lines

Many transformed cell lines will grow both as monolayers and in suspension culture. The CHO-S cell line is one such line having been selected for suspension growth by Thompson from the original K1 CHO cell line isolated by Puck. Because CHO cells are transformed, they do not require as much serum as normal diploid fibroblasts, and we routinely culture them in 7.5% (v/v) FCS. Despite the relative ease with which transformed cells can be cultured, however, unlike normal diploid fibroblasts, they do not enter a stationary phase of long-term viability. In this phase, they rapidly lose viability, and therefore must be subcultured during the exponential phase of growth and cannot be maintained as arrested cultures in reduced serum.

1. CHO cells are stored frozen at ~4 × 10^6 cells/ml at –135 or at –176°C (liquid nitrogen) in growth medium containing 7.5% FCS and 10% (v/v) DMSO. A single vial is removed from the frozen stock, rapidly defrosted in a 37°C water bath, and the cells regained by centrifugation at 300g at room temperature for 3.5 min.
2. The supernatant is discarded and the pellet resuspended in 1 mL of prewarmed medium and placed into a 25-cm^2 flask or a 60-mm diameter dish containing 4 mL of growth medium.
3. Approximately 2 d later, the cells should be almost confluent and ready for subculture. They are trypsinized as described for the primary diploid fibroblasts. After cell detachment, FCS is added to 10% and the cells resuspended as single cells by pipeting up and down about five times with a 5 mL pipet. An aliquot of this cell suspension (up to a total of 10% of the recipient volume of the medium) can be added directly to a new tissue culture vessel containing growth medium and returned to the incubator until the next subculture Alternatively, the cells may be regained by centrifugation, resuspended, and the concentration/mL determined. Known concentrations of cells may then be subcultured by appropriate dilution. In a 25-cm^2 flask with 5 mL of growth medium, CHO cells should yield about 25 × $10^5/cm^2$ but yields are variable depending on serum batch and media used.
4. At this stage, cells maybe transferred to a magnetically stirred spinner flask (commercially available) containing pregassed (95% air/5% CO_2) growth medium; usually a 250-mL spinner flask IS seeded to give a density of ~8 × 10^4 cells/ml. These flasks are then placed in a warm room or in a temperature-regulated water bath (Heto), and stirred at 100 rpm. CHO cells grown in suspension should give ~10^6 cells/mL at saturation density, at which point the medium will be very yellow (acid).

Determination of Cell Number

This can be performed either using an electronic particle counter or a hemocytometer. The former is the more accurate and can be used to count low concentrations of cells (~10^3 cells/mL); the latter requires higher density and is more prone to sampling error, but allows a visual estimation of the "*health*" of the cells and, combined with Trypan blue exclusion, can be used to estimate cell viability.

1. Resuspend cells to give a uniform cell suspension by pipeting up and down against the side of the plastic centrifuge tube.
2. If the cells have been trypsinilzed, 0.2 mL of the cell suspension to 7.8 mL of counting fluid in a 15-mL Falcon snap-cap tube will give a statistically reliable cell count (1000-14,000 particles/0.5 mL counted). Count three aliquots with the Coulter counter set to count 0.5 mL, sum the three counts, divide by 3, and multiply by 40 (for dilution) and 2 to calculate the cells/ mL.
3. The cells can then be appropriately diluted for the experimental setup or subculture.
4. Alternatively, the cells can be counted on a hemocytometer. The cells need to be resuspended at 3-5 $\times$ 10^5 cells/ml. A drop of a cell suspension is added to either side of the hemocytometer, taking care not to overfill it and making sure that the coverslip is firmly in place.
5. Each large square on the hemocytometer (improved Neubauer type) gives an area of 1 mm^2 and a depth of 0.1 mm (i.e., the volume is 10^{-4} mL). Count the cells in the square (usually using the one bounded on each side with triple lines) on either side of the counter, average the counts, and divide by 2 and multiply by 10^4 to give the number of cells/mL. If there are too many cells (>1000), just count the 5 diagonal squares and multiply by 5 to give the number to be multiplied by 10^4. If there are too few cells, count more than one complete square on each side of the chamber, and divide the total cell number accordingly.
6. This procedure can also be used to determine cell viability, since prior to placing the cells in the hemocytometer, they can be diluted 1 1 with 0.5% Trypan blue. The number of cells that can then exclude the stain (i.e., have intact cell membrane) can be determined by counting the cells as described in steps 3-5.

Isolation of Genetically Pure Cell Lines

The isolation of somatic cell mutants is outside the scope of this chapter, and the reader is referred to Thompson for the considerations necessary to isolate such mutants successfully. All cell lines will genetically alter over time, however, and periodically the parental type will need to be purified from variants or revertants. The easiest way to do this is to isolate a single clone. This causes some potential problems, however, since a clone may itself be a variant, and thus several clones will need to be isolated and tested to ensure the phenotype selected is the required one. To overcome this problem of clonal variability, it is usually better to contract the cell population to about 100 cells and then expand this to the mass culture. This contraction should statistically remove any variants from the population. It is worth remembering, however, that any variant that has a growth advantage over the parental type will soon overgrow the whole culture. Once a mass culture is obtained, it should be frozen in a large number of vials (20-50) to provide a base for future experiments. This enables the investigator to grow a culture for approx 3 mo before discarding it, and then to return to the frozen stock for the next set of experiments. This protocol reduces the genetic drift in the culture and avoids the necessity of frequent genetic purification using the following methods.

1. Trypsinize a culture, recover the cells, and determine the cell number.
2. Dilute to 2.5 cells/mL with 20 mL of growth medium.
3. Plate out 0.2 mL/well into a 96-well tissue-culture plate
4. Incubate plates at 37°C in an humidified incubator for 10-12 d. Do not move or disturb the plates, mitotic cells will float off and form satellite colonies.
5. Examine every well with a microscope, and ring those that have a single clone. These may be pure clones but a second cloning ensures that you end up with populations derived from a single cell.
6. Trypsinize two to three of these individual clones with 0.2 mL of trypsin and, once detached, transfer the well's contents into 4 mL of growth medium in a snapcap tube.
7. Pipet this up and down to ensure a single cell suspension, and then plate it again at one-tenth serial dilutions (i.e., 0.4-3.6 mL) and 0.2 mL/well into a 96-well tissue-culture dash.
8. Return these new plates to the incubators. Add medium from a different batch to the trypsinized wells of the old plates, and also return this to the incubator. This provides a backup in case the new plates are contaminated. Again, do not move the plates.
9. After 10-12 d, select individual clones in the new plates, and expand them up to mass culture (remember to always keep a backup culture).
10. Freeze a large stock (20-50 vials), since at this stage, you will have a genetically pure line (except for the mutations that may have occurred during the clone's expansion). Split the frozen stock between a liquid N_2 store (long-term) and a –70 or –35°C store (short-term experimental stock)

11. Alternatively, the mass culture that needs to be genetically cleansed can be plated into 60-mm dishes containing 5 mL growth at 100 cells/dish.
12. Leave these to grow for approx 10 d. Trypsinize the ~100 clones from each plate and expand them together to a mass culture in the normal way.
13. Freeze 20-50 vials of these cultures as described in step 10.

Karyotyping

It is often desirable to karyotype your cells. This chapter, therefore, deals with a simple method, derived from Deaven and Petersen for producing karyotypes of Chinese hamster cells.

1. A culture growing in the exponential phase of growth (i.e., having a high mitotic index) in a 10-mL suspension culture (2 × 10^5 cells/mL) or as a monolayer (106 cells/60-mm plate) is treated with colcemid at 0.06 μg/mL for 2 h to accumulate cells in mitosis.
2. For the monolayer culture, tap the plate and remove the medium containing detached mitotic cells Trypsinize the remaining monolayer, pool with the medium, and proceed.
3. Regain cells by centrifugation at 300g for 3.5 min at room temperature.
4. Resuspend cells in 1 mL of growth medium, add 3 mL of distilled water, and invert to mix (do not pipet because the cells are fragile).
5. Leave for 7 min to allow the cells to swell (this time can be altered if satisfactory spreads are not obtained).
6. Add 4 mL of freshly prepared ice-cold fixative (methanol:acetic acid, 3:1) directly to the hypertonic solution to avoid clumping.
7. Regain the cells by centrifugation at 300g for 3.5 min.
8. Disperse the pellet gently by agitation (do not pipet) in 10 mL of fixative.
9. Repeat this procedure three times. At this point, the fixed cells can be stored for a week at 4°C or slides can be made immediately.
10. Using a Pasteur pipet, drop two to three drops of the resuspended cells onto a chilled slide from about 20 cm. Blow gently onto the surface, and place the slide onto a hot plate at 60-65°C (just too hot to keep the palm of one's hand on the plate).
11. Leave the slide to dry for 5 min and then place in a staining chamber (a Coplin jar) ensuring that the surfaces do not touch.
12. Stain the karyotypes with Giemsa for 3 min.

13. Wash the slides by dipping the slides through three additional Coplin Jars each containing 50 mL of water
14. Dry the slides and count the chromosome number under the microscope, or process for banding.

Serum and Media Testing

Before a new batch of serum or media is purchased, it is advisable to obtain a sample from the manufacturer and test its growth-supporting characteristics. This is particularly important for serum. Usually select two of the most used cell types in the lab—currently these are a human diploid fibroblast strain and CHO cells-to test their growth and plating efficiencies.

1. Make up individual aliquots of growth media, all containing the same media batch, but with the different test sera and including the serum batch currently being used (or vice versa if you are testing media batches).
2. Plate the cells into 15 dishes for each test media at 5×10^5 cells/60 mm tissue-culture dish and containing 5 mL of the media.
3. Every day for 5 d thereafter, Trypsinize the cells from triplicate plates and determine the cell number/plate.
4. Plot a growth curve (log cell number vs time), and calculate the doubling time and saturation density.
5. At the same time as setting up the growth curves, seed in triplicate 60-mm dishes containing 5 mL of the appropriate media with 100 and 200 cells (6 plates/test)
6. After 10-12 d fix the culture for 15 min by flooding with formalin.
7. Tip the media and formalin down the drain, and stain the clones with methylene blue.
8. Leave the stain for 15 min, and then wash it away with water.
9. Leave the plates stacked up against each other to dry in a 37°C room.
10. Count the colonies.
11. The three parameters of doubling time, saturation density, and plating efficiency should allow the section of a serum (or media) that gives optimal growth.

Notes

1. The shelf-life of a powdered medium is several years Once reconstituted, however, this is reduced to 2-3 mo, mainly because glutamine is unstable If older medium is used, the glutamine should

be replenished (292 μg/mL) The pH of a medium, on storage, should not be allowed to rise, and to achieve this, good plastic caps with close-fitting rubber inserts should be used we also find it useful to seal the caps with a strip of Parafilm, since this prevents condensation around the cap rim and, thus, minimizes the risk of fungal contamination. Medium containing HEPES can also be used to avoid bicarbonate buffering. We have never been entirely happy, however, with the cell's long-term growth characteristics in HEPES-containing medium.

2. α-MEM is a rich, multipurpose medium developed by Stanners et al to grow hamster cells. We have not had the experience of any mammalian cell type that will not grow in this medium, including hybridomas. It is slightly more expensive than most media, however, and many cells will tolerate less rich and, therefore, cheaper media.
3. Purified trypsin can also be used and is sometimes necessary, e.g, for macrophage cell lines, but it is much more expensive and usually not necessary. The citrate chelates Mg^{2+} and Ca^{2+} and replaces EDTA (Versene) in the buffer.
4. It is advisable to keep fibroblast cultures from individual animals distinct, since it may be required to distinguish between individuals genetically.
5. If the explant is not necrotic, it is possible to remove it with sterile forceps and transfer it to a new culture flask for further outgrowth of cells.
6. The detailed derivation of the various CHO stains is given in Gottesman. It should be noted that CHO is a proline auxotroph and should always be maintained in proline-containing medium.
7. CHO cells can maintain viability, providing the medium pH does not become alkali, at 4°C for extended periods of time (7-10 d). Cultures in capped bottles can therefore be moved to the cold room to avoid subculture under desperate circumstances.
8. Primary cell cultures may also be grown on microcarriers in suspension culture.
9. The Coulter counter should have a 140-μM aperture and the thresholds set as described in the machine's Instruction Manual. Serum in the PBS prevents cells from aggregating and giving unreliable counts. The counter sometimes gets partially blocked, only experience of the time taken for each count and for the cell's particular display on the spectroscope will indicate problems

with counting. Gentle brushing of the orifice with a camel-hair brush will unblock the counter. The Coulter counter can also give a visual display of cell volumes This, when combined with a pulse height analyzer, can be used quantitatively to measure cell volume or to determine cell viability by estimating the amount of cell debris in a sample.

10. To maintain genetically pure cell lines, it is absolutely essential not to cross-contaminate cultures To ensure this, fresh pipets must always be used at every step. Do not re-enter a media bottle with a pipet that has been near a culture Similarly, never pour from a media bottle into a culture. Splash-backs can occur. If you have more than one culture at a time in a tissue culture hood, only one of these should be opened at any one time. Meticulous attention to these small details will prevent the cross-contamination scandals (e.g., HeLa cells in all cultures') that one so often reads about.
11. It is usual to prepare one slide and check it with phase contrast microscopy so that adjustments can be made on subsequent slides. If there are many nuclei without cytoplasm and a few metaphases, reduce the swelling time. If there are many scattered chromosomes, blow less vigorously. If the metaphase spreads are overlapping, either swell for a longer time (up to 40 min) or blow more vigorously. All these parameters need to be adjusted according to the local environment conditions and cell type.
12. If many cell lines are being used, it is often impractical to test the serum out on all the cell types. Usually the most difficult to grow are chosen for the test, but caution needs to be exercised since we once had a batch of serum that supported the cloning and growth of primary diploid fibroblasts but failed to allow cloning of CHO cells!
13. This procedure need only be performed about once every year Enough serum can then be ordered for the next year, since the serum is stable at –20°C for at least 2 yr. We used to check our serum using [^{3}H]-thymidine incorporation 1 d after seeding the cells, but given the hazard of using radioactive thymidme, we abandoned this procedure. It is less labor-intensive, however, than measuring growth curves and gives perfectly adequate results.

2

DNA IN STEM CELL PATTERN

According to the rules of *Mendelian genetics*, sister chromatids are equivalent, and genes are composed of DNA alone. Violations to both of these rules have been discovered, which explain the stem-cell-like pattern of asymmetric cell division in the fission yeast *Schizosaccharomyces pombe*. In this chapter, we highlight key ideas and their experimental support so that the reader can contrast these mechanisms, which are not based on differential gene regulation, with those discovered in other diverse systems presented in this monograph.

FISSION YEAST AS A MODEL SYSTEM FOR INVESTIGATING CELLULAR DIFFERENTIATION AT THE SINGLE-CELL LEVEL

S. pombe is a haploid, unicellular, lower eukaryotic organism whose genetics has been studied very thoroughly. Its genome comprises only three chromosomes, with DNA content similar to that of the evolutionarily distantly related budding yeast, *Saccharomyces cerevisiae*. This organism has been exploited as a major system for cell cycle studies as well as for studies of cellular differentiation. The single cells of fission yeast express either P (Plus) or M (Minus) mating-cell type and divide by fission of the parental cell to produce rod-shaped progeny of nearly equal size. Yeast cells do not express mating type while growing on rich medium. Only when they are starved, especially for nitrogen, do cells express their mating type and mate with cells of opposite type to produce transient zygotic diploid cells. Normally, the zygotic cell immediately enters into the meiotic cell division cycle and gives rise to four haploid spore segregants, two of P type and two of M type.

The mating type choice is controlled by alternate alleles of the single mating-type locus (*mat1*). Stable diploid lines can be easily constructed by selecting for complementation of auxotrophic markers before the zygotic cells are committed to meiosis. The diploids can then be maintained by growth in rich medium, which inhibits meiosis and sporulation. Once these cells experience nitrogen starvation, they undergo meiosis and sporulation without mating. The sporulation process requires heterozygosity at *mat1*. Strains that switch *mat1* are called homothallic, and those that do not switch are called heterothallic.

Conjugation in cells of homothallic strains occurs efficiently between newly divided pairs of sister cells. Switching occurs at high frequency. The most remarkable feature of the system is that switching occurs in a nonrandom fashion within a cell lineage. Miyata and Miyata (1981) followed the pattern of matings between the progeny of a single cell growing under starvation conditions, on the surface of solid medium. They found that among the four grand-daughters of a single cell, a single zygote was formed in 72–94% of the cases. In no case did they observe two zygotes. The mating mostly occurred between sister cells, whereas non-sister (cousin) cells mated infrequently. It appeared, therefore, that among the four grand-daughters of a single cell, only one had switched. With this procedure, it was not possible to determine switching potential of cells past the four-cell stage since two of them formed a zygote and underwent meiosis and sporulation, so that their future potential could not be ascertained. Subsequent studies used diploid cells instead where one homolog contained a non-switchable heterothallic *mat1* allele, whereas the other contained a homothallic locus. Such a diploid will not sporulate when it is homozygous (*mat1P/mat1P* or *mat1M/mat1M*) at *mat1*, but will stop growing and initiate sporulation once switching produces *mat1P/mat1M* heterozygosity. Diploid cells

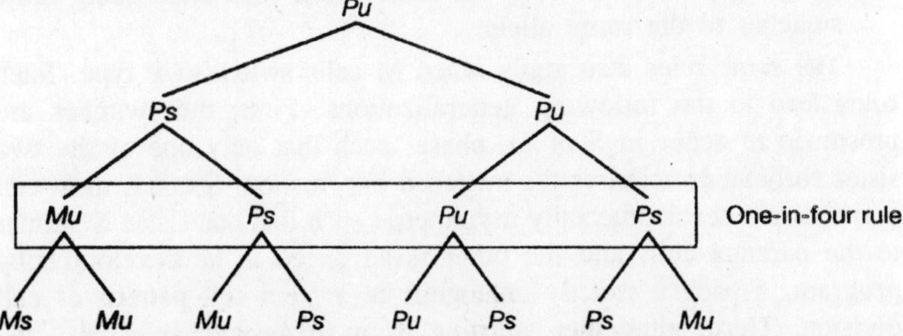

Fig. 2.1. The program of cell-type switching in S. pombe cell pedigrees. The subscripts u and s, respectively reflect unswitchable or switchable cells.

keep on switching regardless of their *mat1* constitution. In such diploid pedigrees, the same rules of switching described for haploids were observed. More importantly, one can determine the competence of switching past the four-cell stage by microscopically monitoring the competence of individual cells to sporulate. Such studies have defined the rules of switching as follows.

Rules of Switching

1. *The single-switchable-sister rule:* In most cell divisions (80–90% of cases) an unswitchable (e.g., Pu) cell produces one Ps (switching-competent) and one Pu unswitchable cell like the parental Pu cell. Thus, both sisters are never switching-competent.
2. *The single-switched-daughter rule:* Switching-competent Ps cell produces one switched and one switching-competent Ps cell in approximately 80–90% of cell divisions. Simultaneous switching of both daughters is never seen.
3. *The recurrent switching rule:* Like the parental cell, the sister of the recently switched cell maintains switching competence in 80–90% of cases. Consequently, chains of pedigree result where one daughter in each cell division is switched.
4. *The rule of switched allele is unswitchable:* To conform to the one-in-four granddaughter pattern, the newly switched allele must be unswitchable, although this notion has not been experimentally established. It is supported by the Miyata and Miyata (1981) observation, since they never observed two zygotes among four granddaughters of a single cell.
5. *The directionality rule:* Since a switchable cell switches to the opposite mating type in 80–90% of cell divisions, it must be that cells show bias in direction of switching such that most switches are productive to the opposite allele rather than undergoing futile switches to the same allele.

The same rules also apply when M cells switch to P type. Such rules lead to the following generalizations. First, the switches are presumed to occur in S or G_2 phase, such that only one of the two sister chromatids acquires the switched information. Second, most cell divisions are developmentally asymmetric such that one sister is similar to the parental cell, and the other is advanced in its developmental program, a pattern exactly analogous to a stem cell pattern of cell division. Third, altogether, starting from an unswitchable cell, two consecutive asymmetric cell divisions must have occurred to produce a single switched cell in four related granddaughter cells.

Switches Result from Gene Conversions at *mat1*

The *mat1* locus is a part of a cluster of tightly linked *mat1-mat2-mat3* genes on chromosome II. The expressed *mat1* locus contains either *mat1P* or *mat1M* allele. Because cells containing a haploid genome are able to express either mating cell type, both cell types must contain sufficient information to interchange *mat1* alleles. The *mat2P* and *mat3M* alleles are silent and are only used as donors of genetic information for *mat1* switching. The *mat2* gene is located approximately 15 kb distal to *mat1*, and *mat3* is located another 11 kb from *mat2,* separated by the sequence called the *K-region*. The P-specific region is 1104 bp long, whereas the nonhomologous M-specific region is 1128 bp long. Very short homologies represented by H1, H2, and H3 sequences flank the indicated cassettes. Each *mat1* allele codes for two transcripts, one of which is induced during starvation. The *mat1* interconversion results from a gene conversion event whereby a copy of *mat2P* or *mat3M* is substituted with the resident *mat1* allele. Consequently, the differentiated state is maintained as a genetic alteration that is subject to further rounds of spontaneous switching.

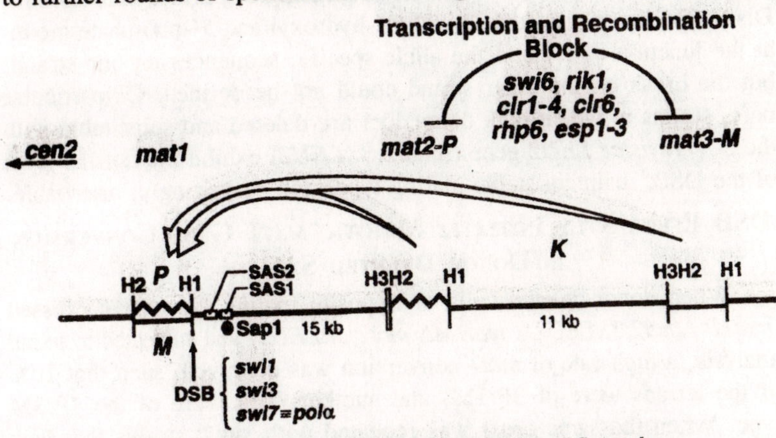

Fig. 2.2. The system of mating-type switching of S. pombe.

cis- and *trans*-Action Functions Required for Switching

Southern analysis of yeast DNA indicated that nearly 20–25% of the *mat1* DNA is cut at the junctions of H1 and the allele-specific sequences. By analogy to the *MAT* switching system where a *trans*-acting, *HO*-encoded endonuclease cleaves *MAT* to initiate recombination, it was proposed that the *double-stranded break* (DSB) at *mat1* likewise initiates recombination. In support of this proposal, several *cis*- and *trans*-acting mutations were isolated that reduce the level of the DSB and, consequently, reduce the efficiency of switching. Interestingly, the amount of cut DNA remains constant throughout the cell cycle, although no study has directly demonstrated that the break actually exists in vivo.

Several *cis*-acting deletion mutations in *mat1* have implicated *mat1*-distal sequences in formation of the DSB. One mutation, C13P11, reduces switching and contains a 27-bp deletion that includes 7 bp of the distal end of the *mat1* H1 region. Another mutation, smt-o, totally blocks switching and contains a larger deletion in the same region as well as two sites, called SAS1 and SAS2, which comprise a binding site for a protein called Sap1p.

Mutations of three unlinked genes, *swi1*, *swi3,* and *swi7*, reduce switching by reducing the level of the DSB. The functions of *swi1* and *swi3* remain undefined, but interestingly, *swi7* encodes the catalytic subunit of DNA polymerase α. This result implicates the act of DNA replication in generation of the DSB. Nielsen and Egel (1989) mapped the position of the break by genomic sequencing of purified chromosomal DNA. The break was defined with 3´-hydroxyl and 5´-phosphate groups at the junction of H1 and the allele-specific sequences on one strand, but the break on the other strand could not be defined. Of particular note, strains in which both donor loci are deleted and substituted with the *S. cerevisiae LEU2* gene (*Δmat2,3::LEU2*) exhibit the normal level of the DSB, maintain stable mating type, and surprisingly, are viable.

DSB Efficiently Initiates Meiotic *mat1* Gene Conversion in Donor Deleted Strains

When donor-deleted cells of opposite mating type were crossed (*mat1P Δmat2,3::LEU2* x *mat1MΔ mat2,3::LEU2*) and subjected to tetrad analysis, a high rate of *mat1* conversion was observed, such that 10% of the tetrads were of 3P:1M, and another 10% were of the 1P:3M type. When the same cross was repeated with *swi3*⁻ strains that lack the DSB, the efficiency of meiotic *mat1* gene conversion was correspondingly reduced. It was suggested that the DSB designed for mitotic *mat1* switching can also initiate meiotic gene conversion such that only one of two sister chromatids is converted, since no 4:0 or 0:4 conversions were observed. This meiotic gene conversion assay tests the switching competence of individual chromosomes and was the key technique in deciphering the mechanism of *mat1* switching in mitotic cells.

Competence for Switching is Chromosomally Borne

Discovering the mechanism by which sister cells gain different developmental fates is central to understanding eukaryotic cellular differentiation. The single-cell assay for testing *mat1* switching, either by mating or by determining sporulation ability as discussed above, suggests that the developmental decision is imparted to sister cells by cell-autonomous mechanisms. It would therefore seem that the switching potential must be asymmetrically segregated to daughter cells either through the nuclear/cytoplasmic factor(s) or via the DNA template. In the first model, essential components, such as those encoded by *swi* genes, would be unequally expressed, differentially stabilized, or asymmetrically segregated to daughter cells.

In the second model, since the DSB seems to initiate recombination required for *mat1* switching and the break may be chromosomally inherited, it may be that only one of two sister chromatids is imprinted in each cell division, thus differentiating sister cells. The term imprinting implies some sort of chromosomal modification such that only one of the two sister chromatids is cleaved to initiate recombination. Any mechanism, however, must explain not only how sisters acquire different development potential, but also how two consecutive asymmetric cell divisions are performed such that only one in four related granddaughter cells ever switches.

Since the level of the DSB is highly correlated with the efficiency of switching, it was reasoned that generation of DSB in some cells, but not in other related cells, is the key to defining the program of switching in cell lineage. Should the observed pattern of switching in mitotically dividing cells be the result of chromosomal imprinting, I hypothesized that the likely candidates to catalyze this epigenetic event are the gene functions involved in generating the cut at *mat1*, such as those of *swi1*, *swi3*, and *swi7*. It has not been possible to directly demonstrate the inheritance of the imprint and correlate it to switching in mitotically dividing single cells. However, testing meiotic *mat1* gene conversion potential of individual chromosomes provided a key test of the model.

Meiotic crosses involving *Δmat2,3::LEU2* strains generate a high rate of *mat1* gene conversion due to *mat1* to *mat1* interaction by which both 3P:1M and 1P:3M asci are produced in equal proportion. Because the spores are haploid and donor-deleted, the recently converted allele is stably maintained in meiotic segregants. We presume that meiotic *mat1* gene conversion events are also initiated by the break resulting from the imprint at *mat1*. With the meiotic gene conversion assay, it became possible to directly test switching potential of individual chromosomes as well as the effect of *swi1*, *swi3*, and *swi7* genotype on switching competence. As *S. pombe* cells mate and immediately undergo meiosis and sporulation, the diploid phase exists transiently. The key result was that a cross between donor-deleted strains *mat1M swi3*$^-$ and a *mat1P swi3*$^+$ generated aberrant tetrads, primarily with 3M:1P segregants, in which only *mat1P* converted to *mat1M*. On the other hand, if *swi3*$^-$ mutation was present in the *mat1P* strain, the *mat1M* changed to *mat1P*.

Similarly, crosses involving a *swi1*$^-$ or a *swi7*$^-$ parent generated meiotic *mat1* conversion in which only the *mat1* allele provided by the *swi*$^+$ parent gene converted. Thus, clearly (1) the competence for meiotic gene conversion segregates in *cis* with *mat1*; (2) the *swi1*$^+$, *swi3*$^+$, and *swi7*$^+$ functions confer that competence; and (3) the presence of these functions in the zygotic cells provided by the *swi*$^+$ parent fails to confer the gene conversion potential to the *mat1* allele that was previously replicated in the *swi*$^-$ background. Those meiotic experiments unambiguously showed that chromosomally imprinted functions are catalyzed at *mat1* by the *swi* gene products at least one generation before meiotic conversion. On the basis of these results, we suggest that the same imprinted event may form the basis of mitotic switching, resulting in the specific pattern of switching in cell pedigrees.

Nonequivalent Sister Cells Result from Inheriting Differentiated, Nonequivalent Parental DNA Chains

If *mat1* switching is initiated by the DSB, it follows that differentiated sister chromatids must be the reason that only one of the sister cells becomes switching-competent or ever switches. Restated another way, How is it that only one of four descendants of a chromosome switches? To explain the one-in-four granddaughter switching rule, we imagined that one of the decisions to make a given switch must have occurred two generations earlier in the grand-parental cell (Mu or Pu). Specifically, a strand-segregation model was proposed in which "Watson" and "Crick" strands of DNA are nonequivalent in their ability to acquire the developmental potential for switching. It was proposed that some *swi*$^+$ gene functions catalyze a strand-specific imprinting event, which in the following cycle will cause switching again in a strand-specific fashion.

The proposal was that strand-specific imprinting allows the DNA to be cut in vivo and switching follows. The inherent DNA sequence difference of two strands alone must not be sufficient, because if it were, each cell would produce one switched and one unswitched daughter. To explain the two-generation program of switching, imprinting in one generation and switching in the following generation was imagined. It was hypothesized that the imprinting event may consist of DNA methylation or some other base modification, an unrepaired RNA primer of Okazaki fragments, a protein complex that segregates with a specific strand, or a site-specific single-stranded nick that becomes DSB in the next round of replication.

Several follow-up tests of the strand-segregation model have established this model. First, strains constructed to contain an additional *mat1* cassette placed in an inverted orientation approximately 4.7 kb away from the resident *mat1* locus cleaved one or the other *mat1* locus efficiently, but never simultaneously in the same cell cycle, as imprinting occurs only on one specific strand at each cassette. Second, as opposed to the switching of only one in four related cells in standard strains, cells with the inverted duplication switched two (cousins) in four grand-daughter cells in 34% of pedigrees. Third, the inverted cassette also followed the one-in-four switching rule and switched in 32% of cases.

Clearly, in such a duplication-containing strain, both daughters of the grand-parental cell became developmentally equivalent in at least one-third of cell divisions. Thus, all cells are otherwise equivalent, ruling out the factor(s) segregation model, and the pattern is strictly dictated by inheritance of complementary and nonequivalent DNA chains at *mat1*. It was also hypothesized that the strand-specificity of the imprint may result from the inherently nonequivalent replication of sister chromatids due to lagging versus leading-strand replication at *mat1*. Suggestive evidence for this idea came from the finding that *swi7* implicated in imprinting in fact encodes the major catalytic subunit of DNA polymerase α. This polymerase provides the primase activity for initiating DNA replication; thus, it is inherently required more for lagging-strand replication than for leading-strand replication.

Fourth, more recent observations biochemically established that the imprint is either a single-stranded and strand-specific nick or an alkali-labile modification of DNA at *mat1*. Both of these studies showed that the observed DSB is an artifact of DNA preparation created from the imprint at *mat1*, since DNA isolated by gentle means from cells embedded in agarose plugs exhibited much-reduced levels of the break. Arcangioli (1998) showed that mung bean nuclease treatment of the DNA results in generation of the DSB. This result, combined with the primer extension experiments, led Arcangioli (1998) to conclude that the imprint is a single-stranded nick which persists at a constant level throughout the cell cycle.

In contrast, Dalgaard and Klar (1999) found both strands at *mat1* to be intact while one of the strands breaks after denaturation with alkali, but not with the formaldehyde treatment. Although these biochemical studies are discordant with each other, nonetheless, both support earlier suggestions and the model. Combining genetic and biochemical results, the strand-segregation model is now clearly established and, henceforth, would be referred to as a strand-segregation mechanism.

Imprinting Mechanism

The DSB was initially discovered when the DNA was prepared with the conventional method, which includes a step of RNase A treatment. All the biochemical studies can be reconciled should the imprint consist of an RNase-labile base(s). Arcangioli (1998) concluded that the imprint must be a single-stranded nick, since mung bean nuclease treatment produces the DSB. It should be noted, however, that this nuclease also has RNase activity, in addition to DNA-cleaving activity at the nick. The alkali-labile site discovered by Dalgaard and Klar (1999) is also consistent with the idea that the imprint is probably an RNA moiety left unrepaired from an RNA primer that has been ligated to form a continuous DNA-RNA-DNA strand.

It was previously suggested that lagging- versus leading-strand replication may dictate imprinting. Dalgaard and Klar (1999) directly tested this idea by proposing an "*orientation of replication model*" where it was shown that when *mat1* is inverted at the indigenous location, it fails to imprint/switch. A partial restoration was obtained if origin of replication was placed next to the inverted *mat1* locus. Furthermore, *mat1* was shown to be replicated unidirectionally by centromere-distal origin(s) by experiments defining replication intermediates with the two-dimensional gel analysis.

These results, combined with the earlier finding that *swi7* encodes DNA polymerase α, led Dalgaard and Klar (1999) to suggest that the imprint is probably an RNA base(s) added only by the lagging-strand replication complex. Alternatively, it may be some other base modification conferring alkali lability to one specific strand. Both these biochemical studies suggest that the DSB is an artifact of the DNA preparation procedure, yet both studies suggest that the imprint leads to transient generation of the DSB at the time of replication of the imprinted strand by the leading-strand replication complex. It is proposed that such a transient DSB initiates recombination required for switching *mat1*. Because meiotic *mat1* conversions are only of 3:1 type and only one member of a pair of sisters switches, recombination must occur in S or G_2 such that only one sister chromatid receives the converted allele. Even the transient DSB fails to cause lethality in donor-deleted strains. In principle, the intact sister chromatid may be used to heal the break.

Since recombination-deficient (*swi5*⁻) strains can also heal the break, the yeast probably has the capacity to heal the break without recombination. Two *mat1 cis*-acting sites located near the cut site and the cognate binding factor encoded by *sap1* somehow dictate imprinting at *mat1*. One possibility is that these elements promote maintenance of the imprint by prohibiting its repair. In summary, the biochemical results provide evidence for the notion that DNA replication advances the program of cellular differentiation in a strand-specific fashion.

It remains to be determined exactly how the imprint is made. Dalgaard and Klar (1999) found DNA replication pausing at the site of the imprint. Analysis of DNA replication intermediates around *mat1* revealed another element located to the left of *mat1* where replication terminates in one direction and not in the other to help replicate *mat1* only unidirectionally. This study showed that swi1p and swi3p factors act by pausing the replication fork at the imprinting site as well as by promoting termination at the polar terminator of replication. One possibility is that pausing at the fork helps imprinting by providing sufficient time to lay RNA primer at the imprinting site. Using DNA density-shift experiments, Arcangioli (2000) showed that 20–25% of *mat1* DNA is replicated such that both strands are synthesized de novo during S phase. This work also showed directly that the newly switched *mat1* does not have the imprint (i.e., nick), further supporting the strand segregation model.

Silencing of the *mat2-mat3* Regions is Caused by an Epigenetic Mechanism

A mechanistically very different imprinting event has been shown to keep the donor region silent from expression and from mitotic as well as meiotic recombination. Even when another genetic marker, such as *ura4*, was inserted in and around the *mat2-mat3* region, its transcription was highly repressed. Starting with such a Ura$^-$ strain, several *trans*-acting factors of *clr1-4* (*clr* for cryptic loci regulator) were identified, mutations of which relieve silencing and recombination prohibition of this interval. Two other previously defined mutations in *swi6* and *rik1* loci likewise compromise unusual properties of this region.

Several other newly identified genes, *esp1-3*, *rhp6*, and *clr6*, have also been implicated in silencing. Molecular analysis of these *trans*-acting factors and sequence analysis of the 11-kb K-region between *mat2* and *mat3* loci have suggested that this region is silenced due to organization of a repressive heterochromatic structure making this region unaccessible for transcription and recombination. First, 4.3 kb of the 11.0 kb region between *mat2* and *mat3*, called the K-region, shows 96% sequence identity with the repeat sequences present in the chromosome II centromere. A similar silencing occurs when *ura4* is placed in centromeric repeat sequences. Second, *swi6*, *clr4*, and *chp1* and *chp2* encode proteins containing a chromodomain motif thought to be essential for chromatin organization. Third, *clr3* and *clr6* encode homologs of histone deacetylase activities that are certain to influence organization of chromatin structure. Fourth, accessibility of *mat2* and *mat3* loci to in vivo expressed *Escherichia coli dam*$^+$ methylase is influenced by the *swi6* genotype.

Interestingly, when the 7.5-kb sequence of the K-region was replaced with the *ura4* locus (*KΔ::ura4* allele), the *ura4* gene expressed in a variegated fashion. Remarkably, both states, designated *ura4-off* and *ura4-on* epistates, were mitotically stable, interchanging only at a rate of approximately 5.6×10^{-4}/cell division. Even more spectacularly, when cells with these states were mated and the resulting diploid was grown for more than 30 generations and then subjected to meiotic analysis, we found that each state was stable and inherited as a Mendelian epiallele of the *mat* region. Thus, the epigenetic state is stable in both mitosis and meiosis as a Mendelian, chromosomal marker.

To explain this kind of inheritance, we advanced a chromatin replication model in which silencing occurs on both daughter chromatids by self-templating assembly of chromatin in the *mat2/3* region. The proposal is that preexisting nucleoprotein complexes presumably segregated to both strands of DNA promote assembly of chromatin on both daughter chromatids to clonally propagate and deliver a specific state of gene expression to both daughter cells. Two recent studies provide support to the chromatin replication model. First, transiently overexpressing *swi6*$^+$ in cells with *ura4-on* state efficiently changes them to *ura4-off* state; once changed, overexpression is not required to maintain the altered state. Second, transiently exposing the *ura4-off* cells to histone deacetylase inhibitor trichostatin A efficiently changes them to *ura4-on* state.

In both of the change-of-state experiments, changes were genetically inherited at the *mat* region and were correlated with the changes in the recruitment of swi6 protein to the *mat* region chromatin. Thus, in this case, the committed states of gene expression are inherited epigenetically rather than through variations in DNA sequence.

Strand-segregation Mechanism for Explaining General Cellular Differentiation

Two important lessons learned from the fission yeast system are that (1) by the process of DNA replication developmentally nonequivalent sister chromatids can be produced, and (2) stable patterns of gene expression can be inherited chromosomally over the course of multiple cell divisions akin to the general phenomenon of imprinting so prevalent in mammals. The question arises as to whether the first of these mechanisms is only applicable to yeast. It is impossible to answer this question because in multicellular systems it is not feasible to experimentally test such models because developmental potential and segregation of differentiated chromatids cannot be ascertained at the single-cell level in mitotically dividing cells.

In principle, however, it is possible to imagine that the act of DNA replication may modulate activities of developmentally important genes in a strand-specific fashion. It is not necessary to expect that such modulation occurs only through DNA recombination as found in yeast; it could rather be due to differential organization of chromatin structure of sister chromatids from both homologs in diploid organisms. (we never liked the idea of DNA methylation being the primary mechanism of imprinting and gene regulation.) Once established, these states may be maintained through multiple cell divisions akin to the epigenetic control operative in the K-region of *mat2/3* interval.

To produce the stem-cell-like pattern, we then propose that the differentiated chromatids from both homologs have to be segregated nonrandomly to daughter cells by yet another mechanism such that one daughter cell will inherit chromosomes with the developmentally important gene in an active state, while the other cell inherits an inactive state. Which daughter will get which sets of chromosomes will have to be influenced by other axes of the developing system, such as a dorsoventral axis. Such a proposal has been made to explain the left–right axis determination of visceral organs of mice. It is proposed that the *iv* gene (for situs inversus) product functions for nonrandom segregation of sister chromatids to daughter cells at certain cell division during mitosis whenever the left–right decision is distributed during embryogenesis. Interestingly, the *iv*$^-$ mutant produces randomized mice such that half of the mice have the heart located on the left side, and the other half have situs inversus such that the heart is on the right side of the body.

Recently, it was found that the *iv* gene encodes dynein, which is a molecular motor that functions to move cargo on microtubules. Of course, the alternate, accepted but not yet proven, model to explain the behavior of the *iv*$^-$ mutant mice is that the mutation causes random distribution of a hypothetical morphogen-producing center, which in *iv*$^+$ mice is localized only to one side of the body. However, the nature of the morphogen, the mechanism of its graded distribution, and the localization of the morphogen production to only one side remain undefined.

Consequently, the morphogen model is only descriptive, because it does not suggest experimental tests to scrutinize its validity. This is not to say that the opportunity for a morphogen-like mechanism does not exist elsewhere in biology. For example, there is ample evidence that such a mechanism operates in the rather unusual development of *Drosophila*. Because the *Drosophila* egg is very large compared to most cells, the graded distribution of egg constituents is required to ensure such a mechanism. In most other developmental systems, decisions are probably made right from the first zygotic cell division such that the sister cells are nonequivalent in their developmental potential. New decisions for regulating developmentally important genes may be made at each cell division.

Clearly, investigation of more model systems is needed to ask fundamental questions of specification and distribution of developmental decision in multicellular systems. Another case where such a mechanism may be operative is development of human brain laterality such that in most individuals the left hemisphere of the brain is specified to process language, while the right hemisphere processes emotional information. It is speculated that a genetic function, analogous to that of the *iv*$^+$ function for mice visceral specification, may have evolved for nonrandom segregation of Watson and Crick strands of a particular chromosome. Thus, chromosomal rearrangements or defects in the hypothesized *RGHT* gene may predispose individuals to develop bilaterally symmetrical brains, causing psychiatric disorders such as schizophrenia and manic-depressive disease. In circumstantial support for the strand-segregation mechanism, segregation of sister chromatids in embryonic mouse cells and in mouse epithelial cells is shown to be nonrandom.

Program of Cell-type Switching of Budding Yeast Compared with the Fission Yeast System

The stem cell pattern of cell type change is also observed with the evolutionarily distantly related yeast *S. cerevisiae*. Analogous studies with this system have yielded a wealth of knowledge regarding mechanisms of silencing, recombination, cell-type determination, and cell-lineage specification. Both of these yeast systems have become models to address fundamental questions of cellular differentiation. The budding yeast system, in fact, has become a classic textbook case. Most interestingly, the details of the molecular mechanisms of both systems vary in fundamental ways at every level; lessons learned from both systems should be taught to future biologists.

The budding yeast cells inherently divide asymmetrically by budding in which the older (mother) cell pinches off a small (daughter) cell. The daughter cell gains in size by growing in the longer G_1 phase before it starts its division cycle, while the mother cell initiates the next cycle right away. The two sexual types of *S. cerevisiae* are designated *a* and α, which are correspondingly conferred by the *MATa* and *MAT*α alleles of the mating-type locus.

These two cell types efficiently interchange, and the changed cells of opposite mating type establish a *MATa*/*MAT*α diploid phase in which further switching is prohibited by heterozygosity at *MAT*. Cells of the diploid phase under starvation conditions undergo meiosis to produce two *a* and two α spore segregants, which will repeat the switching process to establish diploid colonies. Thus, budding yeast exists primarily in diploid phase, while fission yeast predominantly exists as a haploid culture.

MAT switching also occurs by a gene conversion process where the resident *MAT* allele on chromosome III is replaced by a copy of the donor locus from *HML*α or from *HMRa*. The donor loci are located more than 120 kb away, one to the left and the other to the right of

MAT, on opposite arms of chromosome III. Only *MAT* is expressed, while both *HM* loci are kept unexpressed by several *trans*-acting factors encoded by *MAR/SIR* loci.

As with any other feature of this system, the program of switching of *S. cerevisiae* is drastically different from that found in *S. pombe*. Notably, only mother cells switch in G_1, with each mother producing both switched daughters. The recombination event is initiated by a transient DSB at *MAT* by the expression of *HO*-gene-encoded site-specific endonuclease only in mother cells. Many *trans*-acting factors are required for expression of *HO*. One such factor is *ASH1* message, which is differentially localized to the daughter cells where it acts as a negative regulator of *HO* expression. Thus, totally different strategies are used by these yeasts to control the program of cellular differentiation; the fission yeast uses a *mat1 cis*-acting strand-specific imprinting mechanism, whereas the budding yeast uses the more conventional differential regulation of the *trans*-acting *HO*-endonuclease gene to initiate recombination required for switching. Likewise, silencing mechanisms are also quite different in these yeasts.

The overall strategy of both yeasts involves DNA recombination, but mechanisms are very different and complementary. Since the sequences of mating-type loci are very different, it is not surprising that these yeasts have evolved very different molecular mechanisms for switching and silencing. We suspect that Darwinian evolution is not only based on divergence of DNA sequence; it may also be based on evolution of biological principles. For example, in the case of evolution of the mating-type system in both yeasts, first duplication of unrelated sequences in different yeasts is required. Once that happens, evolution of any mechanism promoting site-specific initiation of recombination in one and silencing of the other duplicated segment would create the opportunity for a process such as mating-type switching. Once additional model systems are investigated, more strategies will be discovered. For example, haploid cloned lines of malaria parasites produce both male and female haploid gametocytes. Is sex switching going on there similar to the phenomenon of sex change of yeast?

Concluding Remarks

In both yeasts, an individual cell serves as a somatic as well as a gametic cell. Thus, it is expected that developmental decisions operative in these systems in both mitosis and meiosis can be investigated with the application of sophisticated tools at the single-cell level. In both yeasts, the program of cellular differentiation is due to very different but cell autonomous controls. Furthermore, the mechanism of silencing is best understood in these systems. From the studies of *S. pombe*, it can be stated that mitotic chromosome replication does not always produce identical daughter chromosomes. This is not to say that Mendel's law of segregation of genes or the law of gene assortment is violated. Rather, Mendel's laws apply only to chromosome and gene segregation during meiotic division, but production of nonequivalent sister chromatids during replication occurs in mitotically dividing cells of fission yeast. We could consider this as the Law of Nonequivalent Sister Chromatids. Unlike many other systems reported in this monograph, it is worth stressing that production of nonequivalent chromatids or maintenance of specific epigenetic state through cell division does not require differential gene regulation of upstream regulators. Such mechanisms are likely to be prevalent in other systems of cellular differentiation.

3

EPITHELIAL STEM CELLS

The inner lining of the colon and small intestine is a simple columnar epithelium that is constantly renewed by cellular migration from pockets of proliferative activity or crypts. Small intestinal cells leave the crypt and migrate onto villi that protrude into the gut lumen. Only at the point of exit from the crypt do villus epithelial cells become fully mature with respect to biochemical markers of differentiation. These cells are principally epithelial columnar cells, goblet cells, and enteroendocrine cells. Within the crypt, immature proliferative forms are morphologically identifiable as being precursors to these mature cell types. Near the crypt base, immature cells lacking the morphological characteristics of mature types are found and coexist with a fourth differentiated cell type, the Paneth cell. These features present as a gradient with progressively higher positions being associated with a more developed morphological appearance. Cell kinetic experiments have determined the cycling characteristics of epithelial cells within the crypt, the time of appearance of identifiable differentiated cells from proliferative precursors, and the turnover times of crypt and villus populations. Studies of this kind, in both small intestine and colon, analyzed as a function of cellular position within the crypt (from position 1 at the crypt base), have led to the central tenet of gut stem cell biology: that cells originate from, and therefore stem cells reside near, the crypt base. Proliferating cells at higher locations in the crypt may have the potential to divide 1, 2, 3, 4, and possibly 5 or 6 times, with the actual number being determined by distance from the crypt base stem cell pool. Consequently, these cells are an amplifying transit population capable of up to 6 transit divisions.

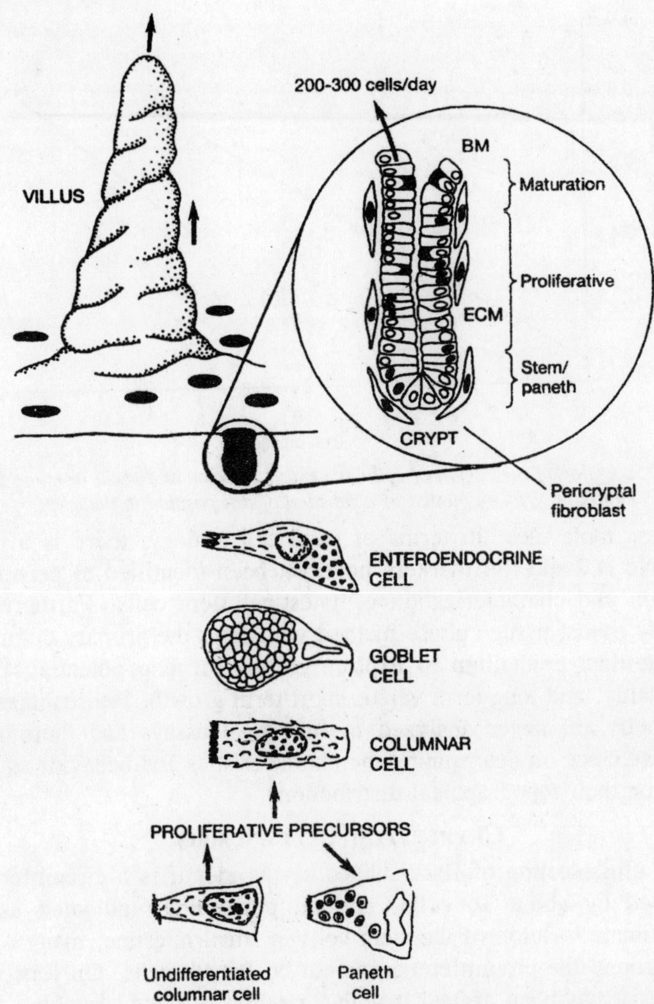

Fig. 3.1. Schematic representation showing clockwise, from top, the relationship between small intestinal villi and crypts; the organization of crypts into discrete functional compartment with the likely cellular and extracellular and basement membrane regulators of epithelial function; the four principal intestinal cell type.

Many aspects of gut biology support the concept of a graded change with progressive loss of proliferative and regenerative potential accompanying the acquisition of a differentiated phenotype. This organization creates tremendous opportunities to analyze pathways of differentiation and to examine the consequences of perturbation within proliferating populations, for example, by manipulating signaling or

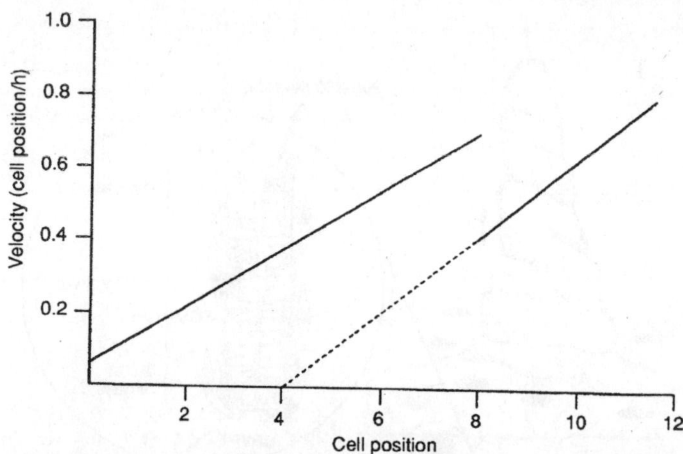

Fig. 3.2. Location of the region of cellular genesis in colon and small intestine by cell migration velocity plotted as a function of cell position in the crypt.

adhesion molecules. In terms of stem cell biology, there is a down side. No cell-specific markers have yet been identified to permit the isolation and characterization of intestinal stem cells. Furthermore, no truly robust tissue culture method yet allows the primary culture of the intestinal epithelium to explore colony forming potential, pluripotentiality, and long-term versus short-term growth. Hence, intestinal stem cells are never analyzed in functional assays and there is an undue reliance on determining the characteristics and behavior of cells based on their topographical distribution.

Crypt Stem Cell Number

A cross-section of flask-shaped crypts identifies a circumference occupied by about 16 cells. At cell position 4, indicated as the approximate location of the stem cells in small intestine, many of the cells around the circumference would be Paneth cells. On functional grounds it has been argued that in a column of cells stretching from the crypt base, any resident stem cell would reside above the highest Paneth cell position and could therefore occupy any position between 2 and 7. If this argument is questioned, an additional 10 or so undifferentiated columnar cells are mixed with the Paneth cells. Thus, the number of undifferentiated crypt base columnar cells available to contain the stem cell population is in the range of 20–30 per crypt with 16 in an "*undulating annulus*" centered on cell position 4 and the remainder mixed with Paneth cells. Integration of cell kinetic data within a mathematical simulation (which assumes the annulus model) suggests that around 4–6 stem cells per crypt could most easily maintain the epithelium.

Other evidence comes from clonality studies in which clones of marked cells are induced following mutagen exposure. Mutation affects one allele of a polymorphic locus. Two detection systems have been used extensively. One involves loss of lectin-binding ability for *Dolichos biflorus agglutinin* (DBA) following mutation of the *Dlb-1*^{b} allele (Dolichos lectin-binding locus) which in *Dlb-1*^{b}/*Dlb-1*^{a} heterozygous mice determines the expression of the lectin-binding site in intestinal epithelia. The other clonal marker involves detection of loss of *glucose 6-phosphate dehydrogenase* (G6PD) activity in mice functionally hemizygous for a wild-type *G6PD* allele. Such hemizygosity arises in male mice or in females homozygous for the wild-type allele after X-inactivation. These studies demonstrate that with time following chemical mutagenesis, intra-crypt epithelial clones are first detected, and subsequently, whole mutant crypts. In the small intestine, the *Dlb-1* assay indicates that the time taken for such monoclonal crypts to appear is up to 8–12 weeks, whereas in the colon using the G6PD assay the equivalent time is around 4 weeks. The differences between the two tissues are probably real, but direct comparison is impossible due to the differences in the mutagens used and the likely cytotoxic effects.

In mice, the total transit time of cells from crypt base to villus tip is around 5 days. Analysis of *Dlb-1* mutant clones at 10 days (and therefore assaying a progenitor population from within the stem cell region of the crypt) demonstrates that the vertical clones of mutant cells present on the villus epithelium following migration from the crypt are some 2 cells wide. This equates to 20–25% of the average cellular output of the crypt and indicates that 4–5 stem cells per crypt have survived mutagenesis. Similar numbers are indicated by a gain-of-function version of the *Dlb-1* assay. For the colon, the estimate is 2–3 stem cells. Whether the cell death induced by mutagen exposure makes this calculation an underestimate is not known at present. These results indicate that there is a mechanism whereby several stem cells in each crypt are supplanted by the progeny of only one. The progression to crypt monoclonality in adult animals has been confirmed by the accumulation of spontaneous whole crypt mutations in small intestine and colon in mice and humans. Thus, stem cell turnover is a physiological process. The mechanism by which an estimated 4–6 stem cells per small intestinal crypt and 2–3 per colonic crypt are replaced has been the subject of some debate.

Extrapolation of survival curves for regenerating foci of small intestinal epithelium (microcolonies) following irradiation has allowed estimates of the number of clonogenic cells to be made under different experimental conditions. Although the method can be criticized for the assumptions made and a high degree of observed variability, a general picture of the radiosensitivity of the stem and proliferative cells has emerged from split dose protocols and different dosage regimens. The steady-state stem cells may be radiosensitive and rarely contribute to regeneration. An additional population of around 6 clonogenic cells per crypt with intermediate sensitivity regenerates the epithelium at low doses (less than 9 Gy). Above 9 Gy, a larger population of 16–24 radioresistant clonogens contributes to regeneration. Adaptation of the method to the colon gives qualitatively similar findings. The results imply a gradient of radiosensitivity among stem cells and their first, second, and third generation progeny which are normally destined to enter, or are actually in, the transit amplifying population. Apparently cells normally destined for differentiation can be called back to regenerate a functional stem cell compartment, although from these acute studies it is unclear how long they can continue to fulfill this role.

Modes of Division

There is no unequivocal evidence that indicates whether intestinal stem cells divide asymmetrically or symmetrically. The progression to monoclonality described above implies different fates for stem cells within the crypt. However, there is no evidence that this asymmetry in fate is determined prior to division. Different models for self-renewal of the stem cell population and which might explain the progression to monoclonality have been proposed. In one model, a probabilistic maintenance of the steady-state stem cell pool arises through purely symmetrical division with two stem cells or two committed cells arising at each stem cell division. This might create monoclonal crypts through progressive clonal expansions and extinctions. In this model, problems associated with complete stem cell extinction in a proportion of crypts (for which there is little evidence) due to small stem cell numbers are resolved by postulating that increased stem cell numbers in other crypts provide a signal for crypt replication through fission. Certainly there is evidence for a crypt cycle, and this might provide a route for the propagation of stem cells. However, it seems speculative to conclude that fluctuations in stem cell numbers are the driving force for crypt fission. The relative contribution of clonal expansion due to the nature of stem cell division and of segregation of clones into crypts by fission in explaining the monoclonality of the crypt epithelium is unknown. Bifurcating crypts (intermediate fission forms) are present in the steady state, and their numbers are increased by mutagen/cytotoxic treatments. Hence, there may be a substantially greater contribution from fission events following chemical mutagenesis. This model denies any regulatory role for control of stem cell divisions through feedback mechanisms. In contrast, a multifactorial mathematical model indicates that a working crypt could be maintained by mainly asymmetrical stem cell division (95% of mitoses) with occasional division (5% of mitoses) being symmetrical and generating two stem cell daughters. Here again, it is the symmetrical divisions that allow clonal expansion (and extinctions) and might therefore account for monoclonality.

It is generally agreed that, in contrast to invariant asymmetrical divisions, symmetrical stem cell divisions are regulatable by environmental factors, because the probability of self-renewing divisions can be controlled. Indeed, the gut epithelium is greatly influenced by environmental signals, and the ability of its stem cells to regenerate a normal epithelium following irradiation amply demonstrates that symmetrical self-renewing divisions can be triggered under conditions of regeneration. However, if the converse applies and there is little need to modulate the size of the stem cell pool in the steady state, in theory asymmetrical division could be the norm. There is no published evidence in intestinal stem cells for an asymmetrical cellular distribution of mammalian homologs of molecular determinants of asymmetry as described in other tissues and species.

Consequently, asymmetrical divisions in the intestinal epithelium may be dictated immediately subsequent to mitosis by the relative vertical displacement of the two daughters. If so, this is not due to an inbuilt planar orientation of each mitotic event in the crypt as a whole, which can be at almost any orientation except at right angles to the epithelial sheet, but even very small vertical displacements might determine fate. Supporting evidence for nonrandom cell deletion, and therefore asymmetrical divisions in intestinal epithelium, comes from computer simulations of the amount of genetic diversity affecting microsatellite repeats in small samples of around 200 epithelial cells in mismatch-repair-deficient mice.

Higher levels of genetic diversity are predicted when one daughter of stem cell division is selected for extinction than when both daughters have the opportunity to be retained. Nonrandom cell death is compatible with the different fates of two daughters being determined prior to division (intrinsic asymmetry) or as a consequence of cell position.

NICHE

The concept of a unique microenvironment or niche maintaining the intestinal stem cell population first came from cell marking studies in which crypt base columnar cells that had taken up phagosomes after exposure to [^{3}H]thymidine were tracked over time. These experiments demonstrate that prior to the acquisition of differentiated characteristics, stem cells first leave the crypt base. For example, Paneth cells colocalize within the stem cell zone, but their precursors are only recognized immediately above the stem cell zone and subsequently migrate downward. This implies that undifferentiated cells first have to leave the stem cell zone to enter an environment permissive for differentiation.

The nature of the niche in terms of its composition and in the aspects of stem cell behavior it regulates is still not understood. The above observation seems to require that niche determinants are localized to the crypt base, and consequently, attention has focused on the Paneth cells themselves and on molecular determinants with the appropriate distribution.

Paneth cells are long-lived and could therefore maintain a stable environment in the crypt base. They produce a number of factors that could regulate proliferation and differentiation in neighboring cells, including epidermal growth factor, tumor necrosis factor α, guanglin, and matrilysin, as well as other factors thought to be involved in the regulation of bacterial populations. However, ablation of the Paneth cell population in mice transgenic for an attenuated diphtheria toxin expressed from an upstream Paneth cell promoter (cryptidin 2) causes no change in the rate of crypt to villus migration or in the ordered differentiation of enteroendocrine, columnar, or goblet cells, or any change in the relative numbers of these cells. Furthermore, species lacking Paneth cells, such as pigs and dogs, have in other respects a similar crypt architecture and organization to that found in mice and humans.

There is ample evidence that the maintenance of a functional gut epithelium results from extensive regulation by and interaction with components of the *extracellular matrix* (ECM). Retention and loss of stem cells from the crypt base may be best achieved by regulating their adhesion to the ECM. Regulation of proliferation and commitment by the ECM is also likely to be of major importance. Gene deletion experiments in mice have demonstrated that both positive and negative regulators of growth are provided to the intestinal endoderm during development. *Hlx*, a divergent murine homeobox gene, is expressed in the mesenchyme underlying the mid- and hindgut and is required for gut elongation, and is a positive regulator of growth. In contrast, *Fkh6*, a winged helix transcription factor with a similar mesenchymal distribution, is a negative regulator. Mice deficient in *Fkh6* show intestinal overgrowth and severe dysregulation of proliferation both during development and in the adult. In the adult, the number of dividing cells is increased fourfold and the crypt morphology is altered.

The distribution of components of the ECM underlying the intestine in both mouse and humans has been investigated by immunohistochemistry, and clear patterns of differential expression have been recognized. The composition of the basement membrane immediately juxtaposed to the epithelium could be particularly important in regulation. One major component of the basement membrane that shows high molecular variability is the laminin (LN) group. These are heterotrimeric proteins composed of α, β, and γ chains. LN1 (α1, β2, γ1)and LN2 (α2, β1, γ1) are of particular interest because in humans they show a reciprocal pattern of villus and crypt expression, respectively. LN1 appears to be involved in triggering intestinal differentiation; LN2 is concentrated around the base of the crypt, suggesting a possible role in regulating the stem cell population. However, *dy* mice lacking the α2 chain (due to a natural mutation) appear to have no obvious intestinal abnormality, leaving the role of LN2 equivocal.

Integrin receptors of ECM components can also show spatially distinct patterns of intestinal expression. Integrin α2β1 and α3β1 show reciprocal expression in the crypt and villus, respectively. Determining the nature and consequence of interactions between the crypt epithelium and the ECM can be complicated by varying degrees of binding specificity. In this respect, α2β1 is promiscuous and binds both laminins and collagen. Not surprisingly, α2β1 has been implicated in diverse effects, including gland branching in breast, adhesion, motility, maintenance of differentiation, and cell cycle progression. More detailed analysis of variant molecules can reveal underlying spatial differences. For example, the β4 chain is present in the intestinal basement membrane throughout the crypt-to-villus axis, but in the crypt epithelium it is proteolytically cleaved to remove the cytoplasmic tail. This modification renders integrin α6β4 functionally inactive in terms of

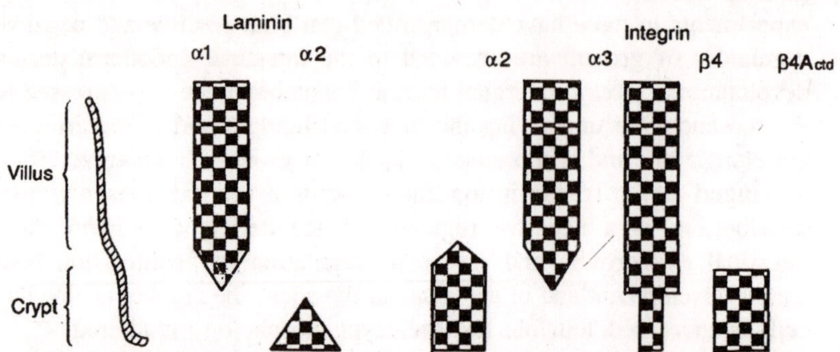

Fig. 3.3. Schematic of crypt:villus pattern of expression of laminin 1 and laminin 2 (subchains α1 and α2) and of integrin α2, α3, and β4 chains in human intestine.

its ability to bind LN5. Again, the functional significance of this modification is unclear, but in addition to conformational changes on adhesion, could affect the ability of α6β4 to signal through the Ras/map kinase pathways and to modulate cyclin-dependent kinase inhibitors.

No unequivocal molecular determinant of the stem cell niche has yet been identified, but there is an enormous potential for cross-talk between crypt stem cells and the ECM. It is likely that an effect at a given cell position will be mediated by multiple determinants/signals, and consequently, the composition of the niche will be complex. More problematic is discriminating whether an effect, such as the intestinal overgrowth and dysregulation observed in *Fkh6* null mice, involves or requires a stem cell defect or whether changes in the transit population are sufficient. Similarly, with integrin-mediated interactions, the regulated function may involve a general effect on proliferation or differentiation.

Subepithelial fibroblasts and myofibroblasts may play a crucial role in defining the stem cell niche. Isolation of cloned sublines has shown that these are heterogeneous and differ in their ability to induce phenotypic changes in intestinal endoderm from day 14 fetal rats. In particular, one line stimulates proliferation, and development of an undifferentiated glandular structure. Another induces differentiation markers and supports the development of a normal crypt/villus structure. Additional lines isolated from different geographical positions along the longitudinal axis of the gut (proximal and distal small intestine and proximal colon) appear to retain the characteristics of their site of origin. Similar heritable differences in phenotype in myofibroblasts localized along the vertical axis of the crypt could help maintain a favorable environment for intestinal stem cells.

Secreted Factors

Systemic or localized production of growth factors and cytokines must regulate intestinal stem cells, and their localized sequestering may limit availability at the niche. Their effects may extend to altering ECM–epithelial interaction through an effect on stromal fibroblasts. Due to difficulties in maintaining normal cells, the effects of candidate cytokines and growth factors have mainly been assayed using colon cancer lines. Molecules with general stimulating or inhibitory effects on cell proliferation in such systems include interleukins (IL-2, IL-4, IL-10, IL-11), *epidermal growth factor* (EGF) and EGF family members, *insulin-like growth factor* (IGF), and prostaglandins.

In some cases, in vivo stimulatory effects are indicated by association with gut pathologies (e.g., the association of IL-4 and IL-10 expression in inflammatory bowel disease). Many of the molecules may be primarily involved in wound healing. It is unclear even within well-characterized gene families such as the EGF family whether a given factor is regulating specific aspects of intestinal epithelium behavior such as differentiation, survival, or cell proliferation. Equally, which cell compartment—stem or proliferative cells—is responding is unclear. Only for a few molecules has an effect on the intestinal stem cell population been demonstrated either by a change in stem cell survival following radiation or by measurement of cell kinetics by cell position following in vivo administration. Thus, transforming growth factor β (TGFβ), which is thought to be a negative regulator of epithelial proliferation, is expressed within the crypt.

Furthermore, its administration reduces cell proliferation, alters cell cycle characteristics within the stem cell region, and protects stem cells from radiation induced-death as determined by the microcolony assay. Exposure to keratinocyte growth factor (a member of the FGF family) stimulates proliferation and affects crypt fission and also protects stem cells from radiation-induced death. *Stem cell factor* (SCF) and FGF-2 have also been shown to protect intestinal stem cells from the effects of ionizing radiation, and subepithelial intestinal myofibroblasts express the SCF receptor c-kit. Whether radiation protection arises directly from an effect of the cytokine/growth factor or indirectly through stimulating other cell populations is unclear. For example, keratinocyte growth factor is induced by serum factors in cultured dermal fibroblasts by two distinct pathways involving protein kinase C or cyclin-dependent kinases. Furthermore, KGF is inducible from such fibroblasts by treatment with IL-6, IL-1, or TNFα. Thus, production of growth factors and cytokines may regulate the stem cell compartment. Paneth cells, although not essential for stem cell maintenance, may still play a role in more subtle ways and certainly produce TNFα. Lymphoid cells may also be able to regulate stem cell behavior by the localized production of cytokines.

Tcf-4

Recently, it has been shown that germ-line deletion of the high-mobility group transcription factor Tcf-4 causes a failure to lay down the adult pattern of stem (and proliferative) cells in small intestinal epithelium. Proliferating cells in the intervillus region (from which crypts will form after birth) of Tcf-4 null mice are completely absent from El6.5 stage embryos onward. In colon, where Tcf-4 is also expressed, the phenotype is not evident, presumably due to the activity of another member of the Tcf family. The relevance of this finding is that it implicates by association several of the molecular partners of Tcf-4, which are involved in the neoplastic transformation of colonic cells, in the establishment of intestinal stem cells. Thus, β-catenin complexes with Tcf-4 to generate an active transcription factor but is normally found in low concentrations associated with the product of the *adenomatous polyposis coli* (APC) gene. APC acts to suppress signaling via the Tcf-4/β-catenin complex.

Key events in transcriptional activation include: signaling through the wingless/Wnt pathway; stabilization of β-catenin as a cytoplasmic monomer that no longer forms a complex with APC; and translocation of β-catenin to the nucleus where it associates with Tcf-4. Constitutively active Tcf-4/β-catenin complexes are present in $Apc^{-/-}$ colon carcinoma cell lines and in $Apc^{+/+}$ cell lines containing a dominant mutation affecting the amino terminus of β-catenin. One interpretation of these observations is that neoplastic transformation as a consequence of activation of Tcf-4 results in a maintenance of stem cell characteristics: In effect, cells that should undergo terminal differentiation continue to divide and to persist.

The link between colon cancer and stem/proliferative behavior has been taken further: CD44, which is often overexpressed early in the development of colorectal cancers, is restricted to the crypt epithelium in normal mice. This family of glycoproteins is believed to link ECM components to the cytoskeleton and, through an interaction with heparin sulfate, to bind growth factors that may promote receptor signaling. CD44 expression appears to be controlled by the Tcf-4/β-catenin pathway.

The role of Tcf-4 in maintaining functional stem cells in adult epithelium is unknown. In situ hybridization indicates expression of Tcf-4 message throughout the crypt (CD44 has a similar pattern of expression), although no detailed expression studies have determined the pattern of expression of the protein. Other factors may serve to maintain the stem cell zone in conjunction with the Tcf-4-mediated pathway. The requirement for Wnt signaling implicates the mesenchymal cells underlying the crypt epithelium as the probable source. Identification of a crypt-specific Wnt receptor might yet define a stem-cell-specific marker. Overall, the need for cross-talk between cells of the ECM and developing and presumably established stem cell population is demonstrated both by the requirement for Wnt signaling and the action of its downstream target CD44.

HOMEOSTASIS

In the steady stage, the small intestine shows a background of apoptosis which in the crypt is coincident with the stem cell position. Up to 10% of crypt base columnar cells may be dying at any time. It has been suggested that such spontaneous apoptosis acts to remove excess stem cells after symmetrical stem cell division. If correct, this appears to be a small-intestine-specific strategy, as spontaneous apoptosis in the colon is not associated with the stem cell region. This may relate to the expression of Bcl-2 in the latter and not the former. Currently, there are no functional data to indicate the nature of the deleted cells.

Apoptosis as a homeostatic mechanism is supported by experiments causing crypt hyperplasia. Following *small bowel resection* (SBR), both proliferation and apoptosis are increased to a similar degree as part of an adaptive response. Similarly, chimeric mice in which one parental component is directed to overexpress an amino-truncated β-catenin in intestinal epithelium show a four-fold increase in cell division and "*spontaneous*" apoptosis in affected crypt epithelium. Qualitatively equivalent findings have been described in the over-grown crypt epithelium in *Fkh6* null mice. These observations would be in accord with establishment of new steady state with increased numbers of stem cells being deleted. Longer-term study of the proliferation and apoptotic changes accompanying SBR in rabbits shows that 3 weeks post-surgery, crypt cell apoptosis levels are returned to normal while crypt cell proliferation remains stimulated. This dissociation is interesting: Can a steady state be established in which an increased proliferative compartment is maintained by a normal number of stem cells?

Regulation of spontaneous apoptosis by specific gene products has been investigated. p53 and Bax null mice show no difference in the levels of spontaneous apoptosis. (p53 deletion also has no effect on the elevated levels of apoptosis observed following SBR). Bcl-2 null mice show elevated levels of spontaneous apoptosis in colonic epithelium (at the putative stem cell location) but not in the small intestine, and overexpression of Bcl-2 in the latter does not affect the frequency of spontaneous apoptosis. These last observations indicate that Bcl-2 may be involved in homeostasis of normal colonic stem cells but indicate that it is not involved in the small intestine.

Following γ irradiation, there is an increase in the level of apoptosis at the stem cell position that is p53-dependent. In p53 null mice, the peak of apoptosis normally observed at 3–4 hours postirradiation is absent (there is a later p53-independent peak of apoptosis that probably arises due to aberrant mitoses). In colon (and not small intestine) irradiated Bcl-2 null mice show increases in apoptosis over wild-type animals, and again this is linked to cell position 1–2. It has been proposed that the lack of a homeostatic mechanism for spontaneous deletion of excess cells in colon due to Bcl-2 may explain the greater susceptibility of the colonic, as compared to the small intestinal, epithelium to neoplastic induction.

Commitment

There is evidence for proliferative and self-maintaining progenitors committed to maintaining different cell types within the small intestine. Cell marking studies using a gain-of-function modification of the *Dlb-1* mutation assay have identified DBA-positive clones composed only of goblet/oligomucous cells. These are induced by high doses of chemical mutagen and persist for up to 154 days post-mutagenesis. Cell ablation studies in which enteroendocrine cells expressing secretin are deleted have also been supportive of a committed precursor. Secretin is expressed in a subset of small intestinal enteroendocrine cells, which are identified by the coexpression of multiple hormones. Hence, cholecystokinin- and glucagon-expressing cells also express secretin. Gancyclovir, in mice transgenic for a secretin promoter linked to herpes simplex virus thymidine kinase, ablates some 95% of secretin cells (S cells) and reduces cholecystokinin-, glucagon-, and peptide YY-expressing cells to a comparable degree. Cells expressing gastric-inhibitory peptide, substance P, somatostatin, and serotonin are only partially ablated (45–59%), and gastrin cells are reduced by ~13%. Ablation takes about 5 days as mature villus forms are not killed by the treatment. Rather, proliferative precursors within the middle region crypt are susceptible and are observed to apoptose. Furthermore, after a 5-week posttreatment recovery period, there is complete recovery of all cell types, indicating that earlier progenitors (or stem cells) are spared because the secretin promoter is not active. The relative reductions in numbers reflect the closeness of the lineage relationship with S cells.

Commitment of S-cell precursors seems to require BETA2, a basic helix-loop-helix protein. BETA2, in association with the coactivator p300, can coordinate transcription of the secretin gene, cell cycle arrest, and apoptosis. Mice deficient for BETA2 due to gene targeting lack enteroendocrine cells expressing both secretin and cholecystokinin, providing additional support for the model.

Evidence for selective versus instructional mechanisms for stem cell commitment is lacking. Long-term (e.g., goblet cell) precursors and short-term precursors may be regulated by different mechanisms, presumably because they occupy different crypt positions. Inhibition of upward migration of proliferating cells due to overexpression of E-cadherin does not affect the normal pattern of differentiation. Hence, differentiation is not cell-autonomous and relates to cell position rather than to actual age of cells from "birth" among the stem cells. It is unclear whether this also applies to commitment of stem cells as well as the differentiation process itself.

Little is known about molecular mediators of the commitment process. Members of the Cdx family of homeobox genes may be involved. In the adult gut, Cdx-2 mRNA is present in the undifferentiated crypt base cells, but the protein is expressed at highest levels in the villus epithelium. The bulk of work carried out to date suggests that, in fully differentiated intestine, Cdx-2 is primarily involved in the final maturation of enterocytes. Cdx-1 is expressed in small intestine and colon throughout the crypt–villus axis in a graded manner, with the highest levels of expression in the crypt base. In comparison to Cdx-2, rather less is known about the properties of Cdx-1, although it too appears to be able to regulate transcription of intestinal brush-border enzymes and may be able to regulate progression through the cell cycle. However, Cdx-1 is also associated with the transition of normal gastric and esophageal epithelium to intestinal metaplasia, and this may play an important role in maintaining the intestinal phenotype. What, if any, role Cdx-1, or indeed Cdx-2, plays in establishing lineage-committed precursors from the multipotential stem cell pool is currently unknown.

Concluding Remarks

Progress in unraveling the mystery of stemness in intestinal epithelium has been impaired by the limited usefulness of tissue culture systems. As it currently stands, primary cultures can be established from epithelial aggregates and not from completely dissociated cells. Transplantation experiments show that such aggregates can reconstitute normal epithelium. However, the methods do not allow the self-renewing potential of isolated and defined cell populations to be assessed. It is unclear whether this is a resolvable technical problem.

The lack of stem-cell-specific markers is also limiting: Do they exist for intestinal stem cells? It may be that intestinal stem cells are defined by the absence of such markers. Quantitative differences in integrin expression seem to be an important factor in determining the self-renewing capacity of keratinocytes. The emphasis on highly localized crypt base "*stem cell*" markers may be misplaced, and gradients of expression of such molecules may better relate to self-renewal. Considerable thought will be required on how to validate any such markers in the absence of appropriate in vitro assay systems.

Despite these limitations, the intestinal epithelium still attracts due to the clear separation of functional compartments and the ability to detect changes in these following genetic manipulation in transgenic experiments, or to establish what is required for their maintenance in transplantation/restitution experiments. These approaches are powerful, but it is unclear even in the most refined transgenic experiment the extent to which genetic redundancy or adaptive responses in vivo mask important interactions that could be teased apart in a more defined, in vitro, setting. Furthermore, subtle changes in stem cell behavior may not be detected because ultimately in vivo analyses depend on determining the spatial distribution of cells and differentiation markers in descendants of the entire crypt stem cell pool and not in the stem cells themselves.

Given the evident plasticity of a variety of apparently tissue-specific stem cells to trans-differentiate, the question arises: Do intestinal stem cells show this ability? The limitations described above do not allow the establishment of experimental conditions under which such behavior could be defined. However, a number of in vivo observations do imply considerable versatility of small intestinal stem cells. Most dramatic is the ability of crypts adjacent to areas of ulceration to generate a novel *ulcer-associated cell lineage* (UACL). Originating from the stem cell zone of individual crypts, these new cell lineages form complicated networks, change their pattern of gene expression, and establish new proliferative patterns as part of an adaptive response. Crypt epithelium surrounding lymphoid Peyer patches shows adaptation to produce M cells, a minor epithelial cell type found on the mucosal surface overlying the lymphoid follicles. Intestinal metaplasia in the stomach and esophagus is a common pathological observation. Conversely, metaplasia of normally caudal intestine into rostral squamous-type epithelium in haplo-insufficient Cdx-$2^{+/-}$ mice suggests a molecular candidate involved in the programming to generate intestinal phenotypes. Before fully testing the limits of intestinal stem cell plasticity, the requirement for the future is to devise appropriate assay systems for establishing the normal criteria for maintaining stem properties in intestinal cells.

4

Epidermal Stem Cells

The epidermis of mammals forms the outer covering of the skin and comprises both the interfollicular epidermis and the adnexal structures, such as the hairs and sebaceous glands. The major cell type in the epidermis is an epithelial cell called a *keratinocyte*. Interfollicular epidermis is made up of multiple layers of keratinocytes. The basal layer of cells, attached to the underlying basement membrane, contains keratinocytes that are capable of dividing, and cells that leave the basal layer undergo a process of terminal differentiation as they move toward the surface of the skin. The end point of this pathway is an anucleate cell, called a *squame*, which is filled with insoluble, trans-glutaminase-crosslinked protein and provides an effective barrier between the environment and the underlying living layers of the skin. The basal layer of interfollicular keratinocytes is continuous with the basal layer of keratinocytes that form the hair follicles and sebaceous glands; once again, the end point of terminal differentiation is a dead, highly specialized cell, forming the hair shaft or the lipid-filled sebocytes.

If stem cells are defined as cells with the capacity for unlimited self-renewal and also the ability to generate daughter cells that undergo terminal differentiation, then the epidermis is one of the tissues in which a stem cell compartment must be present. Throughout adult life there is a requirement for the production of new interfollicular keratinocytes to replace the squames that are continually being shed from the surface of the skin, and there is also a need to produce new hairs to replace those lost at the end of each hair growth cycle. It seems likely that there is a single, pluripotential, stem cell compartment in the epidermis and that the differentiation pathway selected by stem cell progeny is determined by the microenvironment in which they find themselves. The earliest evidence for this came from wound-healing studies in which it was found that hair follicle keratinocytes could migrate out of the follicle and repopulate interfollicular epidermis. Conversely, when interfollicular keratinocytes are grafted into an empty hair follicle, they can differentiate to produce a normal hair. There is also a report that sweat gland cells can produce interfollicular epidermis; whether this reflects the pluripotential nature of the epidermal stem cell compartment or a process of trans-differentiation remains to be investigated.

Keratinocytes have not featured in the numerous recent accounts of the plasticity of stem cells in a variety of tissues. Two observations suggest, however, that some plasticity exists. First of all, the process of keratinocyte terminal differentiation can be reversed by introduction of a viral oncogene. Second, metaplasia of epithelial cells, including keratinocytes, is not uncommon: This is the formation of one differentiated cell type from another in postnatal life, such as the formation of ectopic intestinal epithelium in the stomach or endocervical epithelium in the vagina.

Proliferative Heterogeneity

Under normal conditions, each epidermal stem cell division results in one daughter to replenish the stem cell compartment and one daughter to undergo terminal differentiation. There is no evidence at present that this is achieved through invariant asymmetric divisions. Rather, there appears to be populational asymmetry, so that although on average each stem cell produces one stem and one non-stem cell daughter, individual divisions can potentially result in production of two stem cells, two terminally differentiating cells, or one stem and one non-stem daughter. This populational asymmetry allows the epidermis to respond to varying physiological need, as when the tissue is damaged through wounding.

Studies of interfollicular epidermis have demonstrated that not all dividing cells are stem cells. Instead, the daughter of a stem cell that is destined to undergo terminal differentiation can divide a small number of times before moving out of the basal cell layer. This dividing population, with low self-renewal capacity and high probability of undergoing terminal differentiation, is known as the *transit amplifying compartment*. In interfollicular epidermis, the prime function of this population is to increase the number of terminally differentiated cells generated by each stem cell division, so that although stem cells have a high capacity for proliferation, they divide infrequently. A second potential attribute of transit amplifying cells, by analogy with the committed progenitor compartment of hematopoietic lineages, is that they have more restricted differentiation potential than the stem cells: In other words, transit amplifying cells may be committed to differentiate exclusively into squames or hair or sebocytes. The increasing use of transgenic mice to study epidermal proliferation and function should soon shed light on this issue. In the meantime, with only proliferative potential as a marker of the transit compartment, it remains possible that instead of discrete populations of stem and transit cells, there are gradients of keratinocyte proliferative and differentiative potential, with transit amplifying cells reflecting an intermediate position between stem cells (maximum self-renewal potential, minimum terminal differentiation probability) and terminally differentiated cells (minimum self-renewal capacity, maximum probability of differentiation).

Assays for Epidermal Stem Cells

To identify epidermal stem cells, it is necessary to have assays to measure the proportion of stem cells in a given population. There are well-established techniques for growing human keratinocytes in culture, and confluent sheets of cultured keratinocytes have been used as autografts, primarily for the treatment of burn victims, since the late 1970s. Long-term follow-up of patients has shown that the progeny of keratinocytes expanded in culture persist and make a normal epidermis for years following grafting. Thus, there is now doubt that stem cells can survive in culture.

In hematopoiesis, stem cells can be assayed by their ability to reconstitute all the blood cell lineages in a lethally irradiated mouse. Although human keratinocytes can reconstitute epidermis when grafted onto immunocompromised mice, it is not feasible to graft individual candidate epidermal stem cells and look for their ability to reconstitute the entire epidermis of the animal. As a result, there has been more reliance on clonal analysis in vitro, making use of the fact that human keratinocytes can be grown at clonal density on a feeder layer of mitotically inactivated mouse 3T3 embryonic fibroblasts.

Although the growth rate of a mixed population of human keratinocytes in culture will undoubtedly be influenced by the proportion of stem cells present, it cannot be used as a quantitative measure of stem cells. This is because the growth rate will depend on the proportion of cells that are dividing and the length of the cell cycle. It will also depend on the proportion of terminally differentiating cells, and, to a lesser extent, on cell loss through apoptosis or in vitro senescence. To evaluate the number of stem cells in a population of keratinocytes in vitro, it is therefore essential to carry out clonal analysis, examining the self-renewal and terminal differentiation potential of individual cells. This has been used both for interfollicular keratinocytes and for keratinocytes of hair follicles.

One of the earliest, and undoubtedly the most thorough, analyses of clonal growth of human keratinocytes in culture was carried out by Barrandon and Green (1987). The importance of the study is that it depended not on the behavior of the clones founded on initial plating of keratinocytes (primary clones), but on the secondary clones that grew following disaggregation and replating of individual primary clones, thereby providing a more rigorous measure of self-renewal capacity. Three types of proliferating keratinocytes were defined on the basis of the type of clone they founded in vitro. When no secondary clones formed or all consisted of terminally differentiated cells (terminal clones) the founder clone was classified as a paraclone. When 0–5% of the clones were terminal, the clone was described as a holoclone. Meroclones were intermediate in their behavior, $>5\%$ and $<100\%$ of the secondary clones being terminal. Holoclones thus have the greatest self-renewal potential and are those most likely to be founded by stem cells. The total life span of a paraclone is no more than 15 generations prior to terminal differentiation and so could be attributable to a transit amplifying cell. Meroclones could be indicative of stem cells that generate transit amplifying cells at higher frequency than holoclones; they should still be attributed to stem cell founders, however, because epidermis from elderly people yields meroclones but few or no holoclones.

Unfortunately, the holo/mero/paraclone assay has not been used extensively, because it is time-consuming and labor-intensive when large numbers of keratinocytes are to be screened. Instead, there has been reliance on analysis of primary clones, in particular scoring terminal or abortive clones (typically, 32 cells or fewer per clone by 14 days after plating, all the cells expressing terminal differentiation markers) and attributing them to transit amplifying cells, the remaining, actively growing, colonies being attributed to stem cell founders. Although the assignment of the abortive clones to transit amplifying cell founders is reasonably uncontroversial, it is worth noting that their proliferative potential is lower than that of the majority of paraclones and, more importantly, that attributing all of the nonabortive clones to stem cell founders is likely to overestimate stem cell numbers.

Clonal analysis, with all its imperfections, remains the only practical way to screen for molecular markers of stem cells and for in vitro studies of factors that regulate exit from the stem cell compartment. As different subpopulations of proliferating keratinocytes become more clearly defined at the molecular level, much of the current uncertainty about the in vitro clonal behavior of the stem cell compartment should disappear.

Epidermal Stem Cell Markers

As in other tissues, there is a clear need for molecular markers characteristic of the stem cell population, preferably cell-surface molecules that allow *fluorescence-activated cell sorter* (FACS) selection. There is a wealth of markers that distinguish basal from differentiating keratinocytes, with changes in keratin expression and the onset of expression of precursors of the cornified envelope that is assembled in squames being most frequently used. However, heterogeneity within the basal compartment has been harder to tackle.

The first surface marker of human epidermal stem cells to be described was β1 integrins, receptors that bind extracellular matrix proteins. β1 integrins are expressed by all cells in the basal layer of the epidermis, but interfollicular keratinocytes with properties of stem cells have two-to threefold higher levels than cells with properties of transit amplifying cells. Elevated expression of β1 integrins is also a marker for stem cells in human hair follicles.

β1 integrins can be used to enrich for stem cells, either in culture or directly from the epidermis, by FACS or differential adhesiveness to extracellular matrix-coated dishes, and to visualize the stem cells using confocal microscopy. Approximately 10% of cells in the basal epidermal layer are thought to be stem cells, and the proportion of basal cells with high β1 integrin levels varies from 25% in palm epidermis to >40% in neonatal foreskin. In vitro, human epidermal stem cells (defined by their ability to found actively growing clones) can be isolated to 90% purity on the basis of their adhesive properties.

Human keratinocytes with the highest expression of the α2β1 integrin (collagen receptor) also express the highest levels of α3β1 and α5β1 integrins (receptors for laminin and fibronectin, respectively), and therefore it is not possible to enrich further for stem cells by using combinations of antibodies specific for these individual integrins. The relationship between log β1 integrin fluorescence and clone forming ability is linear, and this would be consistent with a continuum of keratinocyte behavior rather than discrete subpopulations of proliferative keratinocytes. In mouse epidermis it is also possible to enrich for stem cells on the basis of rapid adhesion to extracellular matrix, although the integrins involved have not been defined.

The α6β4 integrin is a component of hemidesmosomes and is essential for anchoring the epidermis to the underlying basement membrane. Stem cells, like all basal keratinocytes, express α6β4; however, there is no strong correlation between level of expression and proliferative potential, whether assessed in clonogenicity assays or by comparison with the distribution of actively cycling cells in human epidermis. The "α6 bright" population described by Li et al. (1998) as enriched for stem cells appears to correspond to total basal keratinocytes when the FACS profiles are compared with those of Jones and Watt (1993).

Other proposed surface markers include the antigen recognized by mAb 10G7, low surface expression of E-cadherin, and high expression of Delta1. Disappointingly, the combination of E-cadherin or Delta1 with β1 integrins is unlikely to give greater enrichment for stem cells than β1 integrins alone, because the size and location of basal keratinocytes that are defined with each marker are similar.

A number of other proteins have been reported to be markers for the epidermal stem cell compartment. These include keratins 19 and 15, a high level of non-cadherin-associated β-catenin, and p63, a member of the p53 gene family. Because these are intracellular proteins, they are not useful for isolating stem cells; however, they can still provide important information about stem cell properties. p63 is of particular interest, since mice which are homozygous null for the p63 gene lack all stratified squamous epithelia, including the epidermis.

Finally, it is worth pointing out that expression of some of the potential markers of epidermal stem cells may be interdependent. There is good evidence for cross-talk between integrins and cadherins in cells and between the Notch (Delta receptor) and Wnt (upstream of β-catenin) signaling pathways.

Epidermal Stem Cell Patterning

The distribution of stem cells within the epidermis is not random. In hair follicles, the keratinocytes with high proliferative potential are reported to lie in the outer root sheath at the point of insertion of the muscle ("*bulge region*") or lower down. In mouse interfollicular epidermis, it is proposed that a single stem cell lies at the base of a column of suprabasal cells and is surrounded by transit amplifying cells and cells that are committed to terminal differentiation; support for this model comes from more recent lineage marking studies.

β1 integrin staining has revealed a high level of patterning of the stem cell compartment in interfollicular human epidermis. The cells with high β1 levels are found in clusters that lie at the tips of the dermal papillae (where the basal epidermal layer comes closest to the skin surface) in most body sites, but at the tips of the deep rete ridges (where the basal layer projects deepest into the dermis) in palm and sole epidermis. Consistent with this view of the epidermis, the actively cycling cells are concentrated in the areas of low β1 integrin expression, as are the cells that have initiated expression of keratin 10 and are in the process of moving out of the basal layer. Clustering of stem cells implies some lateral migration of cells along the basement membrane, and the high β1 integrin-expressing keratinocytes are indeed less motile than the keratinocytes with lower β1 levels, whether motility is measured in isolated cells or in confluent cell sheets. The simple architecture of the epidermis lends itself to mathematical modeling. It is interesting that from biophysical considerations, adhesiveness of keratinocytes to the basement membrane is predicted to be an important determinant of movement out of the basal layer. The predicted topology of stem and transit cells in some models of normal and hyperproliferative epidermis is consistent with the distribution of clusters of keratinocytes expressing high and low β1 integrin levels.

Regulation of Stem Cell Fate

One of the most potent terminal differentiation stimuli is to place cultured keratinocytes in suspension: Both stem and transit amplifying cells initiate terminal differentiation without any further rounds of division, and by 24 hours the majority of cells are expressing markers of the differentiation pathway. Suspension-induced differentiation can be partially inhibited by ligating β1 integrins with extracellular matrix proteins or anti-integrin antibodies. Recent experiments suggest that the integrin signal is "do not differentiate," transduced by occupied receptors, rather than a positive "differentiate" signal transduced by unoccupied receptors. Ligand binding is required both for integrin-mediated adhesion and integrin-regulated differentiation, but mutagenesis of the β1 cytoplasmic domain has established that the sequences which are required for differentiation control are distinct from those which are required to support extracellular matrix adhesion.

Integrins regulate not only the onset of overt differentiation, but also morphogenesis, down-regulation of integrin function and expression ensuring selective migration of committed cells from the basal epidermal layer. In addition, high levels of β1 integrins are required for keratinocytes to remain in the stem cell compartment in vitro. Introduction of a dominant negative β1 integrin into human keratinocytes in culture increases the proportion of clones attributable to transit amplifying cells. The dominant negative mutant interferes with β1 signaling to MAPK, although MAPK activation in response to growth factors or α6β4 ligation is not impaired. Constitutive activation of MAPK rescues keratinocytes expressing the dominant negative integrin, decreasing the proportion of abortive clones to control levels; conversely, a dominant negative MAPKK1 construct reduces MAPK activation and increases the proportion of abortive clones.

As described above, stem cells have higher levels of non-cadherin-associated β-catenin than transit amplifying cells in vitro. Expression of a dominant negative β-catenin mutant in human keratinocytes promotes the formation of abortive, transit amplifying colonies, whereas expression of stabilized, amino-terminally truncated β-catenin increases the proportion of putative stem cells to almost 90% of the proliferative population. Interestingly, expression of stabilized amino-terminally truncated β-catenin in the basal layer of transgenic mouse epidermis causes keratinocytes to revert to a pluripotent state in which they can differentiate into hair follicles or interfollicular epidermis. Over-expression of stabilized β-catenin leads to ectopic formation of feather buds in developing chick skin.

One of the genes that is regulated by β-catenin signaling is c-Myc. Since c-Myc promotes entry of keratinocytes into the transit amplifying compartment in vitro, it is possible that there is a feedback loop involving the two proteins that controls the ratio of stem to transit amplifying cells in the epidermis. Further targets of β-catenin and c-Myc in keratinocytes remain to be identified. It is, for example, possible that c-Myc down-regulates integrin expression. Just as there may be cross-regulation of epidermal stem cell markers, the same is likely to be true for the molecules that regulate stem cell fate.

In addition to factors that act cell-autonomously to regulate epidermal stem cells, there is good evidence for a role of cell-cell interactions. One of the signaling pathways that is implicated is the pathway downstream from the transmembrane receptor Notch, which binds to transmembrane ligands such as Delta on neighboring cells. Notch is expressed in all the layers of postnatal human epidermis; the Notch ligand Delta1 is confined to the basal layer and is most abundant in the clusters of cells known to express high levels of β1 integrins. High Delta1 expression blocks the responsiveness of epidermal stem cells to Notch signals and may enhance cohesiveness of stem cell clusters, thereby discouraging intermingling with neighboring transit amplifying cells. Furthermore, Notch activation in cells at the edges of the stem clusters stimulates them to become transit amplifying cells.

In surveying the molecules that are important for regulating epidermal stem cell fate, it is striking that the key signaling molecules, integrins, β-catenin, and Notch, play similar roles in diverse tissues and organisms.

It is possible to anticipate several advances that will both increase our understanding of epidermal stem cells and result in further therapeutic applications of keratinocytes. The use of proteomics and gene arrays should yield additional markers of the stem and transit amplifying compartments, enabling us to isolate the different cell populations to greater purity than is possible at present and providing new information about how stem cell fate is regulated. There will be increased use of lineage marking, both in vitro and in vivo, to monitor the behavior of keratinocytes in response to signals from their neighbors. There will be increased exploitation of transgenic mice to study stem cell renewal and differentiation, particularly making use of inducible transgene expression and inducible gene deletions, and this will address the issue of whether transit amplifying cells have more restricted differentiation options than stem cells. Finally, the combination of improved retroviral vectors and optimized transduction protocols will allow the exploitation of epidermal stem cells for gene therapy; for example, to treat inherited skin blistering diseases that result from mutation of genes encoding proteins that anchor the epidermis to the underlying dermis.

Epidermal Lineage

The epidermis is a stratified squamous epithelium that provides the protective layer of the skin. The mammalian epidermis is derived from embryonic ectoderm, and this layer eventually gives rise to a very early epithelial cell that further commits to become epidermal tissue. A single layer of proliferating cuboidal cells (stratum germinatium) represents the putative epidermis in the mouse up to E8.5–E12.5 d of gestation. The stratum germinatium resides on a basement membrane and expresses markers characteristic of simple epithelial cells (i.e., keratins 8 and 18 [K8/K18]). At E12.5–E14.5, an intermediate layer (the stratum intermedium) develops, which is relatively undifferentiated and is able to proliferate. However at E15.5, the stratum intermedium begins to differentiate into the spinous layer and starts to lose its proliferative ability with all mitotic activity disappearing at about E16.5, at which time the granular layer appears followed by the stratum corneum at E16.5–E17.5. By E17.5–E18.5, the mouse epidermis is fully differentiated.

The role of keratins as markers of epidermal development and differentiation from mature K5/K14 positive basal cells to the terminally differentiated epidermal cells of the cornified layer has been widely researched. Despite such intense study, there is still very little understood about the cell fate selection of stem cells towards the epidermal lineage. Equally so, there is a limited understanding of the developmental program of epidermal differentiation; such as, its regulation or the signals necessary to result in the differentiation of epidermal progenitors. While this is partly owing to a lack of appropriate markers, the most important stumbling block has been the absence of an appropriate in vitro model system in which to study the early events leading to commitment.

The ability to generate differentiated epidermal progeny from a continuously growing stem cell population in vitro would provide a unique system for the study of stem cells and very early progenitors. Such a cell line would also make possible a comprehensive analysis of the underlying molecular mechanisms for the onset of embryonic epidermal commitment and differentiation. Similar in vitro approaches have yielded invaluable information on the mechanism of differentiation of other cell types. The development of stem cell models for hematopoietic, neuronal, and muscle lineages, for example, have led directly to important advances in our understanding of terminal differentiation. In the past, several attempts have been made to conduct similar studies on epidermal differentiation. However, the difficulties encountered, in maintaining a stable putative stem cell population in culture, limited the usefulness of these systems. Indeed, until relatively recently, even the maintenance of mature epithelial cells in culture has been problematic.

Embryonic stem cells (ES) have been shown to provide an excellent model system in which to study lineage commitment and progression in vitro for a number of lineages. ES cells are derived from the inner cell mass of 3.5-d mouse blastocysts. When placed upon a suitable fibroblast feeder layer with *leukemia inhibitory factor* (LIF), ES cells proliferate and remain totipotent indefinitely. Removal of ES cells from their feeder layer induces aggregation and differentiation of the cells into simple or cystic ***embryoid bodies*** (EBs). Simple EBs consist of ES cells surrounded by a layer of endodermal cells, while cystic EBs develop an additional layer of columnar ectoderm-like cells around a fluid-filled cavity, morphologically similar to embryos at the 6- to 8-d egg cylinder stage. The expression of markers for mesoderm, endoderm, and ectoderm indicates that cells derived from all three germ layers occur in cystic EBs. We describe protocols here for culture conditions in which ES cells undergo epidermal differentiation in a highly reproducible fashion, providing a powerful system for investigating epidermal lineage, and a route to the isolation and characterization of epidermal stem cells.

Materials

Tissue Culture

1. 1X Phosphate-buffered saline (PBS): prepared from 10X PBS, then aliquoted into 100-mL bottles and autoclaved. To prepare 10X PBS (1 L) combine the following: 11.5 g sodium phosphate dibasic (Na_2HPO_4), 2.0 g potassium phosphate monobasic (KH_2PO_4), 80 g sodium chloride (NaCl), 2 g potassium chloride (KCl) in distilled water. It is important to dissolve the first two ingredients before adding the last two. Make the volume to 1 L. To make 1 L of 1X PBS, mix 100 mL of 10X PBS and 900 mL dH_2O.
2. 7X-OMATIC (4 gallons).
3. Collagen Solution Type 1 from calf skin (20-mL bottle).
4. Dimethyl sulfoxide (DMSO).
5. Dulbecco's modified Eagle medium (DMEM) 1X.
6. Fetal bovine serum (FBS) characterized and screened for ES cell growth.
7. Growth factor reduced matrige basement membrane matrix (BM).
8. Insulin–transferrin–selenium (ITS) 100X.
9. Keratinocyte Growth Media without Calcium.
10. Minimum essential medium (MEM) nonessential amino acids (NEAA) solution 10 m*M*, 100X.
11. MEM sodium pyruvate solution 100 m*M*, 100X.
12. Mitomycin C.
13. Penicillin–streptomycin 100X.
14. Trypsin-EDTA: 0.05% trypsin, 0.53 m*M* EDTA•4Na 1X.

15. Trypsin-EDTA: 0.25% trypsin, 1 m*M* EDTA•4 Na 1X.
16. Corning polystyrene 100-mm tissue culture dishes.
17. Corning polystyrene 60-mm tissue culture dishes.
18. Corning polystyrene 35-mm tissue culture dishes.
19. 100-mm Fisherbrand Petri dishes.
20. 15-mL Polypropylene conical tubes.
21. 5-cc Syringes.
22. 0.2-μm Syringe filters.
23. Cryo tube vials.
24. StrataCooler cryo preservation module.
25. Coverslips.

Media for embryonic fibroblast cells

Embryonic fibroblast (EF) cells are maintained in supplemented DMEM. DMEM is supplemented with 10% heat-inactivated FBS, 1% sodium pyruvate, 1% NEAA, and 1% penicillin–streptomycin. This media is called 10% DMEM. To prepare 100 mL 10% DMEM, combine 10 mL heat-inactivated FBS, 1 mL penicillin–streptomycin, 1 mL NEAA, 1 mL sodium pyruvate, and 87 mL DMEM.

Media for ES cells

R1 ES cells are maintained in supplemented DMEM. DMEM is supplemented with 15% heat-inactivated FBS, 1% sodium pyruvate, 1% NEAA and 1% penicillin– streptomycin. This media is called 15% DMEM. To prepare a bottle of 15% DMEM; to one 500-mL bottle of DMEM purchased from Gibco BRL add the following: 90 mL heat-inactivated FBS, 6 mL penicillin–streptomycin, 6 mL NEAA, and 6 mL sodium pyruvate.

Media for MKC cells

Mouse keratinocyte cells (MKC) cells are maintained on collagen-coated 60-mm tissue culture dishes in supplemented *keratinocyte growth media* (KGM). KGM is supplemented with 1% ITS and 2% heat-inactivated FBS. The supplemented media has been termed KI2. To prepare 100 mL KI2 media combine 97 mL KGM, 1 mL ITS, and 2 mL heat-inactivated FBS.

General comments and required equipment for tissue culturing

As a general rule, all tissue culture protocols must be performed using sterile techniques with great attention given to using clean and detergent-free glassware and all media and solutions must be warmed to 37°C before use.

The tissue culture facility for ES cell culturing requires the following:

1. 37°C water bath.
2. Coulter Cell Counter Z2 series.
3. Glass pipets designated for tissue culture only (10 mL and 25 mL).
4. Humidified incubator at 37°C and 5% CO_2.
5. Inverted microscope with a range of phase contrast objectives (×10 to ×25) equipped with photographic capabilities.
6. Laminar flow cabinet.
7. Liquid nitrogen storage tank.
8. Pipetmen (2, 10, 20, 100, 200, and 1000 μL) designated for tissue culture use only.
9. Refrigerator (4°C) and freezer (–20°C).
10. Tabletop centrifuge.

In situ Hybridization

1. 5-Bromo-4-chloro-3-indolyl-phosphate 4-toluidine salt solution (BCIP).
2. Calf serum (100-mL bottle).
3. Formamide, deionized (500-mL bottle).
4. Glycerol (1 L).
5. 4-Nitro blue tetrazolium chloride solution (NBT) (3 mL).
6. Proteinase K (20 mg/mL stock) (50-mg vial). To prepare a 20 mg/mL stock solution, add 2.5 mL water to a 50-mg vial, mix, and store at –20°C.
7. RNase A (10 mg/mL stock) (100-mg vial). To prepare 10 mg/mL stock solution, add 10 mL water to a 100-mg vial, boil for 5 min, and store at –20°C.
8. Digoxigenin (DIG) RNA labeling mix.
9. RNA polymerase.
10. 10X Transcription buffer.
11. DIG-labeled control RNA.
12. QIAquick gel extraction kit.
13. Heparin, porcine, sodium salt.
14. Glutaraldehyde: 8% aqueous solution.
15. Levamisole hydrochloride.
16. 0.2 *M* EDTA: add 4 mL 0.5 *M* EDTA to 6 mL diethyl pyrocarbonate (DEPC) water.

17. Alkaline phosphatase (AP) buffer; for 10 mL, mix 1 mL 1 *M* Tris, pH 9.5, 500 μL 2 *M* NaCl, 500 μL 1 *M* $MgCl_2$, 10 μL 0.1 *M* levamisole, 100 μL 10% Tween-20, and 7.89 mL DEPC-H_2O.
18. Bovine serum albumin (BSA) (10 mg/mL).
19. DEPC-H_2O.
20. Glutaraldehyde (0.2%)/paraformaldehyde (4%) in PBS-Tween (PBS-T). For 4 mL add 100 μL of 8% glutaraldehyde from –80°C stock to 4 mL of 4% paraformaldehyde. (Glutaraldehyde solution is aliquoted upon receipt and stored at –80°C. Aliquots are thawed only once.)
21. Glycine 20 mg/mL (dissolve 0.04 g of glycine in 2 mL PBS-T and store at –20°C) and 2 mg/mL (add 100 μL of 20 mg/mL stock into 1 mL PBS-T).
22. Heparin 50 mg/mL; dissolve 50 mg (0.05 g) of heparin in 1 mL of 4X standard saline citrate (SSC). Store at –20°C.
23. Hybridization buffer. For 50 mL, combine 25 mL formamide (deionized), 12.5 mL 20X SSC, 50 μL yeast tRNA (50 mg/mL stock), 50 μL heparin (50 mg/mL stock), 500 μL Tween-20 (10% Tween-20 stock), and 11.9 mL DEPC-H_2O. Store at –20°C.
24. Levamisole hydrochloride (0.1 *M*).
25. MeOH/PBS-T series; 25% Methanol/PBS-T, 50% Methanol/PBS-T, and 75% Methanol/PBS-T.
26. $MgCl_2$ (1 *M*).
27. NaCl (2 *M*).
28. NBT plus BCIP staining solution; for 2 mL, mix 6.75 μL NBT, 7 μL BCIP, and 1.97 mL AP buffer.
29. Paraformaldehyde (4%); must be prepared fresh in the following manner: measure 2 g paraformaldehyde into 35 mL warmed DEPC water (55°–60°C), and add 8.5 μL 1 N NaOH to dissolve, and stir. Add 5 mL 10X PBS (RNase Free) and top volume to 50 mL. Check that the pH is 7.2–7.5 (using pH paper).
30. PBS (1X).
31. PBS-T is 1X PBS plus 0.1% Tween-20.
32. PBS-T containing 5% calf serum. For 1 mL of solution, add 50 μL calf serum to 950 μL of PBS-T.
33. Proteinase K (10 μg/mL) must be made fresh. For 2 mL, add 1 μL 20 mg/mL proteinase K stock to 2 mL PBS-T.
34. SSC (20X); for 1 L, mix 175.25 g NaCl, 88.25 g sodium citrate, and adjust the final volume with DEPC-H_2O.

35. SSC (2X), 0.1% Tween-20; for 15 mL, mix 1.5 mL 20× SSC, 150 μL 10% Tween-20, and 13.35 mL DEPC-H_2O.
36. SSC (2X), 0.1% Tween-20 containing 20 μg/mL RNase A; for 3 mL, add 6 μL of 10 mg/mL RNaseA stock to 3 mL of 2X SSC, 0.1% Tween-20.
37. Tris (1 *M*), pH 9.5.
38. Yeast tRNA(50 mg/mL); dissolve 50 mg yeast tRNA in 1 mL of 4X SSC. Store at –20°C.

Ayoub Shklar Staining

1. 37% Formaldehyde (500 mL).
2. Acid fuchsin (100 g).
3. Aniline blue (25 g).
4. Orange G (100 g).
5. Phosphotungstic acid (100 g).
6. 95% Ethanol.
7. Filter paper and funnels.
8. 1X PBS.
9. 10% Neutral-buffered formalin; for 100 mL, combine 10 mL 10X PBS, 10 mL 37% formaldehyde, and 80 mL dH_2O, then store at 4°C.
10. 5% Acid fuchsin; for 100 mL, add 5 g acid fuchsin to 100 mL dH_2O, and stir at least 3 h or overnight. Filter and store at room temperature for 6 mo. Filter before use.
11. Aniline blue–orange G; for 100 mL, mix 0.5 g aniline blue, 2 g orange G, and 1 g phosphotungstic acid in 100 mL dH_2O. Stir at least 3 h or overnight. Filter and store at room temperature for 6 mo. Filter before use.

Immunofluorescence

1. Methanol (4 L) with a quantity stored at –20°C.
2. Humidified chamber.
3. Nalgene centrifuge tubes.
4. 50°C Water bath.
5. Disposable pipets.
6. 1,4-Diazabicyclo[2. 2. 2]octane (DABCO).
7. 1X PBS.
8. 0.2 *M* Tris, pH 8.5.
9. Mowiol 4-88 (500 g). For 20 mL, use the following procedure:

(a) Add 2.4 g Mowiol 4-88 to 6 g of glycerol and stir for 2 h at room temperature.
(b) Add 6 mL dH_2O and stir for several hours at room temperature.
(c) Add 12 mL 0.2 *M* Tris, pH 8.5, and heat to 50°C for 10 min with occasional mixing.
(d) Using a disposable pipet, transfer solution to a centrifuge tube and spin at 5000*g* for 15 min.
(e) Add 2.5% DABCO to reduce fading. Stir to dissolve.
(f) Aliquot into 1-mL tubes and store at –20°C. As needed, a tube is thawed and may be stored for several weeks at 4°C.

Reverse Transcription Polymerase Chain Reaction

1. Thermal cycler.
2. Agarose gel apparatus and reagents.
3. Trizol reagent (100 mL).
4. Cell scrapers.
5. 17 × 100-mm Polypropylene sterile culture tubes.
6. Chloroform (500 mL).
7. Isopropanol (500 mL).
8. 75% Ethanol.
9. RNase inhibitor (20 U/μL).
10. DNase I (256 U/μL).
11. Phenol, saturated.
12. 25 m*M* $MgCl_2$.
13. 10X PCR buffer II.
14. 100 m*M* dNTP set.
15. RNase inhibitor (20 U/μL).
16. MuLV reverse transcriptase (50 U/μL).
17. Random hexamers (50 m*M*).
18. *Taq* DNA polymerase (5 U/μL).
19. DEPC-H_2O.
20. 10X DNase buffer (200 m*M* Tris-HCl, pH 8.4, 20 m*M* $MgCl_2$, 500 m*M* KCl); for 100 mL, combine 50 mL 1 *M* KCl, 1.5 mL 1 *M* $MgCl_2$, 10 μL gelatin, and 10 mL Tris-HCl (pH 8.4).
21. 10 m*M* dNTPs are required for reverse transcription (RT). Dilute stock dNTPs 1:10 with DEPC-H_2O.
22. 2.5 m*M* dNTPs are required for polymerase chain reaction (PCR). Combine 250 μL of each stock dNTP (100 m*M*) into a

sterile microtube and mix (equals 25 m*M*). Aliquot 100 μL into each of 10 tubes and add 900 μL sterile H_2O and mix (equals 2.5 m*M*).

23. PCR primers diluted to 25 pmol/μL with sterile H_2O.

Methods

Tissue Culture

Freezing cells

As a general rule, freeze cells slowly and thaw them quickly. For long-term storage (indefinitely), cells must be kept under liquid nitrogen.

To prepare 10 mL freezing medium, add 1 mL DMSO to 9 mL of appropriate media and proceed with the following protocol:

1. Trypsinize cells in the exponential phase of growth (approx 3 d in culture) from a 100-mm dish.
2. Pellet cells by centrifugation (700*g*, 2–3 min) and resuspend in an appropriate amount of freezing medium and gently resuspend.
3. Aliquot 1 mL of the cell suspension into freezing vials.
4. Immediately transfer the vials to a precooled (–20°C) StrataCooler and store it in a –70°C freezer for 24 h.
5. Transfer the tubes to liquid nitrogen.

Thawing cells

When thawing vials of cells from the liquid nitrogen, one must work quickly and efficiently in order to maintain the integrity of the cells. Immediately after the cells are thawed, transfer the entire contents of the cryo vial into 10 mL of media in a 15-mL Falcon tube and collect the cells by centrifuging at 700*g* for 2 to 3 min. Remove the media and gently resuspend the cells with fresh growth media and transfer to a tissue culture dish. Allow the cells to adhere overnight in a 37°C incubator and change the media the next day.

Fibroblast feeder layers

ES cells require a feeder layer of mitotically inactivated fibroblast cells or gelatinized tissue culture dishes with the addition of LIF in order to remain pluripotent in culture. The ES cells that we routinely use, the R1 ES cell line, were established on an EF feeder layer, and so we continue to use this substrata. Because of the short culturing life span of murine EFs, it is advisable to prepare stocks of frozen vials that are capable of supporting ES cells. Therefore, every new batch of EF cells must be tested before they may be used as feeder layers. However, there are a number of slight variations of this protocol that have been used very successfully by others.

Method for the isolation of EF cells

1. Sacrifice pregnant mice at about 13–15 days postcoitum (dpc).
2. Clean the underside of the mouse with 70% ethanol and remove the uterus to a sterile Petri dish.
3. Remove the embryos from the uterus and transfer them to a sterile Petri dish.
4. Dissect away the heads and internal organs (liver, heart, kidney, lung, and intestine).
5. Wash 10–12 carcasses in a 50-mL Falcon tube with 50 mL 1X PBS, at least 3 times to remove as much blood as possible.
6. Cut the carcasses into small pieces with dissecting scissors. Use the following rule when cutting the embryos: (no. of embryos)^2 = the number of cuts; for example if there were 12 embryos, cut 144 times.
7. Transfer the tissue pieces into a sterile 50-mL Falcon tube and rinse once with 10% DMEM.
8. Rinse twice with 30 mL 1X PBS to remove blood cells and all traces of media.
9. Add 2 mL of 0.25% trypsin-EDTA per embryo (i.e., for 12 embryos add 24 mL trypsin-EDTA) and incubate at 37°C for 20 min.
10. Pipet up and down 24 times, remove the cells in suspension, and transfer 2 mL/100-mm tissue culture dish with 10 mL 10% DMEM and incubate overnight at 37°C.
11. Add 10 mL fresh 0.25% trypsin-EDTA to the remaining tissue and incubate for 20 min at 37°C.
12. Pipet up and down 10 times and plate 2 mL/100-mm tissue culture dish and add 10 mL 10% DMEM. Incubate cultures overnight at 37°C.
13. The next day, trypsinize and collect all cells to mix the populations. Plate on fresh tissue culture dishes.
14. The next day, change the media. When confluent (2 to 3 days), trypsinize, collect all cells, and freeze 1 × 100 mm = 1 vial, store in liquid nitrogen.

Maintenance of EF cells

When thawing EF cells, the contents of one cryo vial is transferred to one 100-mm tissue culture dish, and the cells are grown to confluency for 2 d and then subcultured 1:4 for use as feeder layers. Generally, a new vial of EF cells is opened on a Wednesday and subcultured 1:4 on Friday, in order to be ready for the following week's cultures. Once initial cultures have recovered, EF cells are maintained on 100-mm tissue culture dishes to confluency and then subcultured 1:2 every 3 to 4 d. To maintain a precise culturing schedule, EF cells are generally subcultured on Tuesday and Friday, in order for them to be on a comparable time course to the other cells. To subculture:

1. Remove media and rinse with 1X PBS.
2. Add 3 mL 0.05% trypsin-EDTA to each 100-mm dish.
3. Return plate to incubator for 2 to 3 min until the cells float with gentle agitation.
4. For 3 × 100-mm dishes, add 5 mL 10% DMEM to rinse the 3 dishes and collect all 14 mL into a 15-ml Falcon tube.
5. Centrifuge at 700*g* for 2 to 3 min to pellet the cells. While spinning, prepare 6 × 100-mm dishes by labeling the cell type and passage number and add 9 mL 10% DMEM to each plate.
6. After the spin has completed, suction off the media with a pasteur pipet, being cautious not to disturb the pellet.
7. Gently resuspend the pellet in 6 mL 10% DMEM, by carefully pipetting up and down 5 times.
8. Plate 1 mL (approx 2 × 10^6 cells) to each of the 6 × 100-mm dishes and gently agitate the plate back-and-forth and side-to-side to evenly distribute the cells.
9. Incubate at 37°C until they reach confluency (3 to 4 d). The media is changed every 2 d. EF cells are maintained only to the 7th passage. At this point, cell growth becomes retarded, and the cells exhibit signs of senescence.

Embryonic stem cells in culture

Mitomycin C treatment of feeder layer for ES cells

ES cells are maintained on 100-mm Corning tissue culture dishes that have a mitomycin C-treated EF feeder layer. When the EF cells are confluent, they are treated with mitomycin C in order to halt the division of the cells while they are still able to condition the media.

1. Add 250 μL mitomycin C to a 100-mm confluent EF dish containing 10 mL of media, and agitate back-and-forth and side-to-side.
2. Incubate at 37°C for 2 to 3 h.

3. Rinse the dishes 3 times with 10 mL of 1X PBS each wash, then replace the media with 15% DMEM.

The EF cells are now ready to be used as a feeder layer for ES cells. The content of one vial is thawed onto one Mitomycin C-treated EF plate, and the media is changed 24 h after plating. At the first subculture, it is important to freeze some cells to assure that the cell line will not be depleted.

Maintenance of ES cells

To subculture ES cells:

1. Gently remove media and rinse with 1X PBS.
2. Add exactly 2 mL 0.25% trypsin-EDTA to a 100-mm dish.
3. Return plate to incubator for 1 to 2 min until the cells float with gentle agitation. While waiting for trypsinization to be complete, prepare Fisherbrand 100-mm Petri dishes by adding 10 mL 15% DMEM and labeling them to keep track of the passage number. These will be used for the generation of EBs.
4. When the ES cells are floating, gently pipet up and down 30 times with a 1-mL pipetman in order to make a single-cell suspension.
5. Distribute 200 μL (approx 2.5×10^6 cells) to each prepared EF cells and Petri dish, making a 110 dilution of the original culture.
6. Gently agitate up-and-down and side-to-side and incubate at 37°C.

The ES cell media must be changed every 2 d, and the cultures must be split every 3 to 4 d. For simplicity, ES cells are split every Tuesday and Friday, thereby producing EB's on the same days. It is important that one does not keep ES cells in culture for long periods in order to maintain pluripotency. Extensive culturing will result in abnormal karyotypes and inconsistent differentiation. It is for this reason that we generally keep ES cells in culture up to the 6th passage before new cells are thawed. Undifferentiated ES cells will stain positive for alkaline phosphatase.

AP histochemistry

1. Rinse cells once with cold 1X PBS.
2. Fix for 15 min with cold formalin (on the bench).
3. Rinse with distilled water and incubate in fresh distilled water for 15 min.
4. Meanwhile, prepare the substrate (must be fresh). For 25 mL, combine 0.0025 g naphthol AS MX-PO_4, 100 μL N,N-dimethyl-

formamide, 12.5 mL 0.2 *M* Tris-HCl (pH 8.3), and 12.5 mL distilled H_2O. Add 0.015 g Red Violet LB Salt and filter with Whatman's no. 1 paper directly onto dishes.

5. Incubate for 45 min at room temperature, then rinse with dH_2O.

Differentiation of ES cells into the epithelial lineage

The three germ layers (ectoderm, mesoderm, and endoderm) are established during gastrulation. The ectoderm has the capacity to give rise to the epidermis and the nervous system. Specifically, the epidermis is derived from ectodermal precursor cells, which make epidermal cell fate selection as a result of inductive interactions during early embryogenesis. Following the fate selection, these cells give rise to putative multipotential epidermal stem cells capable of giving rise to the epidermal lineage. Within this lineage, one predicts, is contained progenitor cells with varying degrees of proliferative potential as well as committed–mature progenitors that are capable of further differentiation along the epidermal lineage. Based on this descriptive work, we envisioned that, in order to generate epidermal cells in culture, it is necessary to assay these three distinct stages to be successful in dissecting out the stages of the lineage. In accordance with this rational, our culture protocol consists of three distinct stages:

1. Ectodermal cell fate selection; this can be achieved during EB formation.
2. The generation and differentiation of putatitive epidermal stem cells; during the epithelial sheet formation from EBs on matrigel-coated substrata.
3. The fate of early progenitors; differentiating epidermal stem cells presumably give rise to progenitors. These progenitors may potentially be maintained and expanded in vitro under appropriate conditions. Our secondary (epidermal progenitor cells cultures [EPC]) support this prediction.

These stages were assayed and followed for distinct markers. For example, EBs are cell aggregates that express K8 on their surface. K8 expression is assessed through whole mount *in situ* hybridization with a DIG-labeled probe for K8. Further differentiation is accomplished by the production of *differentiated embryoid bodies* (dEBs), which are adhered EBs that have an emerging spread area of epithelial cells that are K14 positive as determined by indirect immunofluorescence with antibodies against K14 as well as AS staining. The culturing of secondary cultures from dEBs generates early epithelial cells called EPCs. These

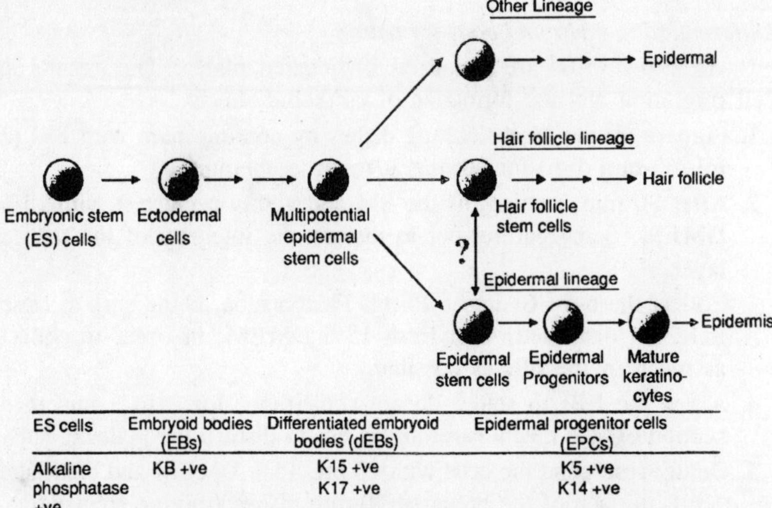

Fig. 4.1. A schematic representing the differentiation of committed epidermal cells from totipotent ES cells and the associated stage-specific markers.

cells progress along the epithelial differentiation pathway to express the markers of differentiation as determined by indirect immunofluorescence and RT-PCR.

Embryoid body (EB) formation

The media is changed on the third day (i.e., Monday for EBs produced on Friday, and Thursday for EBs produced on Monday).

To change the media:

1. Using a 10-mL pipet, collect the entire contents of the Petri dish into a 15-mL Falcon tube.
2. Allow the EBs to settle to the bottom. Do not centrifuge, the settling process takes only a few minutes.
3. Suction off the media and gently resuspend the cells in 10 mL 15% DMEM.
4. Transfer to the original Petri dish. After 6 d in culture, EBs are ready to proceed to the next stage.

This procedure reproducibly generates EBs with an ectodermal layer that can easily be identified by microscopy. The expression profile of K8, which is known to be associated with ectodermal cells during very early development, can be detected by whole mount *in situ* using a K8 riboprobe. Other markers of ectoderm can be used to further characterize differentiation at this stage.

Differentiating embryoid body formation

After 6 d, EBs are plated on BM-coated plates. This results in cell migration and the formation of epithelial sheets.

1. Prepare 60-mm tissue culture dishes by coating them with BM (2 mL/60-mm dish) for 30 min at room temperature.
2. After 30 min, suction off the BM and gently replace it with 15% DMEM, being careful not to disturb the integrity of the matrix layer.
3. Collect the 6-d EBs into a 15-mL Falcon tube, being sure to rinse the Petri dish well, with fresh 15% DMEM, in order to collect as much of the EBs as possible.
4. Allow the EBs to settle, do not centrifuge, for 2 to 3 min then suction off the media carefully as not to disturb the pelleted cells.
5. Gently resuspend the cells with 10 mL 15% DMEM and distribute 1 mL to each of the prepared 60-mm dishes (mixing frequently), thereby making a 1:10 dilution of the original culture. It is important to pipet gently up-and-down while plating the EBs, because they tend to sink quickly due to their mass.
6. Return the cells to the incubator and allow them to grow for 4 d, changing the media on the second day.

Once the cells attach and begin to spread on the plate, they are called dEBs. After a few days, dEBs have cells with epithelial morphology towards the edge of the spread area and, at this point, express keratin 8 (K8), keratin 17 (K17), and keratin 14 (K14) at different stages of the expanding dEB colony. The progression of differentiation is monitored by immunofluorescence, using epidermal-specific keratins. With the onset of K14 expression (approx 4–6 d), secondary cultures may be produced. Detection of the mature epidermal marker K14 is indicative of the progression of differentiation along the epidermal lineage. K8 is also highly expressed in dEBs, indicating that the cells at this stage are "early" cells in the lineage. In addition, the epithelial sheets that are derived from dEBs also contain other early progenitors that express K15, K17, and K19 at varying frequencies during the differentiation process. These progenitors can be further expanded and characterized in secondary cultures by trypsinization.

Epithelial progenitor cells

The proliferative and differentiative capacity of EPCs in vitro can be studied through the establishment of EPC cultures from dEBs by trypsinization. They are early epithelial cells that progress through differentiation along the epithelial pathway expressing K17, K14 and several other epithelial markers in a time- and density-dependent manner. These cultures are suitable for studying the role of growth factors and hormones to devise serum-free culture conditions. These studies are currently ongoing.

1. Prepare 35-mm tissue culture dishes by coating them with Matrigel (1 mL/35-mm dish, approx 0.1 mg) for 30 min at room temperature.
2. After 30 min, suction off the BM and gently replace it with 15% DMEM, being careful not to disturb the integrity of the matrix layer.
3. Trypsinize 4-d dEBs with 2 mL 0.25% trypsin-EDTA per 60-mm dish for 2 to 3 min in the incubator.
4. Collect the cells, washing the dishes well with fresh 15% DMEM, into a 15-mL Falcon tube, then remove 100 μL to count the cells.
5. Centrifuge at 700*g* for 2 to 3 min for pellet formation.
6. While spinning, count the cells using a Coulter Counter.
7. Resuspend the pellet with 15% DMEM and dilute the cells appropriately to plate the desired number of cells.
8. For immunofluorescence, we plate the cells on glass 22 × 22 mm coverslips in 35-mm dishes. EPCs may also be subcultured using a mouse keratinocyte co-culture, thereby enhancing their differentiation.

Mouse keratinocyte feeder layers for low density progenitor cultures

There are two different lines of MKCs that we use for co-culture with EPCs; they are MKC-5 and MKC-6. MKC-5 were isolated from the backskin of K14 knock-out mice, and MKC-6 were isolated from their normal counterpart. They are established cell lines that maintain their basal cell-like characteristics in vitro. Our initial studies suggested that since MKCs are mature keratinocytes, they are conditioning the media, and they are providing factors that are enhancing the survival and the differentiation of EPCs towards the epithelial lineage.

Thawing MKC cells

When thawing MKCs, the contents of one vial are transferred to one 60-mm dish as described above in this chapter. At the first subculture, extra dishes are seeded so that they may be frozen as not to deplete the cell line. Freezing with the cells from one 60-mm dish being distributed to three cryo vials.

Subculturing MKCs

1. Dishes are coated with 100 μL collagen-I solution in 2 mL of 1X PBS overnight in the 37°C incubator.
2. Before use, the dishes are washed 3 times with 1X PBS (2 mL wash), then 4 mL KI2 media is added, and the dishes are appropriately labeled.
3. Trypsinize a 60-mm dish with 2 mL 0.25% trypsin-EDTA for 2 to 3 min in the 37°C incubator until the cells detach with gentle agitation.
4. Collect the cells with 3 mL KI2 into a 15-mL Falcon tube and centrifuge at 700*g* for 2 to 3 min for pellet formation.
5. Carefully suction off the media and resuspend the pellet in 5 mL K12.
6. Gently pipet up and down the entire volume to mix, and then distribute 1 mL to each of the pre-prepared plates, thereby making a 15 dilution of the original MKC-5 plate (approx 2×10^5 cells).

The plates are grown to confluency in about 3 to 4 d, and the media is changed every other day. The MKC-5 cells are subcultured twice a week, on Monday and Friday. At the point of subculturing, extra plates can be set up in order to use them as a feeder layer for EPCs. MKCs are mitomycin C-treated as previously described, using 50 μL for a 35-mm dish containing 2 mL of media.

Analysis of Differentiation

Three distinct differentiation stages, which were described previously, can be analyzed using the techniques that we describe in this section.

In situ hybridization on EBs

This is a powerful technique that allows one to determine the spatial distribution of specific RNA within EBs without disrupting the morphology of the EBs. We have used this procedure to show the spatial distribution of K8 in the developing EB. This process is bone morphogenetic protein (BMP)-dependent, and in the presence of the BMP antagonist noggin, the expression of K8 and Scullin/Claud-6 are inhibited.

In addition, EBs are very suitable for the screening of newly identified genes for their role and association with ectoderm. It is our goal that a battery of new ectodermal genes will be identified.

Preparing a DIG-labeled ribo-probe using the in vitro transcription method

1. Subclone a fragment of the gene of interest into a Bluescript plasmid that has a phage T7 and T3 promoter.
2. Linearize about 20 μg of the plasmid DNA with an appropriate restriction enzyme to produce a 5' overhang at the far end of the fragment from the promoter and run an agarose gel to check for the absolute completion of the digestion.
3. Extract the linearized plasmid using the Qiagen gel extraction kit.
4. For a standard 20 μL of labeling reaction, add the following to a 1.5-mL microtube on ice:
   13 μL DEPC-water
   1 μL linearized plasmid DNA (approx 1 μg)
   2 μL DIG RNA labeling mixture
   2 μL 10X transcription buffer
   2 μL RNA polymerase (T7/T3)
5. Mix and pulse the tube and incubate at 37°C for 2 to 3 h.
6. Add 2 μL of 0.2 *M* EDTA to stop the reaction, and add 30 μL DEPC-water to make the total volume 50 μL.
7. Use a Sephadex G-50 spin column to eliminate the unincorporated NTPs according to the manufacturer's directions:
   (a) Vortex mix the column to resuspend the resin.
   (b) Loosen the cap and snap off the bottom closure.
   (c) Place the column in a 1.5-mL microfuge tube and pre-spin at 1500*g* for 1 min.
   (d) Place the column in a fresh 1.5-mL microfuge tube and add the 50 μL sample to the top-center of the resin bed, being careful not to disturb the bed.
   (e) Spin at 1500*g* for 2 min. This is the purified probe.
8. Estimate the yield of probe and dilute it to 100 μL, then store at -80°C. To estimate the probe concentration, load about 3 μL onto an agarose gel with 1 μL of linearized plasmid as a control and 1 μL DIG-labeled control RNA for yield estimation. Load DNA ladder to estimate the probe size. The 1.6-kb band on the ladder has a concentration of 10 ng/mL.

Whole mount in situ procedure

Day One:

The following steps are performed on ice.

1. Collect EBs and fix with 10 mL 4% paraformaldehyde (4°C) in PBS on ice for 30 min, then change with fresh fixative for 3 h (or overnight), mixing occasionally.
2. Wash twice with PBS-T on ice. Routinely, we use a volume of 0.5 to 1 mL for all washes.
3. Wash with freshly made 25, 50, and 75% MeOH/PBS-T, then twice with 100% MeOH. Dehydrated samples may be stored at −20°C in 100% methanol for 1 mo.

Day Two:

The following are performed on the bench.

1. Take an aliquot of sample (about 100 μL), rehydrate by taking through a MeOH/PBS-T wash series in reverse (100 to 25%) and washing 3 times with PBS-T. Be sure to include a sample for antibody absorption.
2. Digest samples with 500 μL of 10 μg/mL proteinase K at room temperature for 15 min.
3. Wash with 2 mg/mL glycine in PBS-T to stop proteinase K digestion.
4. Wash twice with PBS-T.
5. Refix with fresh 0.2% glutaraldehyde/4% paraformaldehyde in PBS-T for 20 min.
6. Wash 3 times with PBS-T. At this step, one sample will be used for anti-DIG-AP antibody absorption. One absorption reaction is enough for 5 samples in the actual experiment.
7. For antibody absorption, add the following to the sample and incubate overnight at 4°C:

   878 μL PBS-T
   100 μL 10 mg/mL BSA
   20 μL calf serum (or normal sheep serum)
   12 μL anti-DIG-AP
8. Wash the rest of the samples once with 50% hybridization buffer/ 50% PBS-T.
9. Wash once with 100% hybridization buffer at 50–55°C.
10. Prehybridize samples with hybridization buffer (500 μL) for 30 min in the water bath at 55°C, mixing occasionally.
11. Hybridize samples with hybridization buffer containing approx 50 ng/mL of probe. To dilute the probe, 20 μL of the probe reaction was diluted to a final vol of 100 μL. Add 5 μL of diluted probe

into 500 μL of hybridization buffer. Boil it for 5 min. Cool it down to the hybridization temperature and add it to the sample. Hybridize samples (300 μL) in a 55°C water bath overnight.

Day Three:

1. Carefully take out the hybridization buffer containing the probe and store it at -20°C.
2. Wash with 100% hybridization buffer at 60°C for 30 min.
3. Wash with 50% hybridization buffer/50% PBS-T at 60°C for 30 min.
4. Wash with PBS-T at 60°C for 30 min.
5. Wash twice with 2X SSC, 0.1% Tween-20, 15 min each at 60°C.
6. Digest samples in 2X SSC, 0.1% Tween-20, containing 20 μg/mL RNase A at 37°C for 30 min.
7. Wash twice with PBS-T at room temperature, 5 min each.
8. Block for 30 min in PBS-T containing 5% calf serum at room temperature.
9. Incubate samples with absorbed anti-DIG antibody overnight at 4°C. To prepare absorption, spin the absorbed sample for 1 min, take all the supernatant and add it to 3 mL of PBS-T and use it as absorbed anti-DIG-AP antibody. This will eliminate any nonspecific antibody binding.

Day Four:

1. Wash at least 5 times with PBS-T.
2. Wash twice with AP buffer.
3. Stain samples with NBT plus BCIP in AP buffer for 30 min to 2 h or until color development is completed. To stain the samples, place the sample tube on a rack wrapped with foil. Let it sit for about 30 min on a shaker, then check the color development (staining will be blue). Check the color development regularly every 10–15 min until the color turns very blue (up to 2 h or overnight). Be careful not to stain too long as the background increases (a dirty brown color).
4. Stop the reaction by washing twice with PBS-T.
5. Store samples in 100–200 μL of 50% glycerol/PBS-T (containing 1 to 2 μL sodium azide) at 4°C. Samples may be stored without losing staining intensity for extended periods of time at 4°C.
6. Mount the sample in glycerol and photograph.

Indirect immunofluorescence for differentiating cells on glass coverslips

1. Cultures are set up and maintained on coverslips.
2. Fix and permeabilize the cells with -20°C methanol for 10 min at -20°C.
3. Rinse the coverslips 3 to 5 times with 1X PBS to remove the fixative.
4. Place coverslips in a humidified chamber and apply appropriately diluted primary antibody. Incubate for 30 min at room temperature.
5. Rinse each coverslip 3 times in 1X PBS to remove unbound primary antibody.
6. Apply secondary antibody to the coverslips. Again, make sure the cells do not dry out before the antibody is applied. Incubate for 30 min at room temperature.
7. Rinse each coverslip 3 times in 1X PBS.
8. Place Mowiol 4-88 (approx 20 μL for a 12-mm coverslip and 80 μL for a 22 × 22 mm coverslip) on each mounting slide, and carefully place the inverted coverslip on the slide.
9. Observe and photograph.

Ayoub shklar staining for keratin

We routinely use *Ayoub Shklar* (AS) staining as a quick and reliable histological marker to determine the progression of epithelial differentiation of our cultures. Undifferentiated epidermal cells are blue, differentiated epidermal cells are red, and further differentiation is indicated with a color change from orange to yellow to brown. Although the exact stages and extent of differentiation cannot be assessed via this method, AS staining is a good indication of epithelialization of EPC cultures as a function of time.

Staining procedure

1. Aspirate off the medium and rinse the dishes 3 times with 1X PBS.
2. Fix the plates with cold 10% formalin for 30 min at room temperature.
3. Rinse dishes 3 times with 1X PBS.
4. Stain for 3 min with filtered acid fuchsin solution.
5. Aspirate off excess stain, but do not let the dishes dry out.
6. Stain for 45 min with aniline blue–orange G solution.
7. Wash several times with 95% ethanol.

8. Allow the plates to dry inverted. View and photograph cultures in 95% ethanol. Stained plates may be stored indefinitely.

RT-PCR

Trizol RNA extraction

It is crucial that, during this extraction procedure, RNase-free instruments and solutions are used and that gloves are often changed, in order to minimize the risk of RNase contamination.

The following procedure has been derived from the manufacturer's:

1. Lyse cells directly in the tissue culture dish by adding the appropriate amount of Trizol (i.e., for a 35-mm dish use 1 mL, for a 60-mm dish use 3 mL, and for a 100-mm dish use 6 mL).
2. Remove the cells with a cell scraper, then pipet the solution several times to collect all the cells, and transfer the volume to a sterile culture tube. Incubate the samples for 5 min at room temperature.
3. Add 0.2 mL of chloroform (RNase-free) per 1 mL of Trizol used (i.e., for a 35-mm dish use 200 μL, for a 60-mm dish use 600 μL, and for a 100-mm dish use 1.2 mL). Cap the tubes securely and shake vigorously for 15 s, then incubate at room temperature for 3 min.
4. Centrifuge at 12,000*g* for 15 min at 4°C. The mixture will separate into a lower phenol:chloroform phase, an interface, and a colorless upper aqueous phase. The RNA is in the aqueous phase and is about 60% of the original amount of Trizol added.
5. Transfer the aqueous phase into a fresh tube. Precipitate the RNA by adding 0.5 mL RNase-free isopropanol per 1 mL, Trizol originally added (i.e., for a 35-mm dish use 0.5 mL, for a 60-mm dish use 1.5 mL, and for a 100-mm dish use 3 mL). Incubate the samples for 10 min at room temperature.
6. Centrifuge at 12,000*g* for 10 min at 4°C. The RNA forms a pellet on the bottom and the side of the tube.
7. Remove the supernatant and wash the pellet with RNase-free 75% ethanol. Use 1 mL of ethanol for each 1 mL of Trizol used originally (i.e., for a 35-mm dish use 1 mL, for a 60-mm dish use 3 mL, and for a 100-mm dish use 6 mL). Vortex mix the sample and centrifuge at 7500*g* for 5 min at 4°C.
8. Air-dry the pellet and resuspend it in DEPC-H_2O. If it does not dissolve, try incubating for 10 min at 55–60°C.
9. Find the OD at A_{260} and A_{280}. The $A_{260}A_{280}$ should be between 1.6–2.

10. Determine the concentration of the sample using the following equation:

$$[\,]\ \mu g/\mu L = [A_{260} \times 40 \times \text{dilution factor}] \div 1000$$

To remove genomic DNA from RNA samples

1. Digest 30–100 μg RNA so that RNA plus DEPC-H_2O = 87 μL
   Add: 10 μL 10X DNase buffer
   2 μL RNase Inhibitor
   1 μL DNase I
2. Mix and incubate for 15 min at room temperature.
3. Prepare chloroformisoamyl alcohol (24:1): in a –50-mL Falcon tube put 12 mL RNase-free chloroform plus 0.5 mL RNase-free isoamyl alcohol.
4. Add an equal volume of phenol:chloroform to RNA sample, i.e., for a 100-μL sample, add 50 μL phenol and 50 μL chloroform: isoamyl alcohol (24:1).
5. Mix vigorously by inversion until an emulsion forms and spin at maximum speed at room temperature for 5 min.
6. Transfer the aqueous phase (the top phase) to a fresh RNase-free microtube.
7. Ethanol precipitate by adding a 1:10 dilution of RNase-free 3 *M* NaAc, i.e., for a 100-μL sample, add 10 μL 3 *M* NaAc.
8. Add 2X volume 100% EtOH (RNase-free, –20°C), i.e., for a 100-μL sample, add 200 μL 100% EtOH.
9. Mix vigorously and incubate tube at –80°C for 1 h.
10. Spin immediately at maximum speed for 20 min at 4°C.
11. Wash the pellet with 200 μL 70% EtOH (RNase-free, –20°C).
12. Spin again for 10 min at 4°C.
13. Air-dry the pellet (approx 15 min).
14. Resuspend the pellet in DEPC-H_2O.
15. Quantify the DNased RNA (dilute 1 μl in 500 μL DEPC-H_2O) using the following equation:

$$[\,]\ \mu g/\mu L = [A_{260} \times 40 \times 500] \div 1000$$

16. Dilute samples to 1 μg/μL with DEPC-H_2O in order to be used for RT-PCR.

Denature RNA

1. Mix 1 μL RNA with 2 μL DEPC-H_2O in a 100-μL PCR tube. Vortex mix and pulse.

2. Place tubes in a Perkin Elmer thermal cycler and run the RNA at 65°C for 5 min, then hold the temperature at 4°C. The reaction must hold at 4°C for at least 5 min before proceeding to the RT reaction.

Reverse transcription

1. Prepare RT mixture:
   4.0 μL 25 m*M* $MgCl_2$
   2.0 μL 10X PCR buffer II
   2.0 μL 10 m*M* dNTP
   1.0 μL RNase inhibitor
   1.0 μL MuLV reverse transcriptase
   1.0 μL Random hexamers
2. Add 17 μL of the RT mixture to the RNA and mix.
3. Place the tubes in the thermal cycler and run at 42°C for 15 min, 99°C for 5 min and then hold at 4°C. The reaction must hold at 4°C for at least 5 min before proceeding to PCR.

PCR

1. Prepare the PCR mixture (50 μL final volume):
   0.5 μL Template DNA (approx 10 ng)
   5.0 μL 2.5 m*M* dNTPs
   1.0 μL 10X polyelectrolyte complex (PEC) buffer
   1.0 μL 3' Primer (25 pmol/μL)
   1.0 μL 5' Primer (25 pmol/mL)
   0.5 μL *Taq* DNA polymerase
   37.0 μL dH_2O
2. Run samples as per the following schematic:

Various primers for transcription factors/signaling molecules involved in the formation of skin and hair follicles (LEF-1, SHH, and

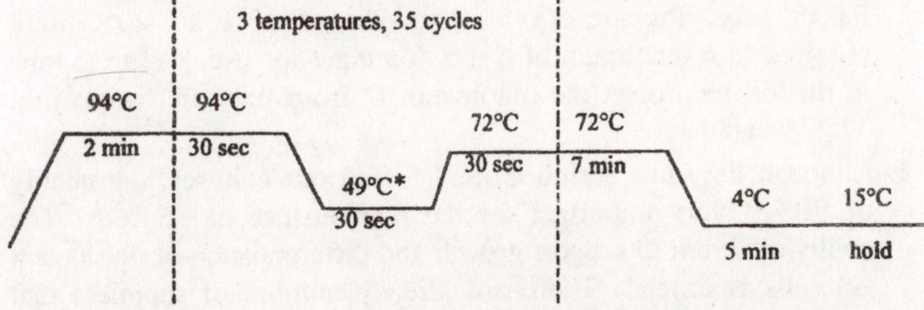

Fig. 4.2. The temperature of the extension step will depend upon the primers being used.

MSX-1) were used. The results indicate that our EPC cultures indeed follow in these lineages. In addition, primers for the endodermal marker AFP were used to prove that our EPC cultures were truly of epithelial and not endodermal lineage. Primers for markers of epithelial terminal differentiation (MK1, involucrin, and loricrin) were used to show that our EPC cultures undergo this process.

Notes

1. 7X-OMATIC is a special soap that we use for all washable tissue culture items. It is important that there is no soap residue on glassware in order to eliminate its potential effects on cultures.
2. This media is high in glucose and contains 4500 mg/L D-glucose and pyridoxine hydrochloride, but no L-glutamine or sodium pyruvate.
3. Serum is heat-inactivated in the following manner: thaw the bottle of serum overnight at 4°C, warm in a 57°C water bath for 7 min with constant mixing, then continue to incubate serum for 30 min, agitating every 10 min. The serum is then aliquoted into sterile 100-mL bottles in the tissue culture hood and allowed to cool before tightening the bottle and freezing. It is important to cool the bottles in order to avoid breakage. Serum may be stored at -20°C for 6–12 mo.
4. Upon receipt of BM, the 10-mL bottle is thawed in a beaker of distilled water, and 1 mL is aliquoted into sterile tubes with a chilled pipet. The 1-mL aliquots are then stored at –20°C until required. When needed, thaw a tube in a beaker of distilled water and add 9 mL of unsupplemented DMEM. At this stage, it may be stored at 4°C. Before use, the diluted BM is further diluted another 1:10 with unsupplemented DMEM to a final working concentration of 0.1 mg/mL.
5. Prepare mitomycin C in the tissue culture hood. Add 5 mL of 1× PBS to a 2-mg vial, recap, and shake vigorously. Empty the vial into a clean weigh boat and syringe filter into a sterile 15-mL Falcon tube. Prepare 500-μL aliquots and store at –20°C until required to a maximum of 6 mo. To thaw for use, wrap the tube in tin foil to protect the mitomycin C from light and warm in a 37°C water bath.
6. Although the same serum is used for all our cultures, the quality of FBS is very important for the maintenance of ES cells. The ability of serum to support growth and differentiation of pluripotent ES cells is crucial. There are already number of suppliers that sell sera that are already tested for supporting ES cell growth.

We use "*ES cell tested*" serum from Hyclone and have found that it is very reliable from batch to batch. Suitable sera batches should then be ordered in large quantities and then bottles may be stored at –20°C for up to 2 yr.

7. To prepare a humidified chamber for immunofluorescence, place a wet piece of filter paper in a 15-cm glass dish with a lid.
8. Usually when freezing cells in the exponential phase of growth, 5 cryo vials can be prepared from each 100-mm dish, and 3 cryo vials may be prepared from each 60-mm dish.
9. To protect oneself from the potential hazard of an exploding cryotube, a face shield and protective gloves should be worn. Prepare an ice bucket (with a lid) with a plastic beaker containing 37°C H_2O. Quickly remove the vial from the liquid nitrogen and place it in the water in the ice bucket and close the lid. Using a pair of forceps, gently invert the vial as to assure even thawing of the cells.
10. We have found that Fisherbrand Petri dishes (made in Canada) are the best dishes for the generation of EBs. In our experience, EBs formed on other plastics adhere to the plate or they aggregate. Fisherbrand Petri dishes reproducibly allow for the generation of EBs that never stick or aggregate.
11. This probe can be used 3 times. Therefore, after each hybridization, carefully take out the hybridization buffer and store it at –20°C. This probe does not need to be boiled again. Just warm it up to the hybridization temperature and add it to the sample right away.
12. Approximately 20–25 μL is required for each 12-mm coverslip and 100 μL for a 22 × 22 mm coverslip. Make sure the coverslips do not dry out before the antibody is applied.

5

Hepatic Stem Cells

The adult mammalian liver contains many different cell types of various embryological origins. Nevertheless, the term liver or hepatic stem cells is used for precursors of the two epithelial liver cell types, the hepatocytes and the bile duct epithelial cells. This terminology also applies to this chapter where only hepatocyte and the bile duct stem cells are discussed.

Organization and Function of Adult Mammalian Liver

Anatomy

The liver is a large parenchymal organ consisting of several separate lobes and representing about 2% of the body weight in the human and 5% in the mouse. It is the only organ with two separate afferent blood supplies. The hepatic artery provides oxygenated blood, and the portal vein brings in venous blood rich in nutrients and hormones from the splanchnic bed (*intestines* and *pancreas*). Venous drainage is into the vena cava. The bile secreted by hepatocytes is collected in an arborized collecting system, the biliary tree, which drains into the duodenum. The gall bladder is part of the distal biliary tree and acts to store bile. The hepatic artery, portal vein, and common bile duct enter the liver in the same location, the porta hepatis.

The main cell types resident in the liver are hepatocytes, bile duct epithelium, stellate cells (formerly called *Ito cells*), Kupffer cells, vascular endothelium, fibroblasts, and leukocytes. Although hepatocytes are responsible for most organismal liver function and represent about 90% of the weight of the liver, they are large cells and only ~60% of total liver DNA is hepatocyte-derived. An adult mouse liver contains about 5×10^7, and an adult human liver about 80×10^9, hepatocytes. Knowledge of the microscopic structure of the liver is essential for understanding hepatic stem cell biology, and two main models for its organization have been proposed. According to one model, the hepatic lobule is the functional unit of the liver. The portal triad consisting of a small portal vein, hepatic artery branch, and bile duct is located on the perimeter. Arterial and portal venous blood enter here, mix, and flow past the hepatocytes toward the central vein in the middle of the lobule. The second model considers liver acini the basic units, with each acinus having the portal triad at the center and the "central" veins at the periphery.

We base our discussions on the lobule model. In both models, liver sinusoids are the vasculature connecting the portal triad vessels

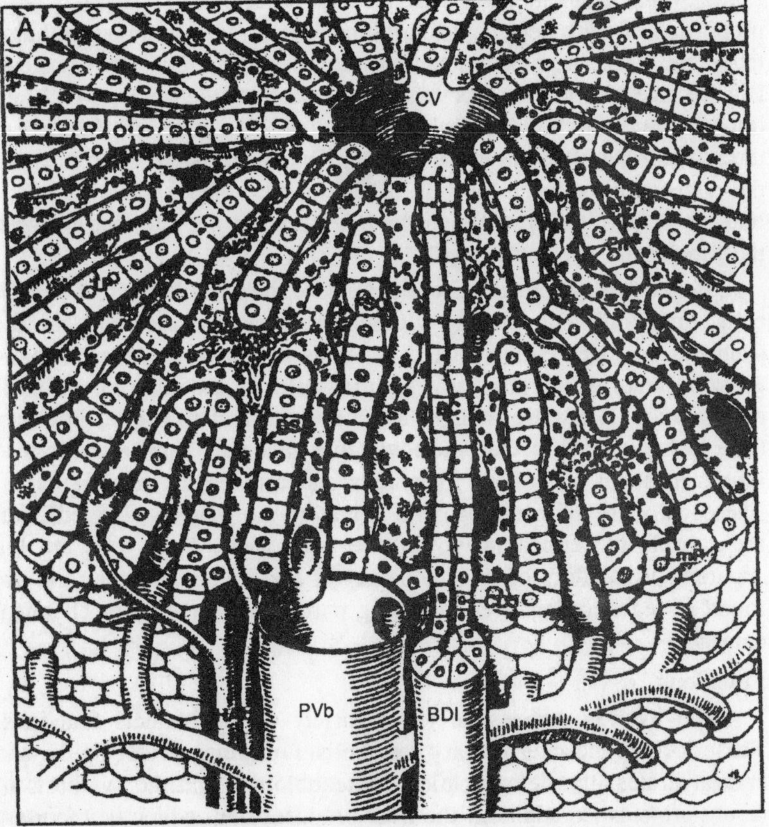

Fig. 5.1. Diagram of vascular supply and sinusoidal structure of the liver lobule.

and the central vein. Unlike other capillary beds, sinusoidal vessels have a fenestrated endothelium, thus permitting direct contact between blood and the hepatocyte cell surface. In two-dimensional images, rows of hepatocytes oriented from portal to central form a hepatic plate. A channel formed by adjacent hepatocytes forms a bile canaliculus that serves to drain secreted bile toward the bile duct in the portal triad. Hepatocytes are large, cuboidal epithelial cells with a basal and apical surface, and the apical surface is also called the *canalicular surface*. Hepatocytes exchange metabolites with the blood on the basal surface and secrete bile at the canalicular surface. Zone-1, -2, and -3 hepatocytes are distinguished on the basis of their relative position within the lobule. Zone-1 hepatocytes are close to the portal triad, zone-2 cells are in the middle, and zone 3 consists of cells directly adjacent to the central vein. Although somewhat variable according to species, a large proportion of adult hepatocytes are binucleated, with some nuclei being tetraploid. Thus, hepatocytes can have 2n, 4n, or 8n total DNA content. Bile secreted by the hepatocytes is collected in bile ducts, which are lined by duct epithelial cells. The smallest bile ducts are located in the portal zone of each hepatic lobule. The canal of Hering represents the connection between the bile canaliculi (the inter-hepatocyte space into which bile is secreted) and the bile ducts, at the interface between the lobule and the portal triad.

Stellate cells represent about 5–10% of the total number of hepatic cells. In addition to storing vitamin A, they are essential for the synthesis of extracellular matrix proteins and produce many hepatic growth factors that play an essential role in the biology of liver regeneration. They are thought to be mesodermal in origin, but very little is known about their turnover and renewal. Kupffer cells also represent about 5% of all liver cells and are resident macrophages. These cells are of hematopoietic origin (bone marrow-derived), but are capable of replicating within the liver itself. Oval cells are the apparent progenitors of liver hepatocytes and epithelial cells, and are found in regenerating liver following partial hepatectomy or chemical damage.

Functions

The liver is responsible for a variety of biochemical functions. These include the intermediary metabolism of amino acids, lipids, and carbohydrates; the detoxification of xenobiotics; and the synthesis of serum proteins. In addition, the liver produces bile, which is important for intestinal absorption of nutrients, as well as the elimination of cholesterol and copper. All of these functions are primarily executed by hepatocytes. The biochemical properties and pattern of gene expression are not uniform among all hepatocytes. The term "*metabolic zonation*" has been coined to indicate the different properties of zone-1, zone-2, and zone-3 hepatocytes. For example, only zone-3 hepatocytes express glutamine synthetase and utilize ammonia to generate glutamine. In contrast, zone-1 and zone-2 hepatocytes express urea cycle enzymes and convert ammonia to urea. Similarly, glycogen synthesis and glycolysis are segregated within the hepatic lobule.

Embryology

In the mouse, the liver develops from ventral foregut endoderm beginning at day 8 of gestation. The first evidence of hepatic differentiation is the induction of albumin and α-fetoprotein mRNA in endodermal cells, even prior to their morphological differentiation. Between days 8.5 and 9.5, the hepatocyte precursors proliferate, and beginning at day 9.5, migrate toward cardiac mesoderm in the septum transversum. Signals from the cardiac mesoderm induce the cells to increase their levels of albumin and α-fetoprotein mRNAs and to form the liver bud. Very recently some of the specific signals produced by the mesenchyme have been identified. Fibroblast growth factors (FGFs) 1, 2, and 8 were sufficient to induce the liver gene expression program in isolated murine foregut endoderm. FGF receptors 1 and 4 are expressed on foregut endoderm cells and are essential for this induction.

At day 10.5, the vascularization of the liver bud begins, followed by a large increase in liver mass. The early cells in the liver bud are positive for both albumin and α-fetoprotein. The more differentiated phenotypes of hepatocytes and bile duct epithelium emerge in mid-gestation. Definitive lineage studies have not yet been reported, but it is generally thought that bile ducts and hepatocytes emerge from common precursors, termed *hepatoblasts*. It is not known whether there is only one type of hepatoblast or whether there is a hierarchy of lineage progression consisting of primitive hepatoblasts and more committed bipotential progenitors. Fetal hepatoblasts can be considered as the equivalent to a fetal liver stem cell.

Several genes that are important for the development of the liver have been identified. Many of the relevant studies have involved mouse knock-out models. The first genes of importance are the transcription factors HNF3β and GATA-4. These genes are essential for the specification of endoderm and also play a role in the development of other epithelial tissues such as the lung and the pancreas. Presently,

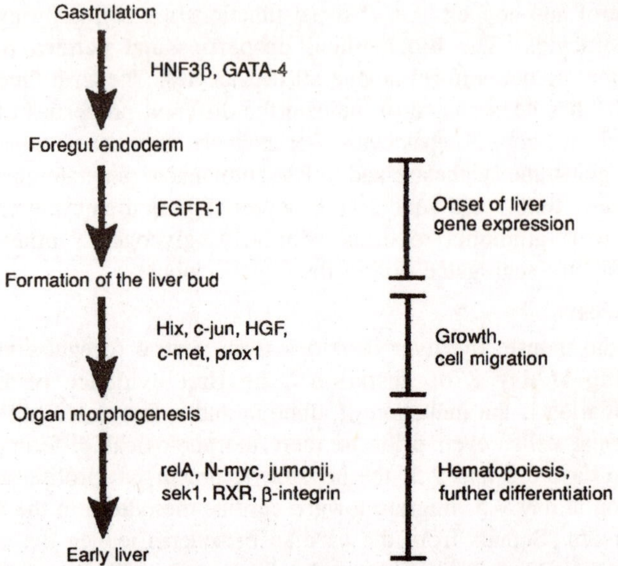

Fig. 5.2. Genes involved in early liver development.

the factor(s) which is responsible for the onset of liver gene expression before formation of the liver bud is not known. This factor would be of obvious interest in the utilization of liver stem cells. In contrast, several factors involved in the formation of the liver bud are already known. c-jun, HGF, and c-met have all been shown to be essential for liver bud formation. Other factors such as β1-integrin act later in the development of the liver.

Overall, our understanding of liver development at the molecular level is still in its early stages. It is likely that many of the yet-to-be discovered mechanisms involved in liver specification during embryogenesis will also apply to progenitor-dependent liver regeneration and liver stem cell biology.

Markers of Hepatic Lineages

Most of the studies pertaining to liver stem cell biology make extensive use of specific markers for different cellular phenotypes, including differentiated hepatocytes, biliary duct epithelium, hepatoblasts, and oval cells. Some markers are based on enzyme histochemical methods, but most rely on the expression of cell-specific antigens, detectable with antibodies. Some markers are expressed only in hepatocytes, some only in bile duct epithelium, some only in oval cells, and some in combinations of these. The best antibodies have been developed against rat marker proteins, and these have been used to map the sequence of antigen expression in liver development and carcinogenesis. Not all markers are expressed similarly in all species, and some monoclonal antibodies do not cross-react. For example, α-fetoprotein is not expressed at high levels in mouse oval cells.

Unfortunately, most of the monoclonal antibodies that have been developed are not against antigens expressed on the cell surface. Hence, *fluorescence-activated cell sorting* (FACS) has not been possible for phenotypic analysis of liver subpopulations.

Hepatocyte-enriched Transcription Factors

Four families of evolutionarily conserved transcription factors are involved in hepatocyte-specific gene expression. These factors are enriched in liver but are not completely restricted to this organ. Because these factors are involved in setting up a hepatocyte program, they are also of interest in liver stem cell biology. Each family is composed of several members displaying similar DNA recognition properties. Current studies suggest a model for liver differentiation based on a regulatory network rather than a genetic hierarchy. The four transcription factor families are (1) the variant homeodomain-containing family of *hepatocyte nuclear factor* 1 (HNF1) proteins, (2) the HNF3 winged helix proteins, (3) the nuclear receptor superfamily, and (4) the leucine zipper C/EBP family.

Two different members belong to the HNF1 homeodomain family: HNF1 and variant HNF1 (vHNF1). Target genes include albumin, α-fetoprotein, and phenylalanine hydroxylase. In adult animals, HNF1 and vHNF1 are present in liver and kidney and intestinal organs. However, vHNF1 is primarily expressed in kidney, whereas HNF1 is abundant in the liver. Both genes are also expressed in a variety of epithelial tissues during embryogenesis. During mammalian development, vHNF1 expression occurs earlier than that of HNF1. vHNF1 can already be found at day 5 of gestation in the mouse in visceral endoderm. It is also expressed at the onset of liver development when ventral foregut endoderm cells proliferate to form the liver primordium. HNF1 is first found in the yolk sac at 8 days but is expressed strongly only at later stages in more differentiated cells of the developing liver. These observations suggest that the vHNF1 may be involved in morphogenesis of organs such as the liver and kidney, whereas HNF1 may be involved in maintaining the differentiated state. Consistent with this hypothesis, HNF1 knockout mice have intact hepatic organogenesis but defects in expression of some liver-specific genes.

The HNF3 winged helix family contains three members, HNF3α, HNF3β, and HNF3γ. The DNA-binding motif of these putative transcription factors has a high degree of similarity to a region of the *Drosophila* gene fork head, which is involved in organogenesis in the fly. The HNF3 family members are expressed during the development of the endoderm as well as in cells of the notochord and ventral neural epithelium. During embryogenesis, HNF3β comes on first, followed by HNF3α and finally HNF3γ. The three genes show different anterior boundaries but an identical posterior boundary (the hindgut) in the developing endoderm. This suggests that they are involved in the regionalization of the definitive endoderm. All three family members are expressed in a variety of epithelial adult organs, including the liver and the intestine. Targeted gene knockouts of some have shed some light on their specific functions. HNF3β knockout mice die at day 10 of embryogenesis because of notochord defects. Recently, a novel transcription factor HNF6 has been found to regulate expression of HNF3β. HNF6 belongs to the cut homeodomain family of transcription factors. HNF6 knockout mice lack a gall bladder and have developmental abnormalities of the intrahepatic bile ducts.

Several orphan nuclear receptors have been identified as hepatocyte transcription factors. These include HNF4, CoupF1, and Arp-1. HNF4 was first identified by its interaction with liver-specific promoters, and its ligand is unknown. HNF4 is involved with diverse metabolic functions, including gluconeogenesis, cholesterol metabolism, and amino acid metabolism. In the adult, HNF4 is expressed at high levels in liver, kidney, and intestine. During mouse development, it is expressed in primitive endoderm at 4.5 days and then becomes restricted to visceral endoderm at day 5.5. GATA6 is upstream of HNF4 and regulates its expression. In embryonic development, HNF4 is essential, as shown by the fact that HNF4 knockout mice die at day 6 of gestation with an endodermal defect. However, the embryos can be rescued later in gestation and thus show that HNF4 is essential for hepatocyte differentiation. CoupTF1 and Arp-1 are negative regulators of hepatocyte gene expression.

The CCAAT enhancer-binding protein family (C/EBP) has four known members; C/EBPα, C/EBPβ, C/EBPδ, and C/EBPγ. These share a highly conserved terminal bipartite domain defined as a basic leucine zipper (bZIP). C/EBPα is expressed predominantly in hepatocytes, intestinal epithelial cells, and fat cells. C/EBPα knockout mice have an interesting phenotype that includes some hyperproliferation of hepatocytes and an absence of brown fat. Like C/EBPα, both C/EBPβ and C/EBPδ are also required for adipocyte differentiation, but all of these factors are also found at high levels in the liver. C/EBPβ has been implicated in regulating genes of the acute-phase response and inflammation.

The PAR subfamily of leucine zipper transcription factors is related to the C/EBP proteins. Three members are currently known, including DBP, *hepatic leukemia factor* (HLF), and TEF. Whereas the C/EBP factors have a more relaxed specificity and bind to PAR recognition sites, PAR proteins are more selective in their recognition and bind only to a subset of C/EBP sites. The PAR subfamily member DBP has the interesting property of being expressed in circadian rhythm-dependent fashion.

Liver Stem Cells

The ancient Greek legend of Prometheus illustrates that the phenomenon of liver regeneration has been known since antiquity. Animals (including humans) can survive surgical removal of up to 75% of the total liver mass. The original number of cells is restored within 1 week and the original tissue mass within 2–3 weeks. This process can occur repeatedly, indicating a very high organ regenerative capacity, which is in contrast to most other parenchymal organs, such as kidney or pancreas. Importantly, liver size is also controlled by prevention of organ overgrowth. Hepatic overgrowth can be induced by a variety of compounds such as HGF or peroxisome proliferators, but the liver size returns to normal very rapidly after removal of the growth-inducing signal. The role of liver stem cells in regeneration has been controversial, but many of the disagreements can be reconciled by considering the different experimental conditions that have been used to study the process. Liver stem cells can be defined in several different ways which are: (1) cells responsible for normal tissue turnover; (2) cells that give rise to regeneration after partial hepatectomy; (3) cells responsible for progenitor-dependent regeneration; (4) transplantable liver repopulating cells; and (5) cells that result in hepatocyte and BDE phenotypes in vitro. Current evidence strongly suggests that different cell types and mechanisms are responsible for organ reconstitution, depending on the type of liver injury. In addition, tissue replacement by endogenous cells (i.e., regeneration) must be distinguished from reconstitution by transplanted donor cells (i.e., repopulation). In the following, we discuss the role of stem or progenitor cells for each of these operant definitions.

Liver Regeneration

Three separate mechanisms for liver regeneration are considered: (1) normal tissue turnover, (2) hepatocyte-driven regeneration after liver injury, and (3) progenitor-dependent regeneration after liver injury.

Liver Regeneration during Normal Tissue Turnover

The average life span of adult mammalian hepatocytes has been estimated to be ~200–300 days. The mechanism by which these cells are replaced has been of interest for some time. One of the main models regarding normal liver turnover was termed the "*streaming liver*". According to this model, normal liver turnover is similar to regeneration in the intestine, with young hepatocytes being born in the portal zone and then migrating toward the central vein. The different patterns of gene expression in zone-1, -2, and -3 hepatocytes were explained by the aging process during this migration and thus represented a typical lineage progression. It has also been noted that the ploidy and size of hepatocytes depend on their location within the lobule. Central (zone-3) hepatocytes tend to be larger and more polyploid than their periportal counterparts. However, recent work has provided strong evidence against the streaming liver hypothesis. First, it was shown in elegant studies that the gene expression pattern in hepatocytes was dependent on the direction of blood flow.

If blood flow was reversed such that portal blood entered the lobule through the central vein and exited via the portal vein, the pattern inverted. Therefore, the lobular zonation is best explained by metabolite-induced gene regulation, not lineage progression. Second, retroviral marking studies provide clear evidence against any hepatocyte migration during normal turnover. Retrovirally marked hepatocytes formed small clones that remained largely coherent and were equally distributed in zones 1, 2, and 3.

These results have been confirmed in elegant studies utilizing the mosaic pattern of X-inactivation in female mice to analyze patterns of hepatocyte growth. Thus, current evidence strongly suggests that normal liver turnover in adult animals is mediated primarily by in situ cell division of hepatocytes themselves and not stem cells.

Regeneration after Partial Hepatectomy (Hepatocyte-driven Injury Response)

Liver regeneration after a 66% partial hepatectomy is relatively well understood in terms of molecular regulation and has been the subject of several excellent reviews. During partial hepatectomy, specific lobes are removed intact without damage to the lobes left behind. The residual lobes grow to compensate for the mass of the resected lobes, although the removed lobes never grow back. The process is completed within one week. Again, as in normal liver turnover, there is no evidence for involvement of, or requirement for, stem cells in this process. Classic thymidine labeling studies show that virtually all hepatocytes in the remaining liver divide once or twice to restore the original cell number within 3–4 days. The earliest labeled hepatocytes are seen 24 hours after partial hepatectomy, with the peak of thymidine incorporation occurring at 24–48 hours, depending on the species. Interestingly, there is zonal variation depending on how much tissue is removed. When only 15% of the liver is surgically removed, periportal (zone 1) hepatocytes divide preferentially, whereas cell division is seen equally in all three zones after 75% partial hepatectomy. Following the hepatocytes, the other hepatic cell types also undergo a wave of mitosis, thereby restoring the original number of all liver cells within 7 days.

Factors involved in liver regeneration after partial hepatectomy

The factors that initiate and control the regenerative response after partial hepatectomy have been the subject of intense study. Obviously, some of these are likely also to be involved in the differentiation of liver stem cells and are therefore of interest in this chapter. Several critical factors for the induction of hepatocyte cell division in this system have been identified. The earliest event, occurring within one minute after the surgical removal, is a large increase in the blood level of *hepatocyte growth factor* (HGF). This rapid increase cannot be explained by de novo synthesis, and it is thought that partial hepatectomy causes remodeling of extracellular matrix in the liver and release of HGF stored therein. HGF then binds to its receptor c-met and activates a signal transduction pathway leading to re-entry of hepatocytes into the cell cycle. Although HGF is a primary mitogen for hepatocytes and is responsible for the early events after partial hepatectomy, other cytokine-receptor interactions are also important in the cascade leading to mitosis. Known factors include interleukin 6 (IL-6), tumor necrosis factor-α (TNFα), transforming growth factor-α (TGFα), and *epidermal growth factor* (EGF). IL-6 and TNFα knockout mice both show significantly delayed regeneration after partial hepatectomy. Although EGF is a primary mitogen for hepatocytes in tissue culture, its role in liver regeneration in vivo is less clear because EGF levels increase only modestly after partial hepatectomy. In contrast,

TGFα mRNA and protein levels increase markedly within hours after partial hepatectomy, and TGFα overexpression can drive hepatocyte replication in vivo. Non-peptide hormones also have a significant role in the regenerative response after liver injury. Triiodothyronine and norepinephrine can stimulate hepatocyte replication in vivo. It is not known whether any of these factors are also important for progenitor-dependent liver regeneration or engraftment and expansion of liver stem cells.

Less knowledge exists about the mechanisms by which hepatocyte cell division and liver regeneration are stopped after the appropriate liver mass has been restored. In particular, the exogenous signals (endocrine, paracrine, or autocrine) involved in sensing the overall liver cell mass and negatively regulating its size are not known. Some evidence suggests that transforming growth factor-β1 (TGFβ1) may be important in terminating liver regeneration. Some endogenous signals are known to participate in the negative regulation of hepatocyte growth. Not surprisingly, they include general tumor suppressor genes. For example, mice lacking p53 or the p53-inducible cell cycle regulatory protein p21 have been shown to have continuous hepatocyte turnover. In addition, some more hepatocyte-specific transcription factors are also known to play a role. A hyperproliferative state was found in C/EBPα knockout mice.

Progenitor Cell-dependent Liver Regeneration

Oval cells

Although neither cell replacement during normal tissue turnover nor after injury by partial hepatectomy requires stem cells for organ regeneration, this is not true for all types of liver injury. In some types of damage to the liver, small cells with a high nuclear/cytoplasmic ratio emerge in the portal zone, proliferate extensively, and migrate into the lobule. These small cells, which eventually become differentiated hepatocytes, are termed oval cells because of their initial observed morphology. Importantly, oval cells are not derived from hepatocytes, but instead are the offspring of cells associated with the canal of Hering. Oval cell proliferation therefore represents an example of progenitor-dependent liver regeneration. The cell that probably resides in the canal of Hering gives rise to oval cells and can be considered a "*facultative liver stem cell*". In the rat, chronic liver injuries caused by chemicals such as DL-ethionine, galactosamine, and azo dyes represent examples of this type of liver damage. The toxic drugs are often combined with surgical partial hepatectomy.

A common feature of progenitor-dependent liver regeneration is that the hepatocytes themselves cannot divide normally. Thus, progenitor-dependent regeneration may be utilized when parenchymal hepatocytes are severely damaged on a chronic basis and/or unable to regenerate efficiently. Oval cells express markers of both bile duct epithelium (CK19) and hepatocytes (albumin). In the rat they also express high levels of α-fetoprotein and are thus similar to fetal hepatoblasts in their gene expression profile. Furthermore, oval cells are bipotential and retain the ability to differentiate into both the bile duct epithelial and hepatocyte lineages in vitro. Because of their similarity to hepatoblasts and their bipotentiality, oval cells have been considered early progenitors by analogy with committed hematopoietic progenitors. Thus, oval cell precursors located in the canal of Hering represent likely candidates for liver-repopulating stem cells.

Several monoclonal antibodies have been raised against rat oval cells and used to study lineage progression. The cell surface marker OV6 has found wide application in a variety of studies. In general, studies with these reagents have confirmed the similarity between oval cells and fetal hepatoblasts. Oval cells have been shown to express both the c-kit tyrosine kinase receptor and its ligand, stem cell factor (SCF, steele factor, MCGF). The SCF/c-kit system plays an important role in stem cell-driven hematopoiesis, melanogenesis, and gametogenesis. This has raised the issue of whether oval cells may respond to similar signals to these other stem cells.

Oval cell proliferation has also been described in a variety of human liver diseases, indicating that progenitor-dependent regeneration can be found in multiple organisms. Oval cells are found in disorders associated with chronic liver injury and are located at the edges of nodules in liver cirrhosis. As in the rat, OV6 is a useful marker for these cells in humans. Interestingly, cells that are c-kit-positive but negative for hematopoietic markers have also been identified in human pediatric liver disease.

Until recently, it has been difficult to induce oval cell proliferation in the mouse and thus take advantage of the powerful genetics in this organism. Using transgenic mice, it would be possible to determine whether factors known to be important in liver regeneration after partial hepatectomy are also required for oval cell-driven regeneration.

Now, however, two protocols have been developed that result in progenitor-dependent hepatocyte regeneration in the mouse. One regimen utilizes cocaine + phenobarbital and the other 3,5-diethoxycarbonyl-

1,4-dihydrocollidine (DDC). Mouse "*oval cells*" differ from their rat and human counterparts in not expressing AFP. The OV6 antibody does also not react with murine oval cells, and to date only one oval-cell-specific antibody, termed A6, has been developed for the mouse. Nonetheless, work on the genetics of oval cell proliferation is now possible. An example is the recent discovery, using transgenic mice, that TGFβ1 inhibits oval cell proliferation.

Non-oval-cell Progenitors

Oval cells are defined by their morphologic appearance in the rat, but there is variability in the marker genes they express at different times after induction of their proliferation. The different induction regimens also result in variability of the phenotype. Therefore, it is not clear whether oval cells are all equivalent or whether different subclasses exist. Recently, another class of hepatocyte progenitors has been described after treatment of rats with retrorsine and partial hepatectomy. Retrorsine blocks the division of mature hepatocytes but does not result in the emergence of classic oval cells, which are α-fetoprotein- and OV6-positive. Instead, foci of small hepatocyte-like cells emerge and eventually result in organ reconstitution. These small cells express both hepatocyte and bile duct markers. At this time, their origin (dedifferentiated hepatocytes, transitional cells in the canal of Hering, bone marrow) is unknown.

Liver Repopulation by Transplanted Cells

The hematopoietic stem cell was defined primarily by its ability to repopulate and rescue hematopoiesis in lethally irradiated hosts. In the 1990s, similar repopulation assays were developed for the liver in several animal models. In liver repopulation, a small number of transplanted donor cells engrafted in the liver expand and replace >50% of the liver mass. Thus, it has now become possible to perform experiments analogous to those done in the hematopoietic system, including cell sorting, competitive repopulation, serial transplantation, and retroviral marking. Hepatic stem cells can now be defined by their ability to repopulate the liver. It should be emphasized that liver repopulation refers to replacement of only the hepatocytes by transplanted cells since efficient repopulation of the biliary system by transplanted cells has not yet been reported.

Models for Liver Repopulation

Three main animal models for liver repopulation studies are briefly described below. The first two are transgenic mice and the third is a chemical liver injury model in the rat.

Urokinase plasminogen activator transgenic mice

In 1991, Sandgren et al. developed mice transgenic for the *urokinase plasminogen activator* (uPA) gene under control of the albumin promoter. These animals displayed liver inflammation with necrosis and a paucity of mature hepatocytes. Most animals died, but some survivors showed spontaneous development of nodular regions of liver containing normal hepatocytes. These nodules were found to be clonal and to have originated from a spontaneous loss of the uPA transgene. The reverted cells had a selective growth advantage and thus repopulated the transgenic liver. Similarly, transplanted wild-type cells (hepatocytes) also have a powerful selective advantage and can repopulate these mice.

Fumarylacetoacetate knockout mice

The enzyme *fumarylacetoacetate hydrolase* (FAH) catalyzes the last reaction in the tyrosine catabolic pathway. FAH deficiency results in accumulation of a hepatotoxic metabolite and causes the human liver disease tyrosinemia type I (HT1). Similar to the uPA transgenic mice, human HT1 patients frequently develop clonal hepatocyte nodules derived from reverted cells, which have lost the disease-causing mutation and express FAH. This finding suggested that transplanted FAH-positive donor cells could be used to repopulate FAH mutant liver. Indeed, it was shown that as few as 1000 donor cells could completely repopulate the liver of an FAH knockout mouse within 6 weeks. In mice, FAH deficiency is lethal in the neonatal period unless the animals are treated with 2(2-nitro-4-trifluoromethylbenzoyl)-1,3 cyclohexane dione (NTBC), a pharmacological inhibitor of tyrosine catabolism upstream of FAH. NTBC blocks the hepatocyte injury, thus allowing the propagation of mutant animals and control of the selective pressure that drives repopulation by transplanted cells.

Retrorsine-treated rats

Both murine models for liver repopulation utilize genetically modified animals. In contrast, Laconi et al. (1995) developed an approach to liver repopulation by chemically blocking the regenerative capacity of host cells using lasiocarpine or retrorsine, structurally similar pyrrolizidine alkaloids. These compounds are selectively metabolized to their active form by hepatocytes, where they alkylate cellular DNA and cause proliferation arrest of hepatocytes in the G_2 phase of the cell cycle. Gordon et al. (2000) have shown recently that partial hepatectomy of retrorsine-treated animals induces proliferation of endogenous, small, hepatocyte-like progenitor cells. Retrorsine followed by partial hepatectomy can also be used to achieve near-total liver repopulation by transplanted donor cells. Typically, genetically marked rat hepatocytes that are positive for the bile canalicular membrane protein dipeptidyl peptidase IV (DPPIV) are injected into the spleen of a congenic strain of mutant rats not expressing DPPIV enzyme activity. Within 2 months, there was 40–60% replacement by transplanted hepatocytes in female rats and >95% replacement in male rats.

Liver-repopulating Cells

The animals described above have been utilized to determine the nature of transplantable liver-repopulating cells and to determine whether undifferentiated stem cells are driving this process. The stem cell hypothesis was strengthened by the observation that liver-repopulating cells could be serially transplanted without loss of functionality. Male wild-type hepatocytes were serially transplanted at limiting dilution through seven rounds of female FAH knockout recipients. Complete repopulation was achieved in each round, and it was estimated that the repopulating cells in the seventh-round recipients had undergone at least 100 cell doublings, similar to what has been seen in serial transplantation of hematopoietic stem cells. Interestingly, the only donor-derived cells in this experiment were hepatocytes. No biliary epithelium or other cell types of donor origin were found, thus raising the possibility of a "*unipotential*" stem cell.

Hepatocytes as Liver-repopulating Cells

In most liver repopulation experiments reported to date, only unfractionated cell suspensions from whole liver were used, and thus, it remained unclear whether the hepatocytes themselves or only a rare subpopulation(s) participates in the process. It would seem reasonable to hypothesize that adult liver cells are not homogeneous in their capacity for cell division and that subpopulations with high repopulation capacity might exist. The highly regenerative capacity of serially transplantable cells in particular has raised the question whether liver stem cells may be responsible for liver repopulation.

In the hematopoietic system, repopulation experiments with purified fractions of total bone marrow were used to identify stem cells. Similar experiments have now been performed with liver cells.

In the FAH mutant mouse model, three sets of experiments were performed to address whether differentiated hepatocytes or putative stem cells were responsible for the repopulation. First, cell fractionation by centrifugal elutriation was used to identify and purify three major-size fractions of hepatocytes (16 μm, 21 μm, and 27 μm). Each fraction was transplanted in competition with unfractionated liver cells carrying a distinct genetic marker, which serves as a baseline reference for liver repopulation. The larger hepatocytes, which represented ~70% of the population, were primarily responsible for liver repopulation. In contrast, small diploid hepatocytes were inferior to the larger cells in competitive repopulation experiments. Second, competitive repopulation was performed between naive liver cells and those that had been serially transplanted up to seven times. Importantly, serial transplantation neither enhanced nor diminished the liver repopulation capacity. If serial liver repopulation were stem cell-dependent, this result would suggest that the ratio of progenitors to differentiated hepatocytes was kept constant during a 10^{20}-fold cell expansion. More likely, this result means that virtually all the original input cells (>95% hepatocytes) were capable of serial transplantation. The third set of experiments involved retroviral marking of donor hepatocytes in vitro and in vivo. Again, no evidence for a rare stem cell responsible for liver repopulation was detected. Together, these experiments strongly suggested that fully differentiated hepatocytes, which constitute the majority of liver cells, were responsible for liver repopulation and have a stem-cell-like capacity for cell division.

Liver Repopulation by Non-hepatocytes

Despite the evidence that hepatocytes themselves are serially transplantable liver-repopulating cells, other cell types are also capable of repopulating the liver. This finding is analogous to the situation in liver regeneration where hepatocytes themselves, as well as undifferentiated hepatocyte progenitors, are capable of reconstituting the organ. In the following, we describe transplantation experiments that demonstrate the capacity of five non-hepatocyte cell types to differentiate into hepatocytes in vivo: (1) fetal hepatoblasts, (2) oval cells, (3) pancreatic liver progenitors, (4) hematopoietic stem cells, and (5) neurospheres.

Fetal Hepatoblasts

During embryonic development, the fetal liver bud contains hepatoblasts, cells that express α-fetoprotein as well as hepatocyte (albumin) and biliary (CK19) markers. These cells therefore may represent fetal liver stem cells capable of hepatocyte repopulation and, potentially, also reconstitution of the biliary system.

Only one report on transplantation of hepatoblasts has been published. This study, using fetal rat liver cells in the retrorsine model, indicated that there were at least three distinct subpopulations of hepatoblasts at ED 12–14. One population appeared to be bipotential on the basis of histochemical markers, and the other two had either a unipotent hepatocyte or biliary epithelial cell phenotype. After transplantation, the bipotential cells were able to proliferate in retrorsine-treated cell transplantation recipients, whereas the unipotent cells grew even in untreated rats. However, none of the fetal liver cell populations proliferated spontaneously, partial hepatectomy or thyroid hormone treatment being required to augment proliferation of the transplanted cells. Nonetheless, fetal liver cells proliferated more readily than adult cells. Finally, the transplanted fetal cells gave rise to both hepatocyte cords and mature bile duct structures. It was not formally proven, however, that both of these cell lineages originated clonally from a common precursor. Together, these results indicated that transplanted fetal hepatoblasts proliferate more readily than adult hepatocytes, and some fetal liver cells may remain bipotential.

Oval Cells

Oval cells are similar to fetal hepatoblasts in that they are also bipotential. These cells have, therefore, been of interest in liver repopulation experiments. Hepatocyte progenitor (or oval) cells can be isolated from the liver of rats treated with D-galactosamine or the pancreas of rats treated with a copper-deficient diet. Copper depletion causes atrophy of pancreatic acini and proliferation of duct-like oval cells expressing genes in the hepatocyte lineage. Transplantation of both hepatic- and pancreatic-derived oval cells has been reported in the rat. Upon transplantation, these cells proliferated modestly and differentiated into mature hepatocytes, even under nonselective conditions. However, because no in vivo selection model was used, their true capacity for liver repopulation was not demonstrated in these experiments. Thus, oval cells can become hepatocytes upon transplantation into the liver, but their proliferative capacity remains unknown.

Pancreatic Hepatocytes

During embryogenesis, the main pancreatic cell types, including ducts, ductules, acinar cells, and the endocrine α, β, and δ cells, develop from a common endodermal precursor located in the ventral foregut. Importantly, the main epithelial cells of the liver, hepatocytes and *bile duct epithelium* (BDE), are also thought to arise from the same region of the foregut endoderm. The hepatic anlage develops ventrally toward the cardiac mesenchyme, which induces the hepatoblast differentiation pathway. The pancreas buds from the same region, with its ventral lobe growing anteriorly in the same direction as the liver and its larger dorsal lobe growing posteriorly. Thus, the ventral lobe of the pancreas is particularly closely related anatomically to the liver. The signals that govern the respective developmental pathways are only beginning to be understood.

This tight relationship between liver and pancreas in embryonic development has raised the possibility that a common hepato-pancreatic precursor/stem cell may persist in adult life in both the liver and pancreas. Indeed, several independent lines of evidence suggest that adult pancreas contains cells which can give rise to hepatocytes. The best-known example is the emergence of hepatocytes in copper-depleted rats after re-feeding of copper. In this system, weanling rats are fed a copper-free diet for 8 weeks, which leads to complete acinar atrophy, and then are re-fed copper. Within weeks, cells with multiple hepatocellular characteristics emerge from the remaining pancreatic ducts. This work has been interpreted to suggest the presence of a pancreatic liver stem cell.

This notion is also supported by the appearance of hepatocellular markers in human pancreatic cancers. More recently, a specific cytokine has been identified as a candidate to drive this process. Transgenic mice in which the *keratinocyte growth factor* (KGF) gene is driven by insulin promoter consistently develop pancreatic hepatocytes. Thus, the existence of pancreatic liver precursors has been shown in several different mammalian species and under multiple experimental conditions. Both adult liver and adult pancreas may continue to harbor a small population of primitive hepato-pancreatic stem cells with the potential to give rise to the same differentiated progeny as during embryogenesis.

Transplantation experiments have verified the pancreatic liver stem cell hypothesis. As mentioned above, pancreatic oval cells induced by copper depletion were shown to give rise to morphologically normal hepatocytes in vivo. To determine whether pancreatic hepatocyte precursors also exist in normal pancreas without the use of toxic induction regimens, repopulation experiments were performed in the FAH knockout mouse model. Pancreatic cell suspensions from adult wild-type mice on a normal diet were transplanted into FAH mutant recipients, and selection was induced by NTBC withdrawal. Extensive liver repopulation (>50%) was observed in about 10% of transplant recipients, and another 35% had histological evidence for donor-derived hepatocyte nodules. Thus, adult murine pancreas contains hepatocyte precursors, even under normal, nonpathologic conditions. It remains to be determined whether these pancreatic liver stem cells also harbor other differentiation potential, particularly toward the pancreatic endocrine lineage.

Bone Marrow-derived Hepatocytes

Recent work has documented that the adult bone marrow of mammals contains cells with a variety of differentiation capacities. The *hematopoietic stem cell* (HSC) giving rise to all blood cell lineages has been known to reside in this compartment for many years. However, bone marrow also contains *mesenchymal stem cells* (MSC) capable of differentiating into chondrocytes, osteoblasts, and other connective tissue cell types. Although HSCs are nonadherent, MSCs adhere to plastic dishes in tissue culture and can be expanded there. A population of primitive, nonadherent cells characterized by their ability to expel the DNA-staining Hoechst dye can give rise to both muscle and hematopoietic cells. Thus, adult bone marrow produces a variety of tissue types of mesodermal origin. Petersen et al. (1999) first suggested that bone marrow also contained epithelial precursors. Cross-sex or cross-strain bone marrow and whole liver transplantation were used to trace the origin of the repopulating liver cells.

Transplanted rats were treated with 2-acetylaminofluorene, to block hepatocyte proliferation, and then hepatocyte injury to induce oval cell proliferation. Markers for Y chromosome, dipeptidyl peptidase IV enzyme, and L21-6 antigen were used to identify liver cells of bone marrow origin, and a proportion of the regenerated hepatocytes were shown to be donor-derived. Next, Theise et al. (2000a) showed that bone marrow-derived hepatocytes also exist in the mouse and that oval cell induction was not required for this phenomenon. Female mice that had received lethal irradiation and bone marrow transplantation from a male donor displayed 1–2% Y-chromosome-positive epithelial cells in their livers. Most recently, two reports demonstrated that donor-derived epithelial cells are also present in human patients who have undergone a gender-mismatched bone marrow transplantation. In all these studies, the epithelial nature of the cells was demonstrated morphologically and by the expression of hepatocyte-specific markers. The nature of the bone marrow cell responsible for repopulation was not shown (adherent versus nonadherent, etc.).

Most recently, the FAH mutant model has been utilized to identify the nature of the bone marrow-derived hepatocyte precursor. Cell-sorting experiments and transplantation of purified hematopoietic stem cells of the KTLS phenotype (c-kit^{high}Thy^{lo}Lin^{neg}Sca-1^{+}) at limiting dilution showed that the primitive HSC is also capable of giving rise to hepatocytes. HSC not only gave rise to cells morphologically similar to hepatocytes, but also formed repopulation nodules and resulted in extensive organ reconstitution.

Thus, it now has been shown that the population of primitive bone marrow cells with the KLTS phenotype contains both hematopoietic and hepatocyte precursors. Although it seems likely, it has not yet been formally shown that the same clonal precursor is responsible for both lineages. The physiologic significance of bone marrow-derived hepatocytes in the response to liver injury is not known at this time. However, it is possible that circulating hepatocyte precursors are an important contributor to progenitor-dependent liver regeneration.

Neurosphere-derived Liver Precursors

In the mouse, neuronal stem cells cultured in neurospheres can give rise to many different cell lineages. Cultured neurospheres can effect repopulation of the hematopoietic system after transplantation. More recently, cultured neurospheres were injected into the blastocyst of a recipient embryo. Upon analysis of adult mice derived from such injections, the donor cells were found in many different tissues, including liver. Donor-derived cells were thought to express the hepatocyte phenotype.

In vitro Models of Hepatic Stem Cells

Several in vitro models for hepatic stem cell growth and differentiation have been developed. Putative liver progenitors from several mammalian species, including mouse, rat, pig, and human, have been isolated and propagated in primary tissue culture. In addition, immortal cell "*liver progenitor*" lines have been developed. Generally speaking, in vitro culture of liver progenitor cells is based on the growth of epithelial cells that are liver-derived but express either no hepatocyte markers or markers of both bile duct epithelium and hepatocytes. These in vitro systems have both medical and scientific goals. The medical purpose is to generate large numbers of hepatocytes in vitro for therapeutic transplantation.

The scientific aims are to understand the factors that control the differentiation of these cells into hepatocytes and biliary duct epithelium. To date, the only cell lines that have had documented therapeutic effects have been cell lines or primary cultures derived from hepatocytes themselves. Despite the availability of a variety of good in vitro model systems, surprisingly little is known about the molecular mechanisms that govern the transition of progenitor cells to hepatocytes and/or bile duct epithelium.

Rat Cell Lines

WB-344 cells

This cell line was clonogenically derived from non-parenchymal rat liver cells and is probably the most intensely studied liver stem cell line. WB3-44 cells are likely derived from canal of Hering cells. Although WB-344 cells can be cultured indefinitely in vitro, they retain the ability to differentiate into morphologically normal hepatocytes after transplantation without forming tumors. To date, WB-344 cells are the only cells to fulfill this stringent criterion necessary to represent a true liver stem cell line. Nonetheless, little is known about the molecular mechanisms regulating the stem cell-to-hepatocyte transition, and liver repopulation with this cell line has not yet been reported.

Oval cell lines

Multiple laboratories have isolated oval cell lines from carcinogen treated rats. Consistent with the proposed role of oval cells in the formation of hepatocarcinoma, these cell lines form tumors upon transplantation into immunodeficient recipients. The isolation, culture, and transplantation of these cells has been well reviewed. Although the phenotypic properties of these cell lines have been described in great detail, the molecular mechanisms regulating phenotypic transitions are not well understood.

Mouse Cell Lines

Although it is normally difficult to establish permanent cell lines from mouse liver, such lines can be established routinely from transgenic mice that overexpress a constitutively active form of c-met. Two morphologically distinct types of cells emerge from such cultures, both of which grow extensively in culture under certain media conditions, but can be induced to differentiate with the appropriate signals. Clonal cell lines with "*epithelial*" morphology resemble hepatocytes and give rise to only hepatocyte-like offspring. In contrast, "*palmate*" clones can give rise to two distinct lineages, depending on the differentiation conditions used. Some general conditions that can cause differentiation of palmate cells in either direction have been discovered. Acidic FGF or DMSO induces hepatocytic differentiation, whereas culture in matrigel induced the formation of bile-duct-like structures. In vivo transplantation and differentiation of these cells has not yet been reported.

HBC-3 cells are a novel bipotential cell line derived from day-9.5 mouse embryonic liver. This clonal cell line can be induced to differentiate into hepatocytes by DMSO or sodium butyrate. Again, matrigel induces the formation of duct structures, but the details of the differentiation process are not understood at the molecular level.

Pig Cell Lines

An interesting cell line was established from cultured pig epiblast. PICM-19 cells are bipotential, similar to the murine cell lines reported above. In contrast to other cell lines, however, PICM-19 cells were derived from a very early embryo prior to the formation of a liver primordium.

Human Cell Lines

Only a single human cell line, AKN-1, which may have liver progenitor characteristics, has been described.

Concluding Remarks

The liver is one of the few organs amenable to repopulation by transplanted cells. Therefore, the subject of liver stem cells is not only of scientific, but also of medical, interest. Current evidence shows that both endogenous liver regeneration and repopulation by transplanted, exogenous cells can be effected by more than one cell type. Hepatocytes themselves have a very high capacity for cell division and can be considered unipotential stem cells, which are used for most tissue repair. In addition, however, facultative liver stem cells distinct from hepatocytes exist and are important in the response to some forms of liver injury. Hepatocyte progenitors have been found in the liver itself, in the pancreas, and most recently in bone marrow. It is unclear whether these are distinct cell types or whether the same cell has been isolated from several different anatomical locations. It also has not yet been conclusively shown in vivo that hepatocyte precursors can differentiate into bile duct epithelium. Oval cells can be viewed as committed progenitors, a transitional phenotype between self-renewing stem cells and differentiated hepatocytes. The molecular mechanisms controlling the phenotypic transitions are not well understood and will be the subject of future research.

6

Heterocrine Stem Cells

Insulin-dependent, or type I, diabetes mellitus (IDDM) is a devastating disease in which affected individuals depend on daily injections of exogenous insulin for regulation of glucose homeostasis and survival. The life-time dependency on insulin invariably leads to a variety of debilitating complications that threaten quality of life and significantly shorten life expectancy of IDDM patients. Insulin-producing cells in the pancreatic islets of Langerhans are the targets for selective inflammatory destruction in IDDM. This loss of the pancreatic islets is a terminal process, since the pancreas lacks any significant ability to regenerate insulin-producing cells. Indeed, pancreatic islet cells have a very low rate of growth in adults, and knowledge of the factors that regulate islet growth is lacking. Increased knowledge of islet stem cells and potential islet regeneration could allow critical advances in the development of new therapies for IDDM. Knowledge regarding pancreatic stem cells has grown dramatically in recent years, but continued progress is still hampered by the lack of in vitro and in vivo assay systems for stem cell activity and function. In this review, we focus on pancreatic stem cells during morphogenesis and regeneration, as well as on the molecular markers associated with these cells.

Ontogeny of Pancreatic Endocrine Cells

During embryogenesis, the pancreas arises from the foregut, initially as two distinct bulges on either side of the duodenum, termed the dorsal and ventral buds. The two buds grow independently, with both endocrine and exocrine tissue components present in each bud. As development proceeds, the stomach and duodenum rotate, bringing the two buds into contact, at which point they fuse to form the pancreas. Endocrine cells in the pancreas are organized into clusters of cells referred to as the islets of Langerhans. The endocrine cells develop from the pancreatic ducts, and stem cells responsible for endocrine cell formation are thought to reside in the fetal pancreatic ducts. The neogenesis of islets from duct epithelial cells occurs during normal embryonic development and in very early postnatal life. By 2 to 3 weeks of age, the pancreatic islets are well defined and distinct from duct structures, as they are in the adult mouse. From this point on, endocrine cells are not observed in the ducts, and the mitotic index of duct and islet cells is very low. Therefore, following the early postnatal period, the normal developmental process that generates islets ceases. Islet growth and the maintenance of islet mass are under strict regulatory controls. Thus, little if any stem cell activity is observed in the normal adult pancreas.

Endocrine gene expression occurs prior to the appearance of the pancreatic diverticulum during pancreatic organogenesis. All four endocrine cell types are thought to arise through progressive differentiation from a common pluripotent stem cell. The endocrine cells first appear as single cells in the duct wall. These cells migrate to form clusters, which grow and become islets during the course of pancreatic development. The glucagon-secreting α cells are typically the first endocrine cell type observed during normal ontogeny. However, the earliest emergence of this endocrine cell type is controversial. In some studies insulin- (β cells), somatostatin-(δ cells), or pancreatic polypeptide-secreting cells (PP cells) are observed prior to or coincidental with the appearance of α cells.

During endocrine cell formation, cells producing more than one hormone are often observed, with cells producing glucagon and insulin arising first. The identification of multiple hormone-expressing cells in the pancreatic ducts during development, as well as in many pancreatic endocrine tumors, has led to the proposal that these multi-hormone-bearing cells represent a transient phase during the differentiation of endocrine cell precursors to the single hormone-expressing phenotype found in mature endocrine cells. However, the derivation of endocrine cell types is not well defined, and controversy remains as to whether or not multi-hormone-expressing cells represent true transitional cell types during endocrine cell formation.

As described above, the pancreas arises from the out-pouching of the gut endoderm in the region of the duodenum during development.

However, heterotopic formation of pancreatic tissue, including mature endocrine cells, can arise from other regions of the gut endoderm, including the stomach and additional regions of the small intestine. Recent studies in chicks have revealed that inhibition of the intercellular signaling protein *Sonic hedge-hog* (SHH) promotes heterotopic formation of pancreatic tissue from regions of the stomach and duodenum. Thus, in addition to "*pancreatic stem cells*" in the region of posterior foregut evagination that normally forms the pancreas, other cells of similar embryonic derivation indeed have the capacity to form pancreatic cells.

Pancreatic Islet Regrowth and Regeneration

In general, adult pancreatic β cells are known to have a poor growth capacity and, in pathogenic states where islet destruction occurs, such as IDDM, regeneration is not detected. In animal models of IDDM, such as the *non-obese diabetic* (NOD) mouse or the *biological breeding* (BB) rat, where the islet cells are lost by presumed immunological mechanisms, islet regeneration is not observed. Furthermore, *streptozotocin* (Sz) injection of adult rats, which results in destruction of β cells and produces a diabetic condition, does not induce islet regeneration. However, under certain conditions of pancreatic regeneration, new islet cells can arise postnatally and later in adult life. For example, limited islet regeneration has been reported under several experimental conditions, such as after partial pancreatectomy of young rats. In this case, new islets have been shown to differentiate from ductal epithelium, recapitulating islet formation during embryogenesis and suggesting that stem cells in the adult pancreas can be activated during regeneration. Some islet cell regeneration has also been reported following ligation of pancreatic arteries, in steroid diabetes, following injection of insulin antibody, after alloxan administration, and after cellophane wrapping. The most impressive regeneration can be found in newborn and neonatal animals following islet destruction. Thus, studies have shown that after treatment of newborn rats with Sz, the pancreas is able to repair itself, and new islet formation is observed.

New islet cells are thought to arise through two distinct processes. One mechanism involves the regeneration of islets through division of existing terminally differentiated islet cells. A second mechanism involves the differentiation of new islet cells from stem cells residing in the pancreatic ducts. In the latter case, quiescent stem cells in the adult are thought to be "*reactivated*" under particular conditions of regeneration, leading to new endocrine cell and islet formation. Clearly, much work needs to be done to define the precise factors ultimately responsible for the induction of stem cell activity during pancreatic regeneration.

Trans-differentiation of pancreatic cells to nonpancreatic cell types is also found under certain conditions, highlighting the plasticity of the pancreatic cell phenotype. In the db/db mouse, proliferating duct cells give rise to ciliated cells, mucous or Paneth secretory cells. In addition, there are instances of human pancreatic metaplasia in which well-differentiated goblet cells and other mucous cells have been reported to have a ductal origin. In rats, hepatocytes develop in the pancreas after maintenance of the animals on a copper-deficient diet, apparently arising from ductal and interstitial pancreatic cells. Additionally, pancreatic hepatocytes are found in a number of other models of experimental destruction of the exocrine pancreas. These latter two observations suggest a possible common stem cell, or at least the potential for plasticity between these two terminally differentiated cell types, implying they may share a closely connected lineage.

Experimental Models of β-Cell Loss and Repopulation of Pancreatic Islets

Many transgenic mouse models for studying IDDM have been described previously. These transgenic mice have, to different degrees, been successful in duplicating various stages of human IDDM and have provided valuable tools for elucidating mechanisms that lead to IDDM. Clearly, the ability of pancreatic stem cells to reactivate islet development in the adult would provide a powerful basis for treatment of IDDM. However, regeneration of islet cells has not been reported among these many transgenic mouse strains, with the exception of a report describing expression of the tumor necrosis factor (TNFα) transgene under control of the insulin promoter. In this case, expression of TNFα caused insulitis, disorganization of islet structure, and proliferation of pancreatic ductules; little regeneration of endocrine cells was observed.

We have developed a unique model system in which new pancreatic growth and islet formation are clearly evident throughout adult life. These Ins-IFNγ mice, who express IFNγ in the β cells of the pancreatic islets, display unusually high proliferative activity in ductal epithelial cells. We have demonstrated that new endocrine cells derive from pancreatic ducts in these transgenic mice, suggesting that a well-defined and functional stem cell population exists in the ducts of the Ins-IFNγ mice. Much of our work has since focused on further defining the stem cell population active in these mice.

The regeneration and differentiation observed in the Ins-IFNγ mouse is very aggressive, with cells showing substantially higher mitotic indices than are normally found in the pancreatic ducts. Interestingly, the pancreases of these mice also display ductal hyperplasia and destruction of islets by infiltrating lymphocytes, characteristic of that seen in type I diabetes mouse models. The most intriguing phenotype of these mice is the balance between regeneration of "new" islets and the destruction of the preexisting islets in the acinar tissue. The newly formed islets bud into the lumen of the ducts, where they are protected from infiltration and destruction. The aggressive expansion of ducts, and the continuous formation of new endocrine cells in the pancreas during adult life, are important characteristics of the remarkable regrowth occurring in the transgenic pancreas. In these animals, new islet cells are formed continuously from duct cells, without a requirement for T cells, B cells, or macrophages.

On the basis of the expression of endocrine, exocrine, and duct cell markers, endocrine cells are thought to derive from duct cells in the Ins-IFNγ mouse through a defined pathway involving a number of transitional cell types, including those bearing a multi-hormone-expressing phenotype. In addition, we have recently demonstrated the expression of a number of molecules known to be critical for normal pancreatic development in the ducts of the Ins-IFNγ transgenic mouse, including PDX-1, HNF3β, Isl-1, and Pax-6. We believe that these PDX-1-expressing ductal cells represent endocrine precursor cells from which new islets derive. Comparison in this case is also made with insulin expression, which is also observed, albeit to a lesser extent than PDX-1, in the ducts of the transgenic mouse. Thus, two distinct populations of PDX-1-expressing cells are observed: one population expressing only PDX-1 and one population expressing both PDX-1 and insulin. These data lead us to speculate that those cells expressing PDX-1 but not insulin represent an earlier precursor cell type than those synthesizing both PDX-1 and insulin.

There are a number of similarities between new islet formation during development and IFNγ-mediated pancreatic regeneration. These similarities include the derivation of new endocrine cells from ductal cells, the presence of multi-hormone-expressing cells in the pancreatic ducts, and the ductal expression of relevant molecules such as PDX-1 and Isl-1 during both processes. These similarities suggest that new endocrine cell formation likely proceeds through similar pathways in both these processes, with new endocrine cells arising from stem cells residing in the pancreatic ducts. Thus, the IFNγ transgenic mouse provides a valuable system for further analysis and characterization of endocrine stem cells. However, it is important to note that the differentiation of pancreatic islets in this model may not be exactly identical to that which occurs during ontogeny. Indeed, it is possible that there is more than one type of islet stem cell and more than one developmental lineage for the differentiation of mature islet cells. For example, Gittes and colleagues have demonstrated that isolated pancreatic epithelium grown under the renal capsule forms pure clusters of mature islets, without any acinar or ductal components. This contrasts with a number of other studies indicating that islets arise from ducts, and it suggests that there might be alternate pathways leading to the formation of mature islets.

Stem Cell-associated Markers

The field of pancreatic islet cell ontogeny enjoyed comparative solitude until the fascinating discovery that factors regulating insulin transcription were required for the differentiation of the mature pancreas. This and related discoveries have fueled the rapid expansion of the pancreas ontogeny field, the discovery of a variety of potential islet stem cells, and new theories on the ontogeny of mature islet cells. Indeed, several proteins associated with pancreatic stem cells have been defined, and studies have shown that a number of transcription factors are important in pancreas differentiation. Significantly, the homeodomain protein PDX-1, which is important in regulating insulin gene expression, plays a critical role in the development of endocrine and exocrine compartments of the pancreas. In mice genetically deficient for the PDX-1 gene, no pancreas develops. HNF3β is an important regulator of PDX-1 gene transcription in the pancreas and is required for endodermal cell lineage development. The exact role of HNF3β in pancreatic islet development has not been resolved, since mice lacking this transcription factor die early in embryogenesis, but given the importance of this molecule in PDX-1 gene expression, HNF3β may be critical for this process as well. We have recently demonstrated striking expression of HNF3β in the ducts of the IFNγ transgenic mouse, suggesting a role for this molecule during endocrine cell formation. Studies using knockout mice have also shown that Isl-1 is required for the differentiation of islet cells. Interestingly, although the dorsal bud does not develop in Isl-1-deficient mice, the ventral bud does form. However, endocrine cells, but not exocrine cells, are lacking in the ventral bud. Thus, whereas Isl-1 clearly contributes to endocrine cell formation, its expression is also important for development of exocrine tissue in the dorsal bud.

Two other homeodomain proteins, PAX 4 and PAX 6, are expressed in the embryonic gut and the adult pancreas as well. Interestingly, gene deletion studies have demonstrated that mature endocrine cells do not develop in the absence of PAX 4 and PAX 6. PAX 4 is essential for the generation of insulin-producing β cells and somatostatin-producing δ cells, whereas PAX 6 is required for the differentiation of glucagon-producing α cells. Analysis of mutants lacking Nkx2.2 has demonstrated a role for this homeobox transcription factor in the development of pancreatic β cells as well.

A number of basic helix-loop-helix (bHLH) proteins have also been implicated in endocrine cell development. For example, neurogenin 3 (Ngn3), a bHLH transcription factor, has recently been shown to be required for the development of all four endocrine cell types. In addition, the bHLH protein BETA2/NeuroD, a transcription factor important in expression of the insulin gene, is involved in proper pancreas development. Mice lacking this protein display significantly reduced numbers of β cells. Reduction in the numbers of other islet endocrine cell types, as well as defective islet and exocrine tissue morphology, is also observed in these mice. Analysis of mice deficient in Hes1 has also demonstrated a role for this bHLH transcriptional repressor in maintenance of the pancreatic precursor pool during development. Hes1 has been proposed to inhibit endocrine cell differentiation mediated by Ngn3. In addition, expression of Hes1 is regulated through the Notch signaling pathway, which has been implicated in endocrine cell development.

Several other proteins are thought to be associated with pancreatic stem cells. For example, studies of the distribution of Nkx 6.1 during early development of the rat pancreas suggest that this molecule may be important in pancreatic development. In addition, the pancreatic and intestinal hormone peptide YY is thought to be expressed by endocrine stem cells during development. Early embryonic endocrine cells in the duct wall also transiently express *tyrosine hydroxylase* (TH), the rate-limiting enzyme of the catecholamine biosynthetic pathway, suggesting that this molecule serves as a marker for early endocrine progenitor cells. In addition, there is a report that acid β-galactosidase expression, which occurs at significantly higher levels in islet-like clusters from the fetal pancreas than in adult islets, might serve as a marker for endocrine stem cells as well. Glut-2 (glucose transporter type 2)-expressing cells have also been proposed as stem cells in the developing pancreas. Additionally, Bouwens and De Blay (1996) have reported on the expression of CK20, vimentin, and Bcl-2 as markers for islet stem cells. We have recently found that Msx-2, a homeobox-containing transcription factor involved in tissue growth and patterning during the development of diverse organs (such as teeth and limb buds), is also expressed at sites of endocrine stem cell activity.

Developmental Lineages

The identification of molecules associated with pancreatic stem cells will be valuable in defining intermediate cell types involved in the developmental progression from multipotent precursor cell to mature endocrine cell types. For example, PDX-1, expressed in the early pancreatic primordia, prior to the formation of the dorsal and ventral buds, is thought to be expressed by the earliest pancreatic stem cells that give rise to both endocrine and exocrine tissues. Development of intermediate progenitor cell types involves the subsequent expression of molecules such as Ngn3 and Isl-1, which are required for the derivation of all four endocrine cell types. Clearly, continued and progressive differentiation of islet cell types would involve the specific expression of a number of additional molecules in the correct temporal sequence, such as Pax4 (required for β and δ cell formation), Pax6 (required for α cell formation), and Nkx2.2 (required for β cell formation).

Concluding Remarks

The identification of islet stem cells, the development of new antigenic markers for these cells, and the establishment of an assay system to monitor stem cell activity will be of critical importance in advancing our knowledge of pancreatic islet stem cell ontogeny. Such knowledge could enable critical advances in the development of new therapies for IDDM, a debilitating disease that severely compromises pancreatic function in affected individuals. For example, new pancreatic islet cells could be differentiated from pancreatic stem cells in vitro for β-cell replacement therapy. In addition, given sufficient knowledge of pancreatic stem cells and the factors that influence their growth and differentiation, it is possible that administration of the appropriate growth factors could allow the induction of new islet cells directly within IDDM patients. The exciting progress and ongoing research in the pancreatic stem cell field are critical if we are to find an eventual means for achieving these important goals.

7

Hematopoietic Stem Cells

Primitive hematopoietic stem cells (PHSC) are the self-renewing precursors that regenerate myeloid and lymphoid cells throughout the life span, and they are required for successful long-term bone marrow transplantation. PHSC are found in *bone marrow cells* (BMC), spleen, blood, and newborn or fetal liver and blood. PHSC will be vital for putative therapeutic measures such as human gene transfer via autologous marrow cell transplantation. PHSC proliferation is required for the long-term success of clinical grafts, yet it cannot be fully defined clinically because 10–20 years may be required to recognize long-term functional cells in human beings.

Fortunately, mouse PHSC appear to provide a good model for human PHSC function, and since mice are short-lived, aging about 30 times faster than humans, PHSC function over much of the life span can be studied directly.

Cells, obtained from bone marrow, cord blood, or mobilized peripheral blood of healthy donors, are clinically useful for transplantation. After transplantation, donor PHSC compete with residual host PHSC for engraftment; thus, high levels of donor PHSC function are essential since host PHSC remain functional in many clinical conditions. Unfortunately, donor PHSC are often damaged by clinical manipulations, so that they repopulate less well than residual host cells. Often the most primitive cells are the most damaged. Competitive repopulation directly models the donor and host stem cell competition for engraftment. One of the most important and demanding problems in hematology is regulating and maximizing the levels of donor PHSC function.

PHSC Repopulate, Self-renew, and Function Long Term

PHSC repopulate both the lymphoid and myeloid systems. The same PHSC produces erythrocytes, platelets, granulocytes, macrophages, B cells, and T cells proportionally, with high correlations between all these cell types.

Demonstrating both lymphoid and myeloid function is not adequate to identify a cell as a PHSC, since most precursors that repopulate both lymphoid and myeloid lineages soon after transplantation are short-lived. The length of time that such short-term multilineage precursors function appears to be proportional to the life span of the species. Short-term precursors disappear 3–4 months after transplantation in mice, but persist for 1–4 years in cats. Human short-term multi-lineage precursors might function for 10–20 years. Thus, long-term functional determination is essential in defining the biology of the PHSC. This makes mouse models more practical than human models for studying PHSC function.

Competitive repopulation assay evaluates primitive PHSC in vivo by testing the defining characteristics of PHSC directly. Donor cells must seed/home and then proliferate and differentiate to continuously form both myeloid and lymphoid cells during large portions of the life span. Function is tested relative to normally functional standard competitor cells. Figure 7.1 illustrates that as donor PHSC numbers increase fourfold, from 2 to 8, with a constant competitor dose of 4, percentages of donor cells only double, from 33% to 66%, if all PHSC repopulate equally. Since use of percentages understates alterations in donor function, we calculate *repopulating units* (RU), where 1 RU is the repopulating ability of 100,000 standard competitor BMC. Numbers of RU from each donor are calculated from the percentage of donor cells where the number of competitor marrow cells used divided by 10^5 = C. By definition, % = 100 (RU/RU + C), so RU = % (C)/(100%).

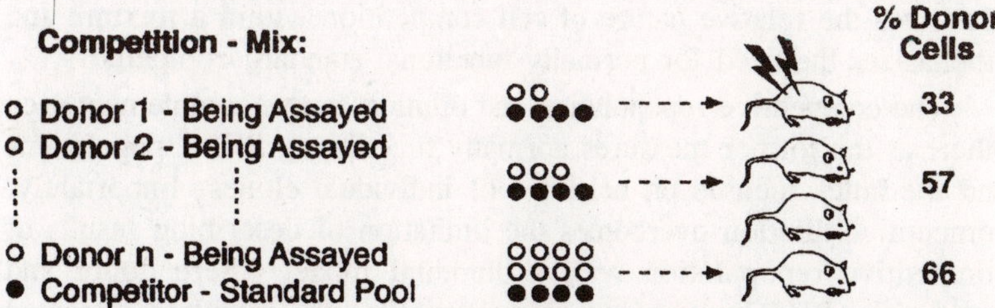

Fig. 7.1. In the competitive repopulation assay, donor cells being tested are mixed with doses of genetically distinguished standard competitor BMC and grafted into irradiated recipients.

PHSC self-renewal is tested most rigorously after long-term serial transplantation. Effects of age on PHSC self-renewal after serial transplantation were shown by the results from competitive repopulation. By measuring retransplanted RUs in carrier BMC, we compare the amount of expansion from PHSC in the original dose of donor marrow cells received by each carrier. Since the same competitor was used with old and young donors of the same strain, the total RU per carrier is directly comparable between each strain's young and old donors.

Competitive dilution assays use limiting dilution with the addition of a standard dose of competitor marrow. This measures relative repopulating abilities of individual genetically marked PHSC without the stress to the cells of enrichment or to the recipients of inadequate numbers of precursors. Low enough numbers of PHSC and standard competitor cells (0.5×10^5 to 5×10^5) are used so that variance in percentage of repopulation depends on relative repopulating abilities of the individual clones, with numbers of clones distributed according to the Poisson equation.

PHSC concentrations per 10^5 B6 marrow cells measured by two different competitive dilution experiments are 0.7–1.1 and 1.0–1.6 (95% confidence limits), which confirm previous results using other models. PHSC numbers in BALB are similar to those in B6 and far less than shown by the *long-term culture-initiating cell* (LTC-IC) assay, with the best estimates for fetal, young, or old cells being 1 PHSC per 100,000 BMC. Unlike B6, BALB relative repopulating abilities change with age; each fetal PHSC repopulates 50–100% better than a young adult PHSC. Old BALB PHSC repopulate about 66% as well as young adult PHSC.

Szilvassy et al. (1990) reported concentrations more than 10 times as high as we find from limiting dilution studies of PHSC. The conflict probably is due to their practice of using defective competitors. This illustrates the relative nature of cell competition within a mixture and emphasizes the need for normally functional standard competitors.

The competitive repopulation and dilution assays complement each other, as the former measures normally functioning PHSC populations and the latter focuses on behavior of individual clones. Importantly, competitive dilution overcomes the limitation of describing results of competitive repopulation with a binomial model where donor and competitor PHSC must contribute equally. To calculate expected percentages of donor cells in each recipient in the competitive dilution, a distribution based on a Poisson model is used to estimate contributions from donor PHSC to the differentiated cell populations, while hypothesizing that donor cell contributions can be a multiple of those from competitor cells.

Ways to Measure Hematopoietic Stem Cells

Differences between short-term and long-term multilineage repopulating precursors are deterministic, since they can be separated using cell surface markers. Spangrude et al. (1988) were the first to use specific cell surface markers to enrich PHSC. By this strategy, PHSC have been enriched very highly, and used in limiting dilution assays, even engrafting a small proportion of lethally irradiated recipients long term using a single donor cell. However, problems with enrichment still exist. It may severely reduce repopulating ability. Some markers are strain-specific, or are effective for fetal but not adult cells. Most seriously, enrichment markers fail to predict PHSC function, since cells carrying all the markers associated with long-term reconstitution proliferate greatly in vivo or in vitro without a proportionate increase in PHSC function.

Spleen colonies in vivo or myeloid cell colonies in vitro are measured to test precursor differentiation. Many types of colony-forming cells can be separated from PHSC, and colony assays often fail to predict PHSC function.

LTC-IC populations contain precursors capable of long-term repopulation in vivo, although they may not compete with normal BMC. The *cobblestone area forming cell* (CAFC) assay measures clusters of tightly packed rounded cells on stromal layers evaluated after 3–5 weeks by a limiting dilution assay. Müller-Sieburg and Riblet (1996) found that long-term CAFC gave results similar to LTC-IC. B6 mice gave the lowest levels of stem cells in marrow, BALB/c were about 5 times higher, and D2 gave the highest levels, 11 times more than B6. However, after serial transplantation, two million BMC from young B6, BALB, or D2 donors expanded to 276, 147, and 98 total RU, respectively. Thus, the strain pattern for self-renewal is the opposite of that for LTC-IC. Furthermore, by competitive dilution, PHSC numbers in BALB are similar to those in B6, far less than in the LTC-IC assay, with the best models having 1 PHSC per 100,000 BMC.

LTC-IC assays may measure an early step in PHSC differentiation. Perhaps B6 cells respond better to stimuli to self-renew, whereas BALB or D2 cells respond better to stimuli to differentiate. This would explain why D2 precursor cells cycle faster than those of B6, as they are responding better to stimuli to differentiate, giving an initial competitive advantage in allophenic mice. However, B6 cells respond better to stimuli to self-renew, giving them the advantage for long-term maintenance.

Summary of Methods Used

Mice

To distinguish donor and competitor (or recipient) cell types, we use mice carrying electrophoretically differing hemoglobin (*Hbb^s* or *Hbb^d*) and *Gpi1* isoenzymes (*$Gpi1^a$* or *$Gpi1^b$*), as detailed below. For the competitive repopulation and self-renewal studies from Chen et al. (2000a), young (2–6 months) and old (20–24 months) mice of normal inbred strains—B6 (*Hbb^s* and *$Gpi1^b$*), BALB (*Hbb^d* and *$Gpi1^a$*), and D2 (*Hbb^d* and *$Gpi1^a$*)— were used as donors. We used congenics differing at the *Hbb* and *Gpi1* loci as competitors, and young mice of the normal inbred strain were used as recipients. Thus, standard competitors were young congenic B6 (*Hbb^d* and *$Gpi1^a$*), BALB (*Hbb^s* and *$Gpi1^b$*), and D2 (*Hbb^s* and *$Gpi1^b$*), respectively. These congenic inbred mice were produced from single stocks that were each back-crossed for more than 18 generations. In all cases, young and old donors were compared with portions from the same pools of standard competitor BMC from young congenics differing from the donors at the *Hbb* and *Gpi1* loci.

For the competitive dilution studies from Chen et al. (2000b), recombinant inbred mice of the CXB-12 line were used as donors. Hybrid CByB6F1 (*$Gpi1^a/Gpi1^b$*) were used as recipients, since the hybrid *Gpi1* band readily identifies recipient cells. Recipients were given sublethal irradiation with 500 rads 14 days before transplantation and intraperitoneally injected with 100 μg/mouse of a *natural killer* (NK) cell antibody (anti-NK1.1, clone PK 136) 3 days before transplantation to deplete NK cells (NK-cell-depleted recipients). In this study, standard competitors were young CByB6F1 (*Hbb^s* and *$Gpi1^b$*) mice, produced by mating congenic BALB (*Hbb^s* and *$Gpi1^b$*) with normal B6.

All mice were produced at The Jackson Laboratory where they had free access to an NIH-31 (4% fat) diet and acidified water and were maintained under standard pathogen-free animal husbandry conditions. Before each experiment, old mice were examined to exclude those with abnormal circulating *white blood cell* (WBC) concentrations (WBC $> 18 \times 10^9$/l or $> 5 \times 10^9$/l), or with grossly visible subcutaneous, intestinal, liver, spleen, lung, lymph node, or thymus tumors.

Competitive Repopulation

BMC were extracted from two femurs and two tibias of each donor or carrier mouse through a 23-gauge needle into 2.0 ml of *Iscove's*

Modified Dulbecco's Medium (IMDM) and filtered through a 100-μm mesh, sterile, nylon cloth to remove debris. Nucleated cells were counted using a model ZBI Coulter Counter. BMC from each donor or carrier were mixed with BMC from a competitor pool in a 1:1 (donor:competitor) ratio, and then iv (*intravenously*) injected into lethally irradiated strain- and gender-matched recipients. Irradiation doses were 11 Gy for B6, 8.5 Gy for BALB, 10 Gy for D2, and were delivered at 1.6 Gy per minute using a Shepard Mark 1 ^{137}Cesium gamma source. Standard competitor BMC were distinguishable from donor BMC by different electrophoretic alleles at the *Gpi1* and *Hbb* loci. In each experiment, the original bone marrow donors were defined as old/young pairs at transplantation. The same competitor pool of BMC was used by each donor pair, so repopulating functions of old and young donors were measured relative to the same competitor standard. This allows the calculation of an old/young RU ratio for each donor pair, so variances can be demonstrated. In each experiment, with inbred strains, a group of control mice received only competitor BMC and were monitored to ensure that irradiated host PHSC did not produce detectable numbers of circulating erythrocytes and lymphocytes.

Serial Transplantation and Competitive Repopulation

To test PHSC self-renewal by serial transplantation, 2×10^6 BMC from each of the young and old B6, BALB, or D2 donors was injected into separate lethally irradiated, strain- and gender-matched, young carrier mice, using five carriers per donor. After 10 months, BMC were pooled from two healthy carriers of each original donor, mixed with BMC from congenic standard competitors as detailed above, and re-transplanted into lethally irradiated recipients. This was done to measure PHSC proliferation in the carriers by competitive repopulation as detailed in the previous section, except that BMC from each carrier were mixed with BMC from its competitor pool in a 5:1 (carrier: competitor) ratio. Carriers differed from donors at the *Gpi1* and *Hbb* markers, but were matched with the competitors used in the assay stage of the experiment. Thus, any host cell function from the carrier would reduce the number of donor cells measured.

Analyses of Donor Engraftment in Carriers

Recipients were bled at time points specified in each experiment. At each bleeding, three microhematocrit tubes of blood (75 μl/tube) were obtained from each recipient through the orbital sinus and were mixed in 2 ml of IMDM with 3.8% sodium citrate. Each sample was then layered onto 2 ml of lymphocyte separation medium and centrifuged at 500*g* for 50 minutes at room temperature to separate lymphocytes (middle layer) from erythrocytes (sediment at the bottom of the tube). Cell lysates were analyzed by cellulose acetate gel electrophoresis to separate isoforms of *Gpi1* and *Hbb*. Percentages of donor-type lymphocytes were calculated based on the proportion of their *Gpi1* after electrophoretic gel bands were quantitatively analyzed using a Helena Cliniscan 2 densitometer. Percentages of donor-type erythrocytes were calculated from the proportion of donor-type *Hbb* in each sample after electrophoretic separation of the two *Hbb* types. These methods have been detailed previously. After each sampling, percentages of donor-type blood cells (mean of donor erythrocytes and lymphocytes) of the four or five recipients for each donor were averaged to calculate RU of donor BMC, using the formula: RU = % donor × C/(100 – % donor). One RU is defined as the repopulating ability of 100,000 standard competitor BMC, where % donor is the mean percentage of donor-type erythrocytes and lymphocytes, and C is the number of repopulating units of competitor BMC, so that for 20 × 10^5 standard competitor BMC, C = 20. The RU:carrier ratio was calculated assuming that the two femurs and two tibias sampled from each carrier contained 25% of its total marrow cells.

Competitive Dilution Using Poisson Modeling

Portions containing 5 × 10^4 CXB-12 donor BMC (*Gpi1*^{a}/*Gpi1*^{a}) and 5 × 10^5 CByB6F1 standard congenic competitor BMC (*Gpi1*^{b}/*Gpi1*^{b}) were injected into each of 38 NK-cell-depleted CByB6F1 (*Gpi1*^{a}/*Gpi1*^{b}) recipients. The Poisson function, $P_i = e^{-N} \times (N^i/i!)$, gives the probability (P_i) that a recipient gets a certain number (i) of PHSC, where N is the average number of PHSC injected to each recipient. The proportion of recipients with 0% CXB-12 type cells 6 months after reconstitution was used to compute the number of PHSC (N) in CXB-12 BMC using the i = 0 term of the Poisson function; $P_0 = e^{-N}$ or $N = -\ln P_0$. This gave N = 1.2 for 5 × 10^4 CXB-12 donor BMC. Then the probabilities (P_i) of having 0, 1, 2, 3, or 4 donor PHSC in a recipient were, according to the Poisson function: 0.3012, 0.3614, 0.2169, 0.0867, or 0.0260, respectively. Since 5 × 10^5 competitor cells were used, N = 5.0 for the competitor, giving probabilities of having 0, 1, 2, 3, 4, 5, 6, 7, 8, or 9 competitor PHSC in a recipient as 0.0067, 0.0337, 0.0842, 0.1404, 0.1755, 0.1755, 0.1462, 0.0653, 0.0363, or 0.0181, respectively. In practice, we calculated two arrays of probabilities, one for the donor and one for the competitor, with i = 0–15 for 16 probabilities in each case. These form a 16 × 16 matrix of 256 combined probabilities, and each is associated with a donor:competitor PHSC ratio that can be used to predict a value representing donor contribution. In the current study, we used the highest 37 probability values since there were 37 recipients available 6 months after transplantation.

We used the 37 corresponding donor:competitor ratios in the Poisson modeling to estimate repopulating ability per cell for CXB-12 PHSC by comparing three hypothesized levels of repopulating abilities (*F*) per cell for CXB-12 PHSC: $F = 1$, $F = 1.4$, and $F = 2$, each relative to a repopulating ability per CByB6F1 standard PHSC of $F = 1$. From the value of *F*, a predicted percentage donor value was calculated for each of the 37 donor:competitor ratios associated with the 37 highest probabilities. This produced three sets of predicted donor percentage values, which were each ranked from low to high and compared to the 37 ranked observed values in a paired *t* test. The hypothesized *F* values that generated predictions significantly different from observed data were rejected. The *F* value whose predictions were not significantly different from the observed data was accepted to represent repopulating ability per cell for CXB-12 donor PHSC relative to those of the CByB6F1 standard.

Effects of strain and age on RU concentrations and total RUs per carrier were analyzed through variance analyses. For Poisson modeling, differences between ranked predicted values and ranked observed data were analyzed through paired *t* tests using the JMP statistical discovery software on "Fit Y by X" and "Fit Model" platforms, respectively.

Results and Discussion

Competitive Repopulation

Competitive repopulation demonstrates that PHSC senescence is strain-specific. We tested PHSC senescence by comparing PHSC function in young and old B6, BALB, or D2 mice through competitive repopulation in vivo. Within each strain, 2×10^6 genetically marked congenic standard competitor BMC were mixed with 2×10^6 BMC from each of four young and four old donors, and each mixture was used to engraft four irradiated recipients per donor. The % donor-type erythrocytes and lymphocytes in recipient peripheral blood measured the functional ability of PHSC from each young or old donor relative to the standard competitor. Throughout this study, percentages of lymphocytes and erythrocytes were closely correlated, so they were averaged in each recipient to calculate RU numbers. By definition, 10^5 competitor BMC produced one RU of functional ability.

RU numbers/10^5 BMC at 8 months after reconstitution were significantly higher in old than in young B6 mice, and were significantly lower in old than in young BALB and D2 mice, indicating that PHSC become senescent in BALB and D2 mice, but not in B6 mice. Measurements at 1, 3, 6, and 8 months all showed similar differences between young and old, so functional abilities of both short-term precursors and long-term PHSC were affected similarly by aging. The old/young RU ratio, the PHSC senescence phenotype, showed that BALB (0.57 ± 0.24) and D2 (0.48 ± 0.07) BMC decline in function with age, whereas B6 (2.52 ± 0.88) increase.

In a preliminary experiment, old/young RU ratios for BALB and D2 were 0.27 and 0.50, respectively, after 9 months. The gain in repopulating ability with age had already been shown in B6 mice, as had the loss with age in BALB mice. Furthermore, these data are very interesting when they are compared to the results of long-term self-renewal of PHSC in other portions of BMC from the same donors that were serially transplanted. They show that PHSC senescence is not tested as rigorously by this competitive repopulation assay as by an assay after serial transplantation and long-term self-renewal.

PHSC Self-renewal

Portions of the BMC samples taken from the donors were also transplanted into lethally irradiated, strain-matched, carrier mice, using 2×10^6 BMC per carrier and five carriers per donor. After 10 months, BMC from two of the five carriers were pooled and re-transplanted to lethally irradiated secondary recipients in competitive repopulation assays to measure RU concentration in carrier BMC. Mixtures of 5×10^6 carrier BMC plus 1×10^6 competitor BMC were re-transplanted to

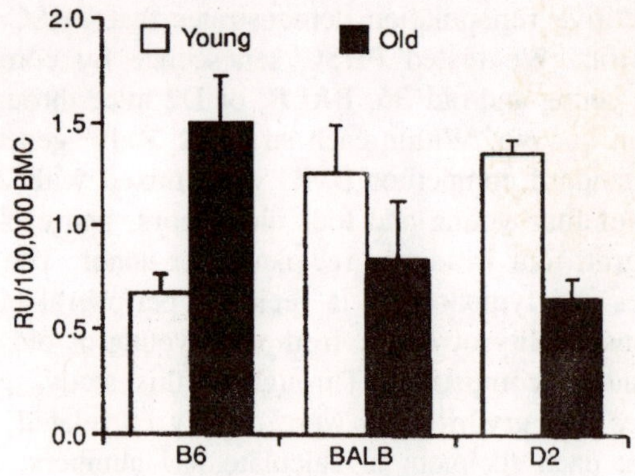

Fig. 7.2. Effects of age on PHSC function.

each of the secondary recipients, which were tested after 1, 3, 6, and 8 months. RU concentrations for carrier BMC were less than 10% of those for fresh donor BMC. Similar results were previously reported for young B6 donors.

The difference between total numbers of donor RU per carrier and numbers in the 2 × 10^6 donor BMC originally injected in each carrier represents RU expansion. However, the initial RU measures apparently included PHSC that failed to expand as well as they functioned in the initial assay. Therefore, the most meaningful measure of functional ability after long-term serial transplantation appears to be the total donor RU per carrier, each produced from a dose of 2 × 10^6 donor BMC. These values can be used to compare expansion in the three different genotypes, giving results in terms of RU produced in each carrier by the two million donor cells. PHSC self-renewal declined with donor age in BALB and D2. The strain differences were large with the following old/young ratios of total RU produced: 0.61 in B6, 0.24 in BALB, and 0.31 in D2.

Strain differences were most obvious comparing old donors. Old B6 donors with 30 RU produced 169 RU, while the 13 and 14 RU from old BALB or D2 donors produced only 35 and 30 RU, respectively. The B6 advantage was evident even comparing young donors: PHSC expansion in young B6 BMC (to 276 RU) was 1.9 times as high as in young BALB BMC (to 147 RU), and 2.9 times as high as in young D2 BMC (to 96 RU). The same genetic differences that retard PHSC senescence in the B6 strain may also increase PHSC self-renewal.

Measuring PHSC Senescence In Vivo

Studies in vivo that continue for a substantial fraction of the life span are required to test PHSC, since their most important feature is long-term function. Multilineage repopulation alone fails to demonstrate PHSC function, since short-term precursors can be multilineage; in fact, most of the multilineage precursors assayed over the short term are more differentiated than PHSC. In the current study, competitive repopulation directly measures how well PHSC produce mature lymphocytes and erythrocytes for about a third of the murine life span. Normal PHSC function in vivo is tested, while avoiding potentially confusing effects from unnatural environments in vitro or from cell manipulations used for enrichment.

PHSC Senescence is Genetically Regulated

The decline in PHSC repopulating ability with increasing donor age seen in BALB and D2 mice suggests that PHSC from these strains become senescent. This agrees with results from inbred CBA mice and B6↔D2 allophenic mice and fits well with the classic idea that cellular function decreases with donor age. Our observation that PHSC from old B6 donors repopulate recipients better than those from young B6 is also consistent with previous reports.

Effects of age on PHSC self-renewal were shown by the results from competitive repopulation after serial transplantation. By measuring re-transplanted RUs in carrier BMC, we compare the amount of expansion from PHSC in the original dose of donor marrow cells received by each carrier. Since the same competitor was used for old and young donors of the same strain, the total RU per carrier is directly comparable between each strain's young and old donors. In BALB and D2 mice, both repopulation and self-renewal declined with age, although measures of self-renewal showed a greater decline. In B6 mice, self-renewal declined slightly, but not significantly, with age, differing from competitive repopulation where B6 PHSC repopulation increased with age.

We suggest two possible mechanisms to explain the genetic differences. First, old PHSC may have a homing defect, and this defect may be much more severe in BALB and D2 than in B6 cells. Serial transplantation followed by competitive repopulation requires original donor PHSC to home twice, so effects of the homing deficiency would be amplified. Second, B6 PHSC may have a proliferative limit but at a much higher level than the limits of BALB and D2 PHSC. Serial transplantation requires extra proliferation and thus shows more of a functional decline with age. Such a decline is consistent with results reported in human hematopoietic stem cell candidates.

B6 Genes are Dominant in PHSC Senescence

The old/young RU ratio for the BALB×B6 F1 was larger than 1, as with B6. Old/young RU ratios calculated from data in Harrison (1983) for the B6xCBA and WBxB6 F1 hybrids are also larger than 1, whereas ratios for the CBA/CaJ parent strain are less. Thus, in different F_1 hybrids between B6 and a parent strain showing PHSC senescence, the B6 pattern is dominant.

Genetic Effects on PHSC Aging

The strain comparisons and genetic analyses reported here lead to the following model: PHSC self-renewal is limited in BALB and D2 mice, so PHSC from 20–24-month-old donors show functional defects. Far more self-renewal can occur in B6 PHSC, so that in old age, PHSC function increases to compensate for other aging defects; thus old donors have more PHSC function than young donors. This B6 phenotype is genetically dominant since it is shown by F_1 hybrids. However, when self-renewal is rigorously tested and PHSC from old and young B6 donors are serially transplanted into lethally irradiated carriers, even in the B6 strain there is a small decline with age. The decline with age in BALB and D2 is far greater; these strain differences in RU expansion in the carriers demonstrate an enormous range of PHSC proliferative potentials.

Competitive Dilution

An important step in defining the biology behind increased PHSC function is to test whether the increase is in numbers of PHSC, function per PHSC, or both. CXB-12 mice have more PHSC as well as more repopulating ability per PHSC. Each of the F_1 recipients is given 5×10^4 CXB-12 donor BMC mixed with 5×10^5 F_1 competitor BMC. After 6 months, 11 of the 37 recipients have 0% CXB-12 type (*Gpi1*[a]) blood cells. This gives the average PHSC number as $N = -\ln (11/37)$, so $N = 1.2$ PHSC in 5×10^4 BMC, or 2.4 PHSC in 10^5 BMC in CXB-12 mice.

To compare relative repopulating functions F of individual CXB-12 PHSC, we produce three sets of predictions. Of these, $F = 1.4$ times the function of the standard fits the real data almost exactly, and the differences between predicted and observed values for $F = 1$ or $F = 2$ are significantly larger than those for $F = 1.4$. Thus, each CXB-12 PHSC is 1.4 times as efficient as each F_1 PHSC in engrafting

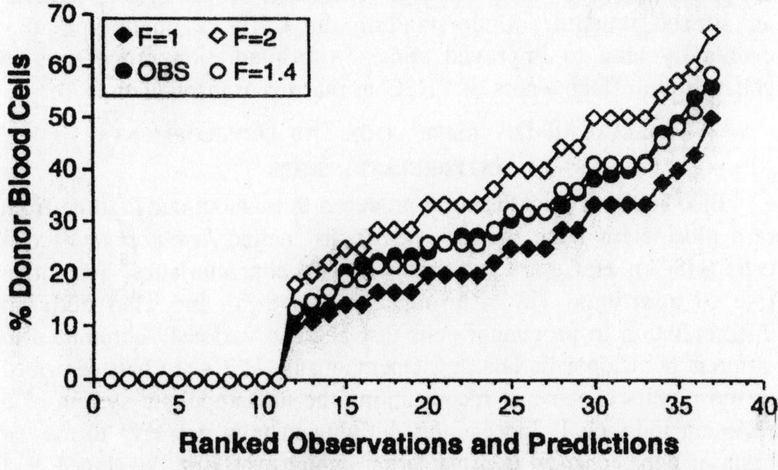

Fig. 7.3. CXB-12 cell numbers and repopulating abilities per cell.

F_1 recipients. Doses of CXB-12 BMC are very low, so hybrid resistance is not saturated, and thus it reduces PHSC numbers and increases the advantage of the F_1 hybrid competitor cells. Concentrations of 2.4 PHSC per 10^5 BMC, and functions of 1.4 times the standard, are minimum estimates.

CXB-12 BMC Do Not Kill Competitor PHSC

Since the model calculates data with the normal condition of 1 PHSC per 10^5 cells for the F_1 competitor, the extremely good fit between the model and real data suggests that there is no killing of standard competitor F_1 PHSC by CXB-12 BMC. Thus, the CXB-12 repopulating advantage does not result from an immune reaction or other destruction of the standard competitor cells. The degree of killing required would have reduced competitor numbers so the model failed to fit with the expected F_1 concentration of 1.0 PHSC per 10^5 competitor BMC. Additionally, if the F_1 standard PHSC N had been reduced by reaction with CXB-12 BMC, we would expect some recipients to have 100% donor-type cells. However, none of the recipients in this experiment had 100% donor-type cells.

The very high number and function of CXB-12 PHSC may be maintained by alterations in a receptor that regulates this cell type, or in its ligand. Increases in numbers of PHSC could result from changes in either ligand or receptor; however, the increased proliferative capacity of CXB-12 cells after transplantation in normal recipients suggests that the change is intrinsic to the PHSC, perhaps an altered receptor. Understanding the CXB-12 phenotype may eventually lead to improved clinical transplantation procedures to enhance the effectiveness of PHSC in marrow transplantation.

Molecular Diversification and Developmental Interrelationships

Blood cells are continuously produced throughout our lifetime from rare pluripotent bone marrow stem cells, called *hematopoietic stem cells* (HSCs). HSCs are endowed with two characteristics: They give rise to additional HSCs through self-renewal and also undergo differentiation to progenitor cells that become variously committed to different hematopoietic lineages. Operationally, HSCs are best described as those cells capable of reconstituting the hematopoietic system of a recipient individual. Indeed, this defining in vivo property forms the basis of bone marrow transplantation, which was first developed as a lifesaving clinical procedure nearly a half century ago.

Despite the recognition decades ago that HSCs exist, many questions regarding their origins, regulation, and developmental potential remain unresolved. These include (1) How are HSCs formed during development? (2) How do they choose between a resting state and self-renewal/ differentiation? (3) How is the remarkable diversity of blood cells (red cells, white cells [neutrophils and monocytes/ macrophages], T- and B-lymphocytes, megakaryocytes, mast cells, eosinophils) established at the molecular level? (4) How can our prior views and understanding of HSCs be reconciled with recent findings of unsuspected plasticity of HSCs and other somatic cells?

Definitions and Markers of HSCs

Characterization of HSCs by their function, that is, by their capacity to sustain long-term multilineage hematopoiesis in a recipient individual, has provided an assay system for cell populations separated by cell-surface markers defined by monoclonal antibodies to surface molecules. In general, the bone marrow of adult animals, most often mice, has been used as the source of potential HSCs. In the mouse, cells of the c-kit$^+$, sca-1$^+$, thy-1^{lo}, lineage-negative phenotype are able to reconstitute recipients at limiting dilution. Extrusion of dyes such as Hoechst 33324 and Rhodamine 123 has also been successfully used to identify populations greatly enriched in HSCs. A preoccupation in the field has been phenotypic description of the HSC. This goal has provided lively discussion among investigators, because it appears that

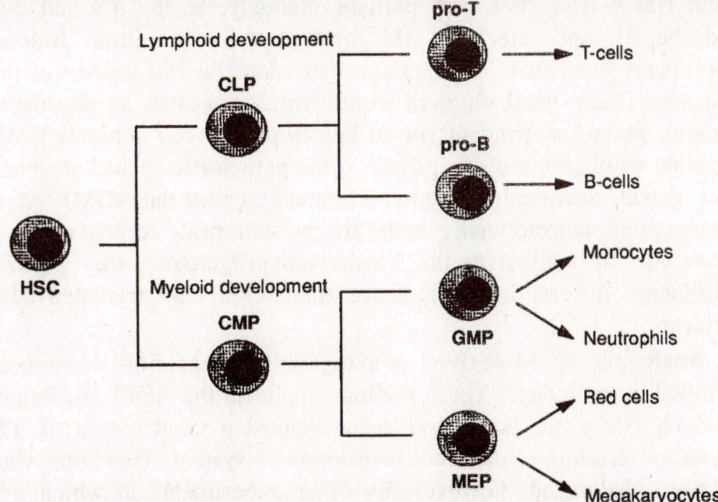

Fig. 7.4. Schematic view of hematopoiesis.

there are subsets of HSCs. For example, the surface marker CD34 was initially believed to be present on all HSCs. More recent evidence points to the existence of rare $CD34^-$ HSCs. Indeed, the $CD34^+$ state may reflect activation of HSCs.

Moreover, separation of cells on the basis of Hoechst fluorescence at two emission wavelengths allows the identification of both replicating and quiescent HSCs. The apparent diversity of HSCs and the failure of in vitro biological assays (at least to date) to define HSCs hamper *direct* molecular characterization of HSCs. Furthermore, study of the parameters controlling activation of HSCs from a resting state and symmetric versus asymmetric divisions has also been impeded. Techniques to expand HSCs in vitro or to immortalize HSCs while retaining their multipotential properties are needed both for improved biological characterization and for clinical application.

Origin(s) of HSCs

The hematopoietic system is an embryologic derivative of mesoderm. In vertebrates, hematopoiesis takes place at successive anatomic sites. Embryonic (or primitive) hematopoiesis occurs in the blood islands of the *yolk sac* (YS) (~E7.5–11 in the mouse). Definitive (or adult) hematopoiesis is transient in the *fetal liver* (FL) (~E11–16 in the mouse) but later sustained throughout life in the *bone marrow* (BM). This temporal pattern is consistent with a simple model in which HSCs arise first, and perhaps uniquely, in the YS and then seed the FL and later the BM through the circulation. Indeed, experiments more than 30 years ago were described in support of this sequence. Chick–quail chimera experiments suggested an alternative scenario, as an independent site of hematopoietic cell generation was identified within the embryo proper in the para-aortic splanchnopleura/aorta, gonad, mesonephros region (commonly called the AGM). AGM precursors of hematopoietic cells are present prior to the onset of circulation. In contrast to the YS-derived progenitors, they provide multilineage differentiation upon transplantation into irradiated adult recipients.

Analogous AGM-derived hematopoietic progenitors have been identified in the mouse. These findings implicate the AGM as a region in which HSCs are born, and argue against a contribution of YS-derived progenitors to the adult hematopoietic system. This latter view has been challenged, however, by other experiments in which YS-derived progenitors are transplanted into *newborn* rather than adult recipients. In this setting, definitive HSCs are observed in the YS, in fact in greater numbers than in the AGM region (at E9). As investigators continue to debate the relative contribution of YS and AGM regions to adult hematopoiesis, a working model posits that multiple, probably independent, origins of HSCs exist in development. The milieu into which HSCs are introduced (e.g., newborn or adult) influences how they are scored by in vivo assays.

A common, emerging theme in the complex origin of HSCs is an intimate relationship between hematopoietic and vascular development. In the vertebrate YS the first blood cells, embryonic red blood cells, arise in close apposition to the endothelial cells of the blood islands. This proximity and the temporal development of the blood and vascular lineages suggested the existence of a common precursor, the hemangioblast. Recent findings, provide experimental support for this hypothesis in the mouse.

Within the embryo proper as well, there is compelling evidence for a close interrelationship between the hematopoietic and vascular programs. Indeed, in this setting, data are more compatible with the emergence of definitive hematopoietic progenitors (presumably HSCs) from a subset of the vascular compartment (dubbed hemogenic endothelium). First, in chick–quail chimera experiments, two subsets of mesoderm have been defined: a dorsal one (the somite) that produces pure angioblasts (angioblastic potential) and a second, ventral one (the splanchnopleural mesoderm) that generates cells with dual endothelial and hematopoetic potentials. In agreement with these findings, $CD34^+$ hematopoietic cells have been observed in the floor of the aorta in the developing mouse and human.

In addition, putative definitive hematopoietic cells ($CD34^+$) are seen in the proximal umbilical and vitelline arteries, two potential sites of HSC emergence. In accord with this, intra-aortic, vitelline, and umbilical hematopoietic clusters are visualized in mice harboring a *lacZ* knock-in of the Cbfa2 (AML1) gene, which is essential for formation of definitive hematopoietic cells.

In vitro studies also argue for a role for endothelial cells in the origin of hematopoietic cells. For example, sorting of endothelial cells of E9.5 YS and embryos for a VE-cadherin$^+$, $CD34^+$, flk-1^+(a receptor for *vascular endothelial growth factor* [VEGF]), $CD31^+$, $CD45^-$, Ter 119^- phenotype yields a population able to generate blood cells of all lineages, including lymphocytes. An unresolved question is the relationship of these "*hemogenic*" endothelial cells to presumptive hemangioblasts.

Regulatory Genes Required for Formation or Maintenance of Stem Cells

Although the anatomic origins of HSCs are not fully resolved, several transcription factors have been demonstrated to be required for either the generation or maintenance/proliferation of HSCs. As with other transcriptional components, their in vivo requirements have been revealed through gene targeting of ES cells in mice. Two factors, SCL/tal-1 and Lmo2, are essential for the generation of any hematopoietic cells, either at the YS stage or later in development. Thus, these factors are necessary for all aspects of primitive and definitive hematopoesis. Moreover, both are required for proper vascular development in the YS, consistent with their expression in the hemangioblast. Remarkably, each factor was first identified through chromosomal translocations in acute T-cell leukemia (T-ALL). Indeed, the SCL/tal-1 gene, which encodes a member of the *basic helix-loop-helix* (bHLH) family, is activated by an upstream deletion or translocation in perhaps 25% of T-ALL, and SCL/tal-1 and Lmo2 expression is more frequently seen in this entity, even without evident chromosomal alterations. Lmo2 protein, a LIM-only polypeptide, is specifically associated with SCL/tal-1 as well as a novel LIM-interacting polypeptide, Ldb1. Moreover, a multi-protein complex containing SCL/tal-1, Lmo2, Ldb1, the ubiquitous bHLH heterodimerization partner E12/47, and GATA-1 has been observed in erythroid and progenitor cells. Whereas SCL/tal-1 (as a heterodimer with E12/47) binds to consensus E-box DNA target sites, Lmo2 does not recognize DNA by itself, but rather appears to provide a bridging function in transcription.

Precisely how SCL/tal-1 and Lmo2 function to promote hematopoietic development is as yet unresolved. The simplest interpretation of available data suggests that these factors are required for the specification of the hematopoietic fate from mesoderm. Expression of SCL/tal-1, Lmo2, and GATA-1 together in *Xenopus* embryos leads to hematopoietic specification in areas of the embryo that would normally exhibit other fates. In addition, injection of SCL/tal-1 RNA into wild-type zebrafish appears to promote both vascular and hematopoietic development, largely by expanding the hemangioblast population. Overexpression of SCL/tal-1 cDNA in zebrafish *cloche* mutant embryos also complements defects in blood and vessel development.

The genes acted upon by SCL-1/tal-1 and Lmo2 that act in transcription to specify hematopoiesis are largely unknown. One possible transcriptional target gene for SCL/tal-1 is that for the membrane tyrosine kinase c-kit, the receptor for stem cell factor (SCF, c-kit ligand), a growth factor for multipotential and mast cells. Defining the roles of SCL/tal-1 in transcription, however, is complicated by recent findings pointing to major DNA-binding-*independent* functions in development. For example, formation of primitive erythroid cells and definitive hematopoietic progenitors takes place without site-specific DNA binding by SCL/tal-1 but requires an intact HLH domain that directs heterodimerization. By inference, therefore, protein–protein interactions of SCL/tal-1, presumably with Lmo2 in this context, are pivotal for HSC generation.

The formation of definitive hematopoietic cells (and HSCs), but not YS erythroid cells, requires AML1 (also known as Cbfa2), a transcription factor related to *Drosophila runt*. As opposed to SCL/tal-1 and Lmo2, which are transcriptionally activated by chromosomal rearrangements, AML1 is expressed as a fusion protein with a variety of partners in diverse translocations in leukemia. In many of these instances, it is believed that the resultant fusion protein functions as a dominant-negative inhibitor of AML1 function. In the absence of AML1, mouse embryos die at mid-gestation without any definitive hematopoietic cells. No progenitors can be scored in colony assays, and homozygous mutants of a *lacZ* knock-in at the locus fail to generate hematopoietic clusters at sites of definitive hematopoietic cell formation, such as the floor of the aorta, and the vitelline and umbilical arteries. Again, the specific target genes regulated by AML1 that are required for HSC formation are unknown. From studies in myeloid cells, it has been demonstrated that growth factor receptors such as *granulocyte colony-stimulating factor* (G-CSF) may be controlled by AML1 in association with other myeloid-restricted transcriptional regulators.

Production of a normal number of HSCs appears to depend on the expression of yet another transcription factor, GATA-2, a member of the GATA family. GATA-2 is highly expressed in immature progenitors and then down-regulated in many, but not all, hematopoietic lineages. Forced expression of GATA-2 inhibits transition from a multipotential progenitor to a committed erythroid precursor. Loss of GATA-2 function in mice is embryonic-lethal due to marked anemia. Although the numbers of YS progenitors are modestly reduced in GATA-2$^{-/-}$ embryos, definitive progenitors are ~100-fold less than wild type. The precise level at which GATA-2 is required is uncertain. During mouse, zebrafish, and *Xenopus* development, GATA-2 expression parallels that of SCL/tal-1 and Lmo2 in regions fated for hematopoiesis. It is conceivable, therefore, that GATA-2 functions within transcriptional complexes, perhaps including SCL/tal-1 and Lmo2, involved in establishing the hematopoietic program. Alternatively, GATA-2 may be required somewhat later in the pathway, perhaps to sustain the viability or proliferative capacity of immature progenitors. The presence of HSCs, albeit at a greatly reduced number in the absence of GATA-2, is consistent with this latter view.

A pivotal decision for HSCs is whether to remain in a resting G_0 state or to commit to self-renewal or production of progenitors. Relatively little is known regarding the regulation of such choices. Recent evidence, however, suggests that the cyclin-dependent kinase inhibitor, p21 (cip1/waf1), may be required to maintain HSC quiescence. In $p21^{-/-}$ mice, HSC proliferation and their absolute numbers were increased under normal homeostatic conditions. However, these HSCs were more susceptible to cell-cycle-specific myelotoxic injury. Increased cell cycling also was observed to lead to stem cell exhaustion. Thus, p21 may constitute a molecular switch controlling the entry of HSC into cycle.

Differentiation and Diversification of Stem Cells

Although our understanding of how HSCs are induced to form is rudimentary, considerably more is known regarding the intrinsic cellular machinery for selection of lineage from a multipotential hematopoietic cell and subsequent differentiation. The growth and differentiation of hematopoietic cells are sustained by growth factors, cytokines that may act on cells of many lineages (e.g., SCF) or principally on a single one (e.g., erythropoietin). A consensus view, not held uniformly, is that signaling pathways activated through the receptor for these growth factors foster viability and are permissive for proliferation, but are not determinative with respect to lineage choice. That is, critical decisions of lineage are executed by nuclear regulatory factors, acting to establish transcriptional programs characteristic of each precursor and lineage. Rather than attempting to provide a comprehensive summary of hematopoietic transcriptional factors, here we illustrate some underlying principles that derive from recent studies.

Dominant Lineage Selection or Reprogramming

The sine qua non for a "*master*" regulator is its potential to alter the differentiation phenotype of a cell into which it is introduced. For ex vivo assessment of function, investigators rely on available cell lines that are not irreversibly "*frozen*" in a given state. In practice, the cell lines suitable for these experiments are limited. Thus, we may underestimate at this time the extent to which some transcription factors are competent to program differentiation of specific lineages. Nonetheless, several clear-cut examples of lineage reprogramming have been reported and serve to illustrate common principles.

The zinc-finger protein GATA-1 is normally expressed at a low level in multipotential progenitors and at higher levels in erythroid precursors, megakaryocytes, mast cells, and eosinophils. Knock-out studies in mice demonstrate that GATA-1 is essential for the maturation of both erythroid and megakaryocyte precursors, but not for their initial generation. It is hypothesized that the related GATA factor GATA-2 shares functions with GATA-1 and, in effect, substitutes for GATA-1 in lineage commitment. This view is consistent with forced expression experiments in two different systems. For example, expression of GATA-1 (or GATA-2 or GATA-3) in a myeloid cell 416B of mouse origin leads to the acquisition of megakaryocytic markers. Similarly, introduction of GATA-1 into chicken progenitors transformed with a *myb-ets* retrovirus reprograms cells to three different fates—erythroid, eosinophil, and megakaryocytic. The level at which GATA-1 is expressed appears to determine the specific lineage that arises. In both systems, lineage reprogramming is accompanied by down-regulation of the myeloid markers of the host cells. Thus, in addition to activating a megakaryocytic/eosinophilic/erythroid program, GATA-1 acts to turn off the program of a "*contralateral*" lineage in which it is not normally expressed. Lineage antagonism, established through critical hematopoietic transcription factors, is emerging as a newly recognized and consistent theme.

Another major hematopoietic transcription factor, PU.1, an ets protein, is able to induce myeloid lineage commitment in multipotential chicken progenitors, and in conjunction with GATA-1 and the basic-zipper protein C/EBP, promotes formation of eosinophils. Again, concomitant with up-regulation of myeloid or eosinophil markers, down-regulation of multipotential markers is observed. In the absence of PU.1 in the mouse, myeloid and lymphoid development is perturbed. In this setting, however, immature PU.$1^{-/-}$ progenitors are formed but are unable to proliferate or differentiate. Thus, as with GATA-1, gain-of-function experiments establish functions in lineage selection, whereas loss of function in vivo does not entirely ablate lineage commitment. This difference most likely rests on compensatory mechanisms available in vivo that are not operative in simpler, test cellular systems.

Combinatorial Positive Actions of Hematopoietic Transcription Factors

Whereas early views of mammalian differentiation highlighted the dominance of single "*master*" regulators, such as the bHLH factor myoD in myogenesis, a more sophisticated appreciation has emerged in recent years. Rather than acting in "*isolation*," key regulatory factors function within a cellular context. In experimental systems the outcome observed depends considerably on the character of the cells in which dominant regulators are expressed. Clear examples of this principle are apparent from studies of hematopoietic regulators.

In probing the mechanisms by which GATA-1 participates in transcription in the context of a GATA-1-negative erythroid precursor cell line, it became apparent that the transcriptional action of GATA-1 might depend on a cell-restricted cofactor. The hypothetical cofactor was isolated by a yeast two-hybrid screen and dubbed "*FOG*" for Friend of GATA-1. Indeed, physical interaction of FOG, a multitype zinc-finger protein, with the N finger of GATA-1 is essential for the completion of both erythroid and megakaryocytic differentiation. Mutations of the N finger of GATA-1 that specifically disrupt interaction with FOG, but retain the DNA-binding contribution of the finger, lead to impaired erythropoiesis and platelet formation and function, both in test systems in culture and in human patients. Similarly, myeloid development relies on the combined presence and action of PU.1 and C/EBP, and eosinophil formation on C/EBP and GATA-1. Although direct experimental verification is not available, one may extrapolate these concepts to suggest that T-lymphoid cells might be GATA-3^+, Ikaros factor$^+$, PU.1^-, GATA-1^-, whereas B cells are PU.1^+, GATA-3^-, GATA-1^-, Pax-5^+. By extension, each lineage can be defined by its transcription factor "*haplotype*", just as cell-surface phenotypes have traditionally been used in immunology.

Suppression of other Lineages as a Mechanism of Lineage Selection

Whereas lineage choice has traditionally been viewed as strictly a positive event, the concomitant down-regulation of markers of one or multiple lineages upon activation of a program, such as that described above for GATA-1, suggests that the manner in which multipotential cells choose a differentiation pathway may be more complex than heretofore imagined. Indeed, experiments that define the role of Pax-5 in B-lymphoid development illustrate how central suppression of alternative fates may be in lineage selection. Prior experiments had shown that Pax-5 is essential for B-cell development. More recently, it has become evident that pro-B cells lacking Pax-5 are not restricted in their lineage fate. Under stimulation with suitable cytokines, these pro-B cells are able to differentiate into macrophages, osteoclasts, dendritic cells, granulocytes, and natural killer cells. Remarkably, Pax-$5^{-/-}$ progenitors reconstitute the T-lymphoid system upon introduction into host mice. Thus, commitment to B-cell development through Pax-5 involves suppression of alternative lineages. Investing complementary actions (activation and suppression) in individual transcription factors provides a parsimonious means of coordinating developmental decisions, particularly where multiple potential fates are available.

Antagonism of Key Hematopoietic Factors and Reinforcement of Lineage Choices

Mechanisms mediating antagonism between lineages are becoming clarified. Recent findings suggest that cross-regulation between lineage-restricted transcription factors is exerted at the level of protein–protein interaction as well as in transcription. GATA-1 and PU.1 serve as the dominant factors for erythroid/megakaryocytic and myeloid development, respectively. Expression of PU.1 in erythroid precursors arrests differentiation and leads to erythroleukemia. Indeed, GATA-1 and PU.1 proteins interact. Association of the amino terminus and ets-domain of PU.1 with the C-finger of GATA-1 (or other GATA factors) blocks DNA binding by GATA-1. Although DNA binding by PU.1 remains unaffected, *trans*-activation by PU.1 is impaired upon interaction with GATA-1, possibly due to displacement of a transcriptional cofactor, proposed to be c-jun. Thus, cross-interactions between these factors serve to favor one lineage (erythroid/megakaryocytic or myeloid) at the expense of the "*contralateral*" option. Unknown mechanisms, either direct or indirect, then serve to down-regulate transcription of GATA-1 in myeloid lineages and PU.1 in erythroid/megakaryocytic cells.

Another example of cross-regulation in lineage choice is provided by recent observations of the roles of C/EBP and FOG in eosinophil selection. Chicken eosinophils generated in the myb-ets system by GATA-1 and C/EBP do not express appreciable FOG. Expression of FOG in these cells leads to down-regulation of eosinophil markers and acquisition of multipotential cell markers. Expression of a regulated C/EBP transgene in multipotential cells leads to rapid transcriptional shut-off of FOG, which acts in this setting as a repressor of the action of GATA-1 at transcriptional targets, such as the gene Eos47. Therefore, induction of eosinophil-specific genes is proposed to require

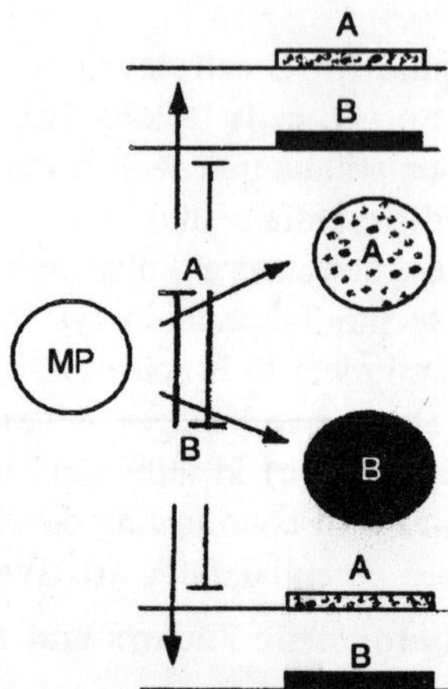

Fig. 7.5. Positive and antagonistic regulatory interactions.

repression of FOG expression, and consequently, relief of FOG repression of GATA-1. In the presence of enforced FOG expression, multipotential cells fail to differentiate into eosinophils. Differentiation reflects a "*collapse*" of the multipotential state, a finding reminiscent of that seen in the absence of Pax-5.

Antagonism established through cross-regulation of critical factors serves not only to execute a lineage choice, but also to reinforce this decision. Although irreversible differentiation might result, the option of reversing these regulatory loops remains, as long as chromatin events do not silence transcription of these factors.

Concentration-deendent Actions of Hematopoietic Transcription Factors

In reprogramming of chicken progenitors to eosinophil, erythroid, and thromboblast lineages by GATA-1, Kulessa and colleagues (1995) observed that the level at which GATA-1 was expressed correlated with the lineage outcome. Intermediate levels were observed in eosinophils and erythroid cells, and high levels in thromboblasts. Of interest, levels of expression differed by only a few fold between lineages. Although these findings might be explained by titration of DNA targets at different GATA-1 concentrations, it seems more likely that assembly (or disassembly) of protein complexes is influenced by subtle concentration changes, particularly given the nature of generally weaker protein–protein interactions. Reducing the expression of GATA-1 in vivo by approximately three-fold due to a targeted mutation of *cis*-regulatory elements results in markedly impaired erythroid cell maturation. Thus, high concentrations of GATA-1 protein are needed to drive cell maturation.

Concentration-dependent influences of PU.1 on differentiation have recently been noted. In particular, reintroduction of PU.1 into PU.1$^{-/-}$ progenitors rescues both macrophage and B-cell differentiation. However, high-level expression favors macrophage development and blocks B-cell development. It is hypothesized that PU.1 in this instance may antagonize the action of Pax-5 via protein interactions.

Coexpression of Lineage Factors in Multipotential Progenitors

Among the lineage-restricted transcription factors, the majority are expressed, at least at low level, in hematopoietic progenitors that have yet to choose a lineage. After lineage commitment, markers of other lineages are extinguished. Furthermore, additional markers of specific lineages, such as DNase I hypersensitivity of immunoglobulin loci, are present in multipotential cells but absent in selected lineages (erythroid, myeloid). Multilineage gene expression has also been documented at the single-cell level by PCR. These findings have led to the hypothesis that multipotential progenitors are "*testing*" various options or poised in a molecular sense to pursue any one of multiple paths. This notion of multipotential cell plasticity is readily reconciled with findings reviewed above regarding cross-regulatory interactions of transcription factors and the assembly of complexes that may be highly dependent on protein concentrations (if not modifications). We envision that immature progenitors may express regulatory factors generally ascribed to specific lineages, perhaps at low level, in bursts, or in a cell-cycle-dependent fashion. Lineage commitment is then associated with increased expression (or activity) of such factors, coupled with antagonistic mechanisms that serve to reinforce an initial, and rather tentative, decision.

Unexpected Plasticity of HSCs and Stem Cells for other Tissues

Cell biologic and molecular findings have provided a working conceptual framework in which to consider the events from HSC to individual hematopoietic lineages. Recent evidence, however, suggests unprecedented plasticity of HSCs. For example, populations of presumptive muscle stem cells (satellite cells) contribute to hematopoiesis upon transplantation into mice. Conversely, purified HSC populations contribute to muscle. Prior evidence also suggested that peripheral blood contains CD34$^+$, presumptive endothelial progenitors, whose relationship to traditional CD34$^+$ hematopoietic progenitors is unknown. The most extreme example of stem cell plasticity thus far described is the generation of hematopoietic cells in the mouse from cultured clonal neural stem cells. Although it has yet to be proven formally in these instances that isolated single cells are endowed with inherent plasticity of the magnitude revealed in these recent studies, it is highly unlikely that low-level contamination of one population with another can account for the results. Rather, it seems that cells are reeducated in some unknown manner by the local microenvironment to adopt new fates, or dedifferentiate into multipotent somatic cells that then re-differentiate along a new path. Although the signals mediating these phenomena are unknown, the suggestion that cell culture conditions may prime muscle stem cells for hematopoietic contribution in vivo hints at experimental approaches to dissecting the relevant components. Determining the environment cues responsible for stem cell plasticity and relating these to the control of critical transcription factors are challenges for the future.

8

Vascular and Muscle Stem Cells

Embryonic stem (ES) cells, the undifferentiated cells of early embryos are established as permanent lines and are characterized by their self-renewal capacity and the ability to retain their developmental capacity in vivo and in vitro. The pluripotent properties of ES cells are the basis of gene targeting technologies used to create mutant mouse strains with inactivated genes by homologous recombination.

ES cells cultivated as embryo-like aggregates, called *embryoid bodies* (EBs), differentiate in vitro into cellular derivatives of all three primary germ layers of endodermal, ectodermal, and mesodermal origin. ES cell lines develop from an undifferentiated stage resembling cells of the early embryo into terminally differentiated stages of the cardiogenic, myogenic, neurogenic, hematopoietic, adipogenic, or chondrogenic lineage, as well as into epithelial, endothelial, and *vascular smooth muscle* (VSM) cells. Terminally differentiated ES cells also show pharmacological and physiological properties of specialized cells: in vitro differentiated cardiomyocytes have characteristics typical of atrial-, ventricular-, Purkinje-, and pacemaker-like cells, and neuronal cells are characterized by inhibitory and excitatory synapses. Neuronal, cardiac, and VSM cells express functional receptors typical for each cell type.

Differentiation protocols for the development of ES cells into cardiomyocytes, skeletal muscle, or VSM cells have been well established. The in vitro differentiation of ES cells allows investigators (1) to analyze developmental processes during the differentiation of stem cells into specialized cell types and early processes of commitment to specific lineages; (2) to study the effects of differentiation factors or xenobiotics on embryogenesis in vitro; (3) to investigate pharmacological effects on functionally active cardiac or VSM cells (which are otherwise not available from in vitro cultivated cells); and of potentially the greatest importance, (4) to establish strategies for cell and tissue therapy.

The following parameters influence the developmental potency of ES cells in culture: (1) the number of cells differentiating in the EBs; (2) the media, quality of *fetal calf serum* (FCS), growth factors, and additives; (3) the ES cell lines used; and (4) the time of EB plating. Genetic manipulation of ES cell lines through either "*gain-*" or "*loss-of-function*" strategies, when used in conjunction with established differentiation protocols, also permits the testing of specific hypotheses related to the development of these cell types. The *gain-of-function* and *loss-of-function* analyses are excellent alternatives and substitutes to in vivo studies with transgenic animals to analyze the consequences of mutations on early embryogenesis and development. Loss-of-function analyses are especially helpful for the investigation of those mutations that result in early embryonal death of homozygous embryos.

In principle, EB aggregates of ES cells develop into many differentiated cell types. To obtain maximal differentiation of a defined cell type, specific cell lines and cultivation conditions have to be used.

In this chapter, we describe methods to differentiate ES cells into functionally active cardiac, skeletal muscle, and VSM cells and to characterize their phenotypes. Furthermore, strategies are presented for the genetic manipulation of ES cell differentiation.

Materials

Cells

The following cell lines have been used for in vitro differentiation into cardiac, skeletal muscle, and VSM cells:

1. Cardiac muscle differentiation: D3, R1, Bl17, AB1, AB2.1, CCE, and E14.1.
2. Skeletal muscle cell differentiation: D3, BLC6, AB1, and AB2.1.
3. VSM cell differentiation: D3, AB1, and AB 2.1.

In addition to ES cells, embryonic germ (EG) cells, i.e., EG-1, or *embryonic carcinoma* (EC) cells, i.e., P19, may be used for differentiation into cardiac and skeletal muscle cells.

Media, Reagents, and Stock Solutions

Solutions for cell culture

1. Dextran T500.
2. 0.01 *M* Tris-HCl, pH 8.0.
3. Activated charcoal.
4. Phosphate-buffered saline (PBS): containing 10 g NaCl, 0.25 g KCl, 1.44 g Na_2HPO_4, 0.25 g KH_2PO_4 × 2 H_2O/L, filter-sterilized through a 0.22-μm filter.
5. Trypsin solution: 0.2% trypsin 1:250 in PBS, filtersterilized through a 0.22-μm filter.
6. Ethylenediaminetetra acetate (EDTA) solution: 0.02% EDTA in PBS, filter-sterilized through a 0.22-μm filter.
7. Trypsin-EDTA: mix trypsin solution and EDTA solution at 1:1.
8. Gelatin solution: 1% gelatin in double-distilled water, autoclaved, and diluted 1:10 with PBS. Coat tissue culture dishes with 0.1% gelatin solution for 1–24 h at 4°C before use.
9. Mitomycin C (MC) solution: dissolve 2 mg MC in 10 mL PBS, filter-sterilize. From this stock solution, dilute 300 μL into 6 mL of PBS (final concentration is 0.01 mg/mL). MC stock solution should be freshly prepared at weekly intervals and stored at 4°C. *Caution*: MC is carcinogenic.
10. β-mercaptoethanol (β-ME): prepare a stock solution from 7 μL of β-ME into 10 mL of PBS (stock concentration is 10 m*M*). Make fresh at weekly intervals and store at 4°C.
11. Cultivation medium I: Dulbecco's modified Eagle's medium (DMEM) (4.5 g/L glucose) supplemented with 15% FCS for feeder layer cells.
12. Additives I: to 100 mL media, add 1 mL of 200 m*M* L-glutamine stock (100X), 1 mL of β-ME stock, 1 mL of *nonessential amino acids* (NEAA) stock (100×).
13. Additives II: to 400 mL medium, add 40 μL of a 3×10^{-4} *M* stock solution of Na-selenite, 10 mL of 7.5% stock solution of *bovine serum albumin* (BSA), and 1 mL of stock solution (4 mg/mL) of transferrin.
14. Cultivation medium II: DMEM supplemented with 15% FCS (selected batches) and additives I for ES cell cultivation.
15. Differentiation medium I: DMEM (or Iscove's modification of DMEM [IMDM]) supplemented with 20% FCS and additives I

for EB differentiation into cardiac, skeletal muscle, and VSM cells.

16. Monothioglycerol (3-mercapto-1,2-propanediol [MTG]): Prepare a stock solution from 13 μL of MTG into 1 mL of IMDM, filter-sterilized through a 0.22-μm filter. Make fresh before use. To 100 mL media, add 300 μL of stock solution (final concentration is 450 μM).
17. Differentiation medium II: DMEM supplemented with 15% dextran-coated charcoal-treated FCS (DCC-FCS) and additives I and additives II to analyze effects of growth factors on differentiation or for EB differentiation into neuronal or myogenic cells.
18. Retinoic acid (RA): prepare in the dark a 10–3 *M* stock solution of all-*trans* RA in 96% ethanol or DMSO. Store aliquots at –20°C and use a fresh sample for each experiment.
19. Dibutyryl-cyclic adenosine monophosphate (db-cAMP): dilute in double-distilled water to a 0.1 *M* stock solution. Store aliquots at –20°C for at least 1 mo.
20. Transforming growth factor β_1 (TGF β_1): prepare a 20 ng/μL stock solution of TGF β_1 in 0.1 *M* acetic acid or 0.05 *M* HCl. Store aliquots in silanized glass tubes at –20°C for at least 2 mo.

Solutions for reverse transcription polymerase chain reaction

1. Diethyl pyrocarbonate-treated water (DEPC-H_2O): add 1 mL DEPC to 1 L double-distilled or Milli-Q water (dH_2O) and stir overnight. DEPC is inactivated by heating to 100°C for 15 min or autoclaving for 15 min.
2. RNA lysis buffer: add 23.6 g of guanidinium thiocyanate to 5 mL of 250 m*M* Na-citrate, pH 7.0, 2.5 mL of 10% sarcosyl, and add DEPC-H_2O to a total volume of 49.5 mL, and mix carefully. Make fresh at monthly intervals. Add 1% β-ME before use.
3. 2 *M* Na-acetate, pH 4.0: dissolve 27.2 g of Na-acetate × 3 H_2O in 0.1% DEPC-H_2O, adjust the pH to 4.0 with glacial acetic acid, and adjust to 100 mL with DEPC-H_2O. Treat the buffer with 0.1% DEPC-H_2O at 37°C for at least 1 h and heat to 100°C or autoclave for 15 min.
4. Acidic phenol: phenol is saturated with DEPC-H_2O instead of Tris. The saturated acidic phenol contains 0.1% hydroxyquinoline (antioxidant, partial inhibitor of RNase, and a weak chelator of metal ions; its yellow color provides a convenient way to identify the organic phase). Store at 4°C for up to 2 mo.

5. Chloroform:Isoamylalcohol (24:1).
6. 75% Ethanol:prepare in DEPC-H_2O.
7. 25 m*M* $MgCl_2$.
8. 10X Polymerase chain reaction (PCR) buffer II: 100 m*M* Tris-HCl, pH 8.3, 500 m*M* KCl.
9. RNase inhibitor: 20 U/μL.
10. Oligo $d(T)_{16}$: 50 μ*M* in 10 m*M* Tris-HCl, pH 8.3.
11. Random hexamers: 50 μ*M* in 10 m*M* Tris-HCl, pH 8.3.
12. MuLV murine leukemia virus reverse transcriptase: 50 U/μL.
13. *Ampli*Taq DNA polymerase: 5 U/μL.
14. 5 m*M* dNTP mixture: 100 m*M* of each dNTP (dGTP, dATP, dCTP, and dTTP) dilute to 20 m*M* with DEPC-H_2O and freeze at –20°C. dNTP (5 m*M*) mixture is freshly made by mixing the equal volumes of 20 m*M* of each dNTP before use.
15. Select PCR primer pairs: dilute synthetic oligonucleotides to 10 m*M* with DEPC-H_2O and freeze at –20°C (a critical step in the PCR).
16. Glycogen: 20 mg/mL.
17. 5 *M* NaCl: dissolve 29.2 g of NaCl in dH_2O, adjust to 100 mL with water and autoclave.
18. TE buffer: 10 m*M* Tris-HCl, 1 m*M* EDTA, pH 7.5, filter-sterilized through a 0.22-μm filter.
19. 6X Loading buffer: 0.25% bromophenol blue, 0.25% xylene cyanole FF, 30% glycerol in dH_2O.
20. 5X TBE: dissolve 54 g Tris-base and 27.5 g boric acid in dH_2O, add 20 mL of 0.5 *M* EDTA, pH 8.0, and adjust to 1 L with dH_2O.
21. Ethidium bromide aqueous solution: 1% (w/v) = 10 mg/mL.
22. Agarose gels: melt electrophoresis grade agarose in 1X TBE by gentle boiling in a microwave oven. Cool to $<60°C$ and pour into an agarose gel mold. Run small gels at around 80–100 V by using bromophenol blue and xylene cyanole FF in the stop mixture as an indicator of migration.

Solutions for immunohistochemical analysis

1. 3.7% Paraformaldehyde (PFA): dissolve 3.7 g PFA in PBS, adjust to 100 mL with PBS, heat the mixture to 95°C, stir until the solution becomes clear, and cool to room temperature.

   Caution: PFA is toxic. Work under the hood and use gloves.
2. Methanol:acetone (7:3) fixative.

3. 10% goat serum or 1% BSA in PBS for blocking unspecific binding of antibodies.
4. Mounting medium: Vectashield.
5. 0.02% Triton-X 100 in PBS.
6. 0.5% BSA in PBS for dilution of secondary antibodies.

Solutions for single cardiac cell isolation

1. Supplemented low-Ca^{2+} medium: 120 m*M* NaCl, 5.4 m*M* KCl, 5 m*M* $MgSO_4$, 1 m*M* EGTA, 5 m*M* Na pyruvate, 20 m*M* glucose, 20 m*M* taurine, 10 m*M* HEPES-NaOH, pH 6.9, at 24°C, supplemented with 1 mg/mL collagenase B and 30 μM $CaCl_2$.
2. KB medium: 85 m*M* KCl, 30 m*M* K_2HPO_4, 5 m*M* $MgSO_4$, 1 m*M* EGTA, 5 m*M* Na-pyruvate, 5 m*M* creatine, 20 m*M* taurine, 20 m*M* glucose, freshly added 2 m*M* Na_2ATP, pH 7.2, at 24°C.

Solutions for introduction of DNA into ES cells

1. Plasmids ideally purified by cesium-chloride gradient ultracentrifugation or by ion-exchange chromatography. DNA should be suspended in H_2O or TE buffer, pH 8.0.
2. Restriction enzymes with appropriate digestion buffers as recommended by the manufacturer.
3. Phenol:chloroform:isoamyl alcohol (25:24:1, v/v/v). Phenol: equilibrated with 0.1 M Tris, pH 8.0, and containing 8-hydroxyquinoline (0.1%).
4. 3 *M* sodium-acetate, pH 5.2, sterilized by autoclaving.
5. 100 and 70% Ethanol for DNA precipitation and washes, respectively.
6. PBS: Dulbecco's PBS, containing 8 g NaCl, 0.2 g KCl, 0.2 g KH_2PO_4, 1.15 g Na_2HPO_4/L, sterilized by autoclaving.
7. Leukemia inhibitory factor (LIF): 10^7 U/mL.
8. Cultivation medium II + 1000 U/mL LIF.
9. Geneticin sulfate (G418) stock solution, puromycin, hygromycin B, gancyclovir. The concentration of each solution needed to kill ES cells should be determined for each batch of selection agent.

Solutions for analysis of transfectants

Rapid DNA isolation from ES cells

1. Cultivation medium I, PBS, 0.1% gelatin in PBS.
2. Proteinase K: dissolve lyophilized proteinase K at 20 mg/mL in dH_2O. Store 125-μL aliquots at –20°C for up to 1 yr in sterile microcentrifuge tubes.

3. DNA lysis buffer (final concentrations): 10 m*M* Tris-HCl, pH 7.5, 10 m*M* EDTA, 10 m*M* NaCl, 0.5% sarcosyl, and 1 mg/mL proteinase K (added fresh).
4. NaCl/ethanol solution: 1.5 μL of 5 *M* NaCl to 100 μL of cold absolute ethanol. The salt will precipitate. Mix to obtain a uniform suspension.

PCR

1. 10X Mg^{2+}-free DNA polymerase buffer: 500 m*M* KCl, 100 m*M* Tris-HCl (pH 9.0 at 25°C), 1% Triton-X 100.
2. 20 m*M* dNTPs stock solution: dilute 100 m*M* stock solution, 1:5 with 10 m*M* Tris-HCl, pH 7.5.
3. 25 m*M* $MgCl_2$.
4. *Taq* DNA polymerase in buffer B.
5. 10 μM PCR primer stock solutions: combine aliquots of primers for PCR and dilute to a concentration of 10 μM in TE. (Primer sets: β-globin as internal PCR control [*sense primer*: AGG TGA TAA CTG CCT TTA ACG A, *antisense primer*: CCC AGC ACA ATC ACG AT]; neomycin sequence primers [*sense primer* TAT TCG GCT ATG ACT GGG CAC AA, *antisense primer*: AGC AAT ATC ACG GGT AGC CAA CG]; for puromycin [*sense primer*: CAG GAA GCT CCT CTG TGT CCT C, *antisense primer*: GCT TAT CCA GTG GAG TGC TGG GTT]).
6. Molecular weight markers: 100 bp DNA Ladder.
7. Strain specific mouse genomic DNA.

Southern blot analysis

1. High concentration restriction enzymes and appropriate digestion buffers as recommended by the manufacturer.
2. RNase A (DNase-free): dissolve lyophilized RNAse A to 10 mg/mL in sterile 5 m*M* Tris-HCl, pH 7.5. Place the preparation at 80°C for 15 min to inactivate any deoxyribonuclease activity. Prepare 100-μL aliquots and store at –20°C.
3. 20X Sodium chloride sodium phosphate EDTA (SSPE): 3.0 *M* NaCl, 0.2 *M* NaH_2PO_4, 0.02 *M* EDTA.
4. 20% Sodium dodecyl sulfate (SDS).
5. Sheared salmon sperm DNA.
6. 50X Denhardt's solution: 1% Ficoll type 400 1% polyvinylpyrrolidone, 1% BSA.
7. Molecular weight markers (λ DNA/*Hind*III fragments).

8. Ready-to-Go DNA labeling (-dCTP) kit.
9. ProbeQuant G-50 Micro Columns.
10. Hybond-N+ Nylon membrane.
11. Prehybridization solution: 50 mL containing 12.5 mL 20X SSPE, 5 mL 50X Denhardt's solution, 1.25 mL 20% SDS (add last), 31.1 mL dH_2O, and 0.15 mL of denatured 10 mg/mL sheared salmon sperm DNA.

Equipment

1. Tissue culture plates: 35 mm, 60 mm, 100 mm.
2. 6-well, 24-well (Falcon) and 96-well microwell plates.
3. Pasteur pipets, 2-, 5-, 10-, 25-mL pipets.
4. Bacteriological Petri dishes: 60 mm for EB mass culture, 100 mm for EB hanging drop culture.
5. Tissue culture plates (60 mm) with sterilized coverslips (n = 4) for immunofluorescence.
6. 2-mL Glass pipets for preparing single-cell suspensions.
7. For feeder layer culture: sterile dissecting instruments, screen or sieve (about 0.5 to 1 mm diameter), Erlenmeyer flasks with stir bars, centrifuge tubes.
8. Tissue culture incubator with 37°C and a 5% CO_2 atmosphere.
9. For pharmacological analysis, the inverted microscope Diaphot-TMD (Nikon) equipped with a 37°C heating plate and a CO_2 chamber is used; the computer-assisted imaging system is coupled via a one-chip *charge-coupled device* (CCD) camera to a computer imaging station (Pentium CPU, 100 MHz) running the LUCIA Laboratory Imaging System including the "HEART" application.
10. GeneQuant RNA/DNA calculator.
11. PCR apparatus: mastercycler gradient or Gene Amp PCR System 9700.
12. 0.5- and 1.5-mL Microtubes and 20-, 100-, and 1000-μL filtertips.
13. Electrophoresis equipment.
14. A CN-TFX darkroom is coupled via GelPrint Workstation to a computer station with digital graphic printer running the PhotoFinish v3.0 program. Alternatively, a Bio-Rad Gel Doc 1000 and Multi-Analyst Version 1.1 software can be used.
15. TINA2.08e software.
16. Inverted Confocal Laser Scanning Microscope (CLSM) LSM-410 equipped with an argon-ion laser.

17. Bio-Rad Gene Pulser II system with capacitance extender and Gene Pulser disposable cuvets, 0.4-cm electrode gap.
18. UV Stratalinker.
19. Multichannel pipets.

METHODS

Cultivation of Undifferentiated ES Cells on Feeder Layer

Feeder layer culture

1. Remove embryos from a mouse pregnant for 15 to 17 d (i.e., NMRI or CD-1 outbred strains. For neomycin-resistant feeder cells, use MTK-neo^R or Zeta mice [NIA] or other available strains), rinse in PBS, and remove placenta and fetal membranes, head, liver, and heart. Rinse the carcasses in trypsin solution.
2. Mince the embryonic tissue in 5 mL of fresh trypsin solution and transfer to an Erlenmeyer flask containing a stir bar.
3. Stir on magnetic stirrer for 25 to 45 min (use longer incubation time if the embryos are older), filter the suspension through a sieve or a screen, add 10 mL of culture medium I and spin down.
4. Resuspend the pellet in about 3 mL of culture medium I and plate on 100-mm tissue culture plates (about 2×10^6 cells/100 mm-dish) containing 10 mL culture medium I, incubate at 37°C and 5% CO_2 for 24 h.
5. Change the medium to remove debris, erythrocytes, and unattached cellular aggregates, cultivate for an additional 1 to 2 d.
6. Passage the primary culture of mouse embryonic fibroblasts: split 12 to 13 on 100-mm tissue culture plates, grow in culture medium I for 1 to 3 d. The cells in passages 2–4 are most suitable as feeder layer for undifferentiated ES cells.
7. Incubate feeder layer cells with MC buffer for 2 to 3 h, aspirate the MC solution, wash three times with PBS, trypsinize feeder cells, and replate to new gelatin (0.1%)-treated microwell plates or to Petri dishes. Feeder layer cells prepared 1 day before ES cell subculture are optimal.

Culture of undifferentiated ES cells

It is important to passage ES cells every 24 or 48 h. Do not cultivate longer than 48 h without passaging, or the cells may differentiate and be unsuitable for differentiation studies. Selected batches of FCS have to be used for ES cell culture.

1. Change the medium 1 to 2 h before passaging.
2. Aspirate the medium, add 2 mL of trypsin-EDTA, and incubate at room temperature for 30 to 60 s.
3. Remove carefully the trypsin-EDTA mixture and add 2 mL of fresh cultivation medium II.
4. Resuspend the cell population with a 2-mL glass pipet into a single-cell suspension and split 1:3 to 1:10 to freshly prepared (60 mm) feeder layer plates.

In Vitro Differentiation

Preparation of growth factor-free FCS (DCC-FCS)

To analyze the influence of growth factors on ES cell differentiation, the medium should be free of high molecular weight proteins, like growth and differentiation factors. We use the DCC treatment of FCS. Prepare DCC-FCS as follows:

1. Dissolve 0.45 g dextran T500 in 1800 mL 0.01 *M* Tris-HCl, pH 8.0, add 4.50 g activated charcoal, and stir the mixture at 4°C in a tightly closed Erlenmeyer bottle overnight.
2. Inactivate FCS by incubation at 56°C for 30 min.
3. Fill 50 mL DCC solution in plastic centrifuge tubes and centrifuge at 2000*g* for 20 min, remove and discard the supernatant, repeat the procedure in the same tube without removing the pellet (= double pellet).
4. Add 50 mL of FCS to the tube with a double pellet, transfer the mixture to a clean glass bottle, and incubate for 45 min at 45°C in a water bath under shaking.
5. Centrifuge the mixture at 2000*g* for 20 min and transfer the supernatant to another centrifuge tube. Repeat steps 3–5.
6. Collect the FCS supernatant in a clean centrifuge tube and sterilize through a 0.22-μmfilter (low protein binding) into sterile flasks. Add at required concentrations to differentiation medium II.

Differentiation protocols

For the development of ES cells into differentiated phenotypes, cells must be cultivated in 3-dimensional aggregates called *embryoid bodies* (EBs) by the hanging drop method, by mass culture, or by differentiation in methylcellulose. The differentiation of cardiac, skeletal, and VSM cells requires different conditions, and these may vary for the particular ES cell line used. In this chapter, differentiation protocols utilizing the hanging drop method are described.

1. Prepare a cell suspension containing a defined ES cell number of 400, 600, or 800 cells in 20 μL of differentiation medium.
2. Place 20 μL drops (n = 50–60) of the ES cell suspension on the lids of 100-mm bacteriological Petri dishes containing 10 mL PBS.
3. Cultivate the ES cells in hanging drops for 2 d. The cells will aggregate and form one EB per drop.
4. Rinse the aggregates carefully from the lids with 2 mL of medium, transfer into a 60-mm bacteriological Petri dish with 5 mL of differentiation medium, and continue cultivation in suspension for 2 to 5 d until the time of plating.
5. Aspirate gelatin solution from pretreated 24-well microwell plates or 60-mm tissue culture dishes and add the appropriate differentiation medium.
6. Transfer a single EB into each well of gelatin (0.1%)-coated microwell plates for morphological analysis, or transfer 20 to 40 EBs per dish onto 60-mm tissue culture dishes containing 4 coverslips (10 × 10 mm) for immunofluorescence, or 15 to 20 EBs onto 60-mm tissue culture dishes for reverse transcription PCR (RT-PCR) analysis of EB outgrowths.
7. Change the medium during EB differentiation every second or third day.
8. To characterize the EB outgrowths morphologically, calculate the percentage of EBs with the specific differentiated cell type (from EBs of at least 48 wells) or calculate the amount of the differentiated cell type as a percentage of the outgrowth area of each EB.

Cardiac muscle cell differentiation

1. Use of 400–600 cells of ES cell lines D3, R1, or CCE for preparation of EBs is optimal for cardiac differentiation.
2. Culture with differentiation medium I.
3. Plate EBs onto gelatin-coated tissue culture plates at d 5 to 7. The first beating clusters in EBs can already be seen in 7-d-old EBs, but maximal cardiac differentiation is achieved after EB plating.
4. For the investigation of early cardiac stages, plate EBs at d 5.

If EC cells are used, they have to be induced to cardiac differentiation by treatment with 1% dimethyl sulfoxide (DMSO) between the first 2 d of EB development and plating at d 5 to 7. EC cells are cultivated without feeder cells and result after DMSO induction in a high number of cardiac cells.

Skeletal muscle cell differentiation

1. Use of 600 cells of ES cell line BLC6 or 800 cells of lines D3, R1, or EG-1 per EB are optimal for myogenic differentiation.
2. Culture EBs with differentiation medium I or II (line BLC6).
3. Plate EBs at d 5. The first myoblasts appear 4 (line BLC6) or 5 to 6 d (D3 and R1, respectively) after EB plating. Skeletal muscle cells begin to fuse into myotubes in the EB outgrowths 1 to 2 d later.
4. A specific differentiation induction of skeletal muscle cells from ES or EC cells is achieved by 10^{-8} *M* RA or 1% DMSO.

Vascular smooth muscle cell differentiation

1. ES cells of line D3 (n = 800) are differentiated as EBs in hanging drops in differentiation medium I.
2. Plate EBs at d 7 and induce differentiation of VSM cells by treatment with 10^{-8} *M* RA and 0.5×10^{-3} *M* db-cAMP between d 7 and 11 after plating (duration and treatment time has to be optimized for each cell line).
3. The first spontaneously contracting VSM cells, which express the vascular-specific splice variant of the VSM *myosin heavy chain* (MHC) gene, appear in the EBs around 1 wk after plating.
4. Change the medium during the differentiation period every day or every second day. A similar VSM cell induction is achieved by cultivating D3 cells (n = 600) as EBs in hanging drops in differentiation medium II containing 2 ng/mL TGF β_1 from d 0 to 5, and plating of EBs at d 5. The first spontaneously contracting VSM cells appear in EBs 10 d after plating, and maximal VSM cell differentiation (60%) is achieved at d 5 + 24 to 5 + 28.

   Alternatively, ES (AB1 or AB2.1) cells (n = 400) were differentiated in hanging drops in M15 (DMEM plus 15% FCS, 0.1 m*M* β-ME, 2 m*M* L-glutamine, 0.05 mg/mL streptomycin, and 0.03 mg/mL penicillin). After plating at d 4.5, the medium is partially exchanged every third day. Maximal VSM cell differentiation (30%) is achieved at days 4.5 + 17 to 4.5 + 19.

Characterization of Differentiated Phenotypes

Semiquantitative RT-PCR Analysis

Preparation of cell samples

The transcripts of genes, which are specifically expressed during ES cell differentiation, are analyzed by RT-PCR with primers of tissue-specific genes. The following steps are used to harvest ES cells or EB outgrowths:

1. Discard the medium and wash twice with PBS.
2. Add 400 μL of RNA lysis buffer per 60-mm culture dish. Allow the lysis buffer to spread across the surface of the dish and transfer the lysate into a 1.5-mL microtube.

The following steps are used to harvest EBs from suspension:

1. Collect EBs by centrifugation at 2000*g* for 3 min.
2. Wash the EBs twice by resuspension in PBS.
3. Add 100 μL of lysis buffer per 10 EBs and lyse the cells completely.
4. Store samples at –20° or –80°C.

Isolation of total RNA

The method described here is based on the use of a chaotropic agent (guanidine salt) for disruption of cells and inactivation of ribonucleases.

1. Thaw lysate (400 μL) and vortex mix for 15 s.
2. Add 40 μL (1/10 vol) of 2 *M* Na-acetate, pH 4.0. Mix carefully.
3. Add 400 μL of acidic phenol and vortex mix vigorously.
4. Add 80 μL of chloroformisoamylalcohol (24:1) and vortex mix again.
5. Store for 15 min on ice.
6. Separate the organic and aqueous phases by centrifugation at 16,000*g* for 10 min at room temperature.
7. Transfer the upper aqueous phase carefully to a fresh tube, add an equal volume of isopropanol and mix well. Store for 1 h at –20°C.
8. Centrifuge at 16,000*g* for 10 min at room temperature. Carefully discard the supernatant.
9. Dissolve the pellet in 300 μL of lysis buffer. If the pellet is difficult to dissolve, heat to 65°C for several minutes. Add an equal vol (300 μL) of isopropanol and mix well. Store at –20°C for 1 h.
10. Centrifuge at 16,000*g* for 10 min at room temperature. Carefully discard the supernatant.
11. Wash the pellet with 500 μL of 75% ice-cold ethanol (made with DEPC-H_2O), vortex mix briefly, recentrifuge at 16,000*g* for 10 min, discard supernatant, and allow the pellet of nucleic acid to dry in the air.

12. Dissolve RNA pellet in 30 μL of DEPC-H_2O and freeze at -80°C.
13. Dilute 1 μL of RNA with 100 μL of DEPC-H_2O, measure OD_{260} and the concentration of RNA using GeneQuant RNA/DNA calculator or a suitable spectrophotometer, adjust all samples to the same RNA concentration (i.e., 0.2 μg/μL) with DEPC-H_2O, and measure again to ensure the same RNA concentration of all samples. The yield of RNA from EBs (n = 20) is in the range of 20 to 100 μg.

Reverse transcription reactions

All RT and PCR solutions are available from commercial suppliers in ready-to-use form. RT reactions are performed in 20 μL of reaction volumes using 0.5-mL microcentrifuge tubes.

1. Label one PCR tube for each sample and appropriate controls. Add the same amount of RNA (0.5–1.0 μg in 3 μL) to each tube.
2. Prepare the following RT-mastermixture for 25 reactions (or a smaller quantity as required) containing: 100 μL of DEPC-H_2O, 50 μL of 10X PCR buffer II, 100 μL of 25 m*M* $MgCl_2$, 100 μL of 5 m*M* dNTPs mixture, 25 μL of RNase inhibitor, 25 μL of specific antisense primers or random hexamers or oligo $d(T)_{16}$ and 25 μL of MuLV reverse transcriptase to a total vol of 425 μL.
3. Add 17 μL of RT-mastermixture to each tube, mix carefully, and centrifuge briefly.
4. Transfer the tubes to a thermal cycler and perform RT reactions for 1 h at 42°C and then heat to 99°C for 5 min.
5. Cool the samples to 4°C or store at -20°C until use.

Polymerase chain reactions

1. Prepare a PCR-mastermixture for 25 reactions (or a smaller quantity as required) containing: 825 μL of ddH_2O, 120 μL of 10X PCR buffer II, 90 μL of 25 m*M* $MgCl_2$, 40 μL of dNTPs mixture, 50 μL of 10 μ*M* 5' sense primer of target gene, 50 μL of 10 μ*M* 3' antisense primer of target gene, 12.5 μL of AmpliTaq DNA polymerase, 12.5 μL of DMSO to a total vol of 1200 μL.
2. Label new PCR tubes and add 2.0 μL of RT reaction product to each tube as template DNA.
3. Add 48 μL of PCR-mastermixture to each tube, vortex mix and centrifuge briefly.
4. Transfer the tubes to a thermal cycler. Amplify the cDNA through 25–40 thermal cycles. Standard conditions are denaturation at 95°C

for 40 s, annealing at 55°–68°C for 40 s and extension at 72°C for 40 s. The conditions depend on the primers and thermal cycler used.

5. Run a parallel reaction containing 2.0 μL of RT reaction product and 48 μL of PCR-mastermixture with primers of the internal standard gene (i.e., β-tubulin or hypoxanthine-guanine phosphoribosyl-transferase [HPRT]) instead of the target gene.
6. Cool the samples to 4°C and store at –20°C.

Post-PCR treatment of samples

1. Transfer the PCR products to 1.5-mL microtubes.
2. Add 2.5 μL of a 1:4 mixture of glycogen: 5 *M* NaCl and 150 μL of ice-cold ethanol to each tube.
3. Incubate at –20°C for at least 1 h and centrifuge at 16,000*g* for 15 min.
4. Dissolve the pellet in 25 μL of TE buffer, add 5 μL of 6X loading buffer, and store at 4°C.

Electrophoresis and quantitative analysis of gene expression

1. Separate one third of each PCR (10 μL) by electrophoresis on a 2% agarose gel in 1X TBE containing 0.35 μg/mL of ethidium bromide at 5–10 V/cm for 70 to 100 min.
2. Illuminate the gel by UV light and obtain a digital image.
3. Quantitate the ethidium bromide fluorescence signals of gels. We use the TINA2.08e software or Multi-Analyst Version 1.1 software to evaluate the relative mRNA levels of the target gene in relation to the internal standard gene.

Immunofluorescence analysis

The formation of tissue-restricted proteins in the EB outgrowths is analyzed by immunofluorescence with a normally equipped fluorescence microscope or with a CLSM.

Single-cell isolation of ES cell-derived cardiac cells

For a better demonstration of the structural organization of intracellular, especially sarcomeric proteins, ES cell-derived cardiomyocytes are isolated as single cells by the following procedure:

1. Isolate the beating areas of EBs (n = 10) mechanically using a microscalpel under an inverted microscope and collect the tissue in PBS Dulbecco in a centrifuge tube at room temperature. Centrifuge at 1000*g* for 1 min and aspirate the supernatant.
2. Incubate the pellet in collagenase B-supplemented low-Ca^{2+} medium at 37°C for 25–45 min dependent on the collagenase activity. For

the isolation of cardiac clusters, shorten incubation time to 10 to 20 min.

3. Aspirate the enzyme solution, resuspend the cell pellet in about 200 μL of KB medium and incubate at 37°C for 60–90 min.
4. Transfer the cell suspension into tissue culture plates containing gelatin-coated slides and incubate in differentiation medium I at 37°C overnight. The KB medium is diluted at least 1:10 with differentiation medium I.
5. Change the medium to differentiation medium I; cardiomyocytes begin rhythmical contractions and are ready for immunostaining after a recovery time of 24 h.

Immunofluorescence for the detection of tissue-restricted proteins

For characterization of ES cell-differentiated phenotypes, monoclonal antibodies (mAbs) against tissue-specific intermediate filament proteins or sarcomeric proteins are suitable.

1. Rinse coverslips containing EB outgrowths twice with PBS.
2. Fix cells onto coverslips with methanol: acetone (7:3) at –20°C for 10 min, or alternatively, with 3.7% paraformaldehyde in PBS at room temperature for 10 min (depending on the antibody used).
3. Rinse coverslips twice with PBS at room temperature for 5 min.
4. Incubate the cells with 10% goat serum in PBS in a humidified chamber at room temperature for 30 to 60 min to prevent unspecific immunostaining.
5. Incubate with the primary antibody at 37°C for 30 to 60 min, or at 4°C overnight.
6. Rinse coverslips with PBS 3 times at room temperature for 5 min.
7. Incubate with the secondary antibody (i.e., dilute Dichlorotriazinyl/Amino Fluorescein [DTAF]-labeled goat anti-mouse IgG 1:100 in PBS with 0.5% BSA; or prepare DTAF-labeled goat anti-rat IgG, at 12 μg protein/mL final concentration, depending on the primary antibody) in a humidified chamber at 37°C for 45 to 60 min.
8. Rinse coverslips twice with PBS at room temperature for 5 min.
9. Rinse coverslips quickly with distilled water at room temperature.
10. Embed coverslips in mounting medium and analyze immunolabeled cells with a conventional fluorescence or CLSM.

Pharmacological analysis of ES cell-derived cardiac cells

Cardiomyocytes differentiated from ES or EC cells develop cardiac-specific physiological properties, as well as cardiac-specific receptors and signal transduction mechanisms. Therefore, ES cell-derived cardiomyocytes are suitable to measure chronotropic effects of cardioactive substances.

1. Plate EBs separately onto 24-well microwell plates at d 5 or 7, cultivate the EB out-growths for further 5 to 7 d until 85 to 100% of the EBs contain clusters of rhythmically contracting cardiomyocytes. Beating cardiomyocytes should comprise 5 to 30% of the EB outgrowths area. Change medium 1 d before measurements and add exactly 1 mL of medium per well.
2. Place the 24-well microwell plate on the inverted microscope equipped with a 37°C heating plate and a CO_2 incubation chamber, localize independently beating areas (n = 20), and measure the spontaneous beating frequency.

   Alternatively, select different areas (n = 20) of pulsating cardiomyocytes by visual control under the inverted microscope coupled via a one-chip CCD camera to a computer imaging station running the LUCIA Imaging System including the "HEART" application. The coordinates (x, y, z) of selected areas are collected by the LUCIA HEART System, and the spontaneous beating activity is automatically determined for each area.
3. Add different concentrations of the test substance and incubate for 3 min before measurement. Determine dose-dependent effects of the beating frequency after cumulative application of increasing concentrations of the test substance by adding positive chronotropic drugs (i.e., BayK 8644 may be used as a positive control) or negative chronotropic drugs (i.e., diltiazem may be used as a positive control) at final concentrations in the range of about 10^{-9} to 10^{-5} *M* (depending on the test substances).
4. Calculate the mean values (≥ standard error of the mean) of beats per minute for each data point from the pulsation rates of ES cell-derived cardiomyocytes with and without addition of drugs. Test for significance by the Mann-Whitney U-test and calculate dose-response curves. Alternatively, the final processing of the data is done by the LUCIA HEART Imaging System resulting in dose-response curves of chronotropic activity.

Genetic Modulation of Differentiation

The existence of ES cells has allowed the creation of almost any kind of mutation in any mouse gene. The study of genetically modified ES cells and their differentiation in vitro has thus provided a novel approach to study the development or study the effects of mutations in the genome. ES cells can be used to study promoter elements instead of using transgenic animals. Additionally, random insertion of exogenous DNA into single sites in the mammalian genome (gene trapping) provides a genome-wide strategy for functional genomics. When performed in tandem with the developmental potential of ES cells to differentiate into distinct cell lineage (expression trapping), such techniques are useful for the identification of novel genes expressed in developing systems. Finally, the Cre recombinase/loxP system has permitted loss-offunction analyses in vitro and subsequent rescue of targeted alleles for examination of individual gene products and their function during development.

Introduction of DNA into ES cells

Preparation of DNA

- *DNA for stable transfections*

1. Super-coiled plasmid DNA, prepared by cesium-chloride centrifugation or ion-exchange chromotography, needs to be appropriately linearized to permit integration into the genome. DNA linearization and preparation should be performed under sterile conditions.
2. Linearize a minimum of 20 μg selection vector and 40–200 μg targeting vector containing a selection cassette in the construct with the appropriate digestion buffer and restriction enzymes (ensure linearization is complete). Bring the final vol to 100 μL with TE buffer.
3. Extract with an equal volume of phenol:chloroform:isoamyl alcohol. Mix until an emulsion appears, separate aqueous and phenolic phases by centrifugation in a microcentrifuge at 4200*g* for 2–5 min at room temperature.
4. Transfer the aqueous (top) phase to a fresh tube.
5. To the tube containing the phenolic phase, add 100 μL of TE buffer. Mix and centrifuge as before. Remove the aqueous phase and combine with what is already in the fresh Eppendorf tube and increase the volume to 0.5 mL.
6. To concentrate the samples, add 1/10 vol 3 *M* Na-acetate and 2.2 vol of ice-cold ethanol. Precipitate overnight at –20°C or for 1 h at –70°C. Centrifuge Eppendorf tubes in a microfuge at 10000*g* at 4°C for 15 min.
7. Remove the supernatant and add 200 μL of 70% ethanol. Mix briefly and centrifuge for 5 min at 10000*g* at 4°C.

8. Remove supernatant and let air-dry in the laminar flow hood. Dissolve the pellets completely in PBS at a concentration of 1–5 mg/mL.

- *DNA for transient transfections*

1. Prepare supercoiled plasmid DNA by standard methods of either cesium-chloride centrifugation or ion-exchange chromotography. Use sterile techniques.
2. To concentrate the solution, precipitate DNA with 1/10 vol 3 *M* Na acetate and 2.2 volumes of ethanol at –20°C overnight.
3. Centrifuge the DNA precipitate in an Eppendorf microfuge at 12500*g* for 15 min at 4°C.
4. Discard the supernatant and add 200 μL of 70% ethanol. Centrifuge as above but for 3–5 min.
5. Discard the supernatant and let air-dry in the laminar flow hood. Dissolve the pellets completely in sterile PBS at a concentration of 1–5 mg/mL.

Electroporation of cells

1. Change the media of the actively dividing ES cells at least 1 to 2 h before harvesting.
2. Under sterile conditions, harvest the ES cells with trypsin-EDTA and centrifuge the suspension at 2000*g*.
3. Wash the cell pellet thoroughly with PBS (three washes with 10 mL PBS each time), centrifuging after each wash.
4. Before the final centrifugation, take an aliquot of cells and count them using a hemocytometer. (From a 60-mm plate, between 3 and 5 $\times$ 10^6 ES cells should be harvested.) Resuspend the cell pellet to 4.0–5.0 $\times$ 10^6 cells/mL in PBS. Cell viability should be 90–95%, and the feeder cells should comprise no more than 5% of the total cell number.
5. Transfer 0.8 mL of the cell suspension to a cuvette.
6. For stable transfections, add 10–20 μg of linearized plasmid DNA to the cuvette. For homologous recombination, the amount of DNA needed is greater because of the larger size plasmid (for every 1000 bp of plasmid, add 2 to 4 μg of linearized plasmid). Alternatively, for transient transfections, add supercoiled plasmid to the cells at a concentration of 3–5 n*M* (for every 1000 bp of plasmid, add 0.8 to 1.5 μg of linearized plasmid). For transient transfections in which loss-of-function or gain-of-function is to be achieved by either deletion of loxP flanked sequences or insertion

of DNA sequences into loxP sites, respectively. We recommend cotransfection of the Cre recombinase containing plasmid with either the loxP flanking positive selection cassettes or loxP flanked gain-of-function constructs. Add the DNA to the cuvettes at a molar ratio of 1:1 (for every 1000 bp of plasmid, add 2 to 4 μg of plasmid), for inducible knock-out or knock-in of a loxP flanked sequence.

7. Once the DNA has been added to the cells, mix the solution thoroughly by gentle trituration. Let the cells and DNA incubate for 10 min on ice.
8. Mix DNA and cells by gentle pipetting and recap the cuvette.
9. Electroporate the cells by delivering a pulse of 250 V at 500 μF. Record the exact voltages, conductance, and pulse duration. After electroporation, the solution will contain viscous material.
10. One minute after delivering the pulse, transfer half of the contents of the cuvette to each of the two culture dishes containing feeder cells and 15 mL of cultivation medium plus 1000 U LIF/mL. Be very gentle during the transfer process, trying not to damage the fragile cells. Let incubate for 40–48 h (37°C, 5% CO_2).

Cultivation and selection of clones

1. Replace medium with fresh medium containing selection drug. Change the medium every 2–4 d during the selection period.
2. After a further 8–12 d, resistant colonies of ES cells should become visible (to the naked eye). Harvest individual colonies and transfer to 96-well plates containing 20 μL trypsin-EDTA at room temperature. Once 20 to 50 individual colonies have been transferred, incubate the plate at 37°C for 5 min. Triturate thoroughly and transfer cells to 96-well plates containing feeder cells, fresh media, and the selection agent.
3. Individual wells should be monitored daily, ensuring that the media does not change color. Expand the cell population by passaging to 24-well plates. Let grow for 24–48 h. Freeze half the cells in the well and use the other half for either continued expansion or preparation of DNA.

Analysis of transfectants

Rapid isolation of DNA from ES cells

1. Plate an aliquot of trypsinized ES cell clones into 4 wells each of a 96-well plate pretreated with 0.1% gelatin.

2. Grow in cultivation media without LIF, changing media frequently to prevent pH changes, to confluence.
3. When the cells are confluent, remove the media by aspiration. Wash the wells twice with PBS.
4. Add 50 μL of DNA lysis buffer to each well. Seal the 96-well plate with parafilm and place in a tupperware box where the bottom is covered in water. Incubate the plate overnight at 55°C in a humid atmosphere.
5. The next day, prepare a mixture of NaCl and ethanol and add 100 μL of the mixture to every well using a multichannel pipet.
6. Leave the 96-well plate on the bench at room temperature for 20–30 min or until the solution becomes transparent.
7. Invert the plate gently onto paper towels to discard the solution. Gently add to the side of the wells 150 μL ice-cold 70% ethanol with the multichannel pipet. Invert the plate, as before, to discard the 70% ethanol.
8. Repeat the 70% ethanol washes, being particular gentle. After the second wash, most of the wells become transparent, and it is possible to see the DNA on the bottom of the well. Some of the wells may still contain salt (white as opposed to translucent). If so, just wash these wells with 70% ethanol again.
9. After the final washing, invert the plate, and let it partially dry for a few minutes.
10. The DNA is now ready to use or can be stored until needed. For storage, seal the plates with parafilm and place at –20°C.
11. For immediate use, add 30 μL sterile H_2O into each well and dissolve the DNA very well.
12. For PCR, 0.5–1 μL of this DNA is usually sufficient for genotyping. For Southern blotting, a vol of 25 μL is needed. Approximately 5 μg DNA can be obtained from a single fully confluent well of a 96-well plate. This quantity is sufficient for digestion with one restriction enzyme and preparation of a Southern blot.

PCR analysis

1. The PCR conditions should be established with a positive control and with a plasmid DNA template with characteristics similar to that predicted from the chromosomal structure after random DNA insertion events or targeted events (homologous recombination or Cre recombinase-mediated insertion into pre-existing loxP sites). Dilute the DNA template to 50 pg/μL in TE.

2. Random insertion events: design a primer set from the gene product of interest. Software such as Primer Express 1.5 is useful for this.
3. Primers for targeted events: in addition to the primer set designed in step 2, design a primer set that has one primer located either upstream or downstream of the targeting DNA sequence and one primer located in the targeting sequence. (This amplification is however dependent on the distance between the primers. If the primer located out of the targeting construct is more than 3 kb away from the internal primer, we recommend Southern blot analysis for genotyping.)
4. Set up parallel PCRs with different Mg^{2+} concentrations and containing primers for β-globin. In a duplicate set of reactions, prepare the same mixes, but omit primers for β-globin. On occasion, β-globin primers interfer with the amplification and a separate primer set is chosen.

10X Mg^{2+} free DNA polymerase buffer	5.0 µL
dNTPs, 20 m*M* dilution	1.0 µL
± β-Globin primer set	1.0 µL
Primer set for gene product of interest	1.0 µL
Template DNA (10–50 pg/µL)	1.0 µL
Mouse genomic DNA (0.5 µg/µL)	1.0 µL
$MgCl_2$	1.8 µL
Taq DNA polymerase (1 U/µL)	0.7 µL
dH_2O	to 50.0 µL

5. Amplify the DNA according to the following cycling parameters.

Initial denaturation:	94°C for 5 min
Primary cycles:	94°C for 45 s
	55°C for 45 s
	72°C for 45 s
	Repeat cycle 24 times
Final step:	72°C for 5 min
	4°C for infinity.

The PCR can be stored at room temperature. Elongation and denaturing times should be increased for longer DNA products (>1000 bp).

6. Volume (1/10) of the PCR should be transferred to a fresh Eppendorf tube containing 17.5 µL of TE buffer and supplemented with 2.5 µL loading buffer.

7. To analyze the PCR products, load the products on a 1.0% agarose gel. Include molecular weight markers to determine the size of the amplified products. Amplification results should have two DNA products: one for β-globin (control product) at 800 bp and one for neomycin–puromycin–or another selection cassette.

Southern blot analysis

1. Digestions of DNA can be performed either in the 96-well plates (if doing all wells) or in microfuge tubes if the contents of the well are transferred before digestions.
2. To the 25 μL of genomic DNA, add 3 μL of 10× restriction enzyme digestion buffer, 1 μL RNase A, and 1 μL of the appropriate high concentration restriction enzyme (40 U/reaction). Mix contents gently. Do not sheer the DNA.
3. Incubate the digestion at 37°C overnight in a humidified atmosphere.
4. The following day, add another 20 U restriction enzyme and incubate at 37°C for another 2–3 h.
5. Prepare a 0.7% agarose gel containing 1X TBE and ethidium bromide.
6. Stop the reaction with the addition of loading buffer. Load the samples and molecular weight markers on the agarose gel and fractionate the DNA products by electrophoresis. For the best resolution, run the gel at low voltage (0.7 V/cm) in 0.5X TBE.
7. Place the gel on an UV transilluminator and photograph. Record the migration distances for the molecular weight markers.
8. Transfer the gel to a glass baking dish and depurinate the DNA by soaking the gel for 5–10 min in 0.2 N HCl. This is important particularly for improved transfer of large DNA fragments (>8 kb in length).
9. Rinse the gel with distilled water. Soak the gel in strong base (1.5 *M* NaCl, 0.5 *M* NaOH) for 45 min to denature the DNA.
10. Rinse the gel briefly in dH_2O for 5 min. Neutralize the solution by soaking it in 1.0 *M* Tris, 1.5 *M* NaCl (pH 7.4) for 30 min.
11. Transfer the genomic DNA to Hybond-N^+ Nylon membrane by capillary transfer.
12. Air-dry the membrane, then fix the DNA to the membrane by using UV Stratalinker (120,000 $\mu J/cm^2$).
13. Put the membrane into a prehybridization solution and hybridize at 65°C for at least 2 h with agitation.

14. Prepare the probe by using 25–50 ng denatured cDNA and Ready-to-go DNA beads (-dCTP) to make a probe with specific activity approx 10^9 cpm/μg DNA, and add the denatured probe (heated to 95°C for 5 min and quenched on ice) to the prehybridization buffer at an activity of 0.5–2 × 10^6 counts per minute (cpm)/10 mL.
15. Hybridize the probe to the membrane DNA overnight at 65°C.
16. Wash membrane as follows:
    (a) Twice with 2X SSPE buffer; 0.1% SDS at room temperature for 20 min.
    (b) Once with 1X SSPE buffer; 0.1% SDS at room temperature for 15 min.
    (c) Once with 1X SSPE buffer; 0.1% SDS at 50°C for 15 min.
    (d) Once with 0.5X SSPE buffer; 0.1% SDS at 60°C for 15–30 min (monitor the radioactivity, if it is still very radioactive, go to next wash).
    (e) Once with 0.2X SSPE buffer; 0.1% SDS at 65°C for 15–30 min.
17. Expose to X-ray film for 1–10 d at –80°C or use a phosphoimager.
18. Once the cells have been expanded, frozen stocks prepared, and genotyping complete (stable clonal lines), analysis of gain- or loss-of-function tests can begin.

Gain-of-function and loss-of-function analysis in vitro

The developmental pattern of EBs differentiated in vitro may be modulated by exogenous factors, i.e., RA, growth factors, or by genetic means: the targeted inactivation of genes (equals loss-of-function) or the overexpression of genes in ES cells (equals gain-of-function).

The in vitro differentiation of mutant pluripotent ES cells has been used as an alternative and supplement to in vivo studies to analyze the phenotypes of mutant cells during early embryonic development. The strategy is especially useful in those cases where the mutation results in early embryonic death in vivo. This has been successfully used in the analysis of cellular differentiation of i.e., desmin- and β_1 integrin-deficient ES cells.

Notes

1. ES cell lines should be cultivated without antibiotics. In some cases (i.e., for selection procedures), the addition of a penicillin–streptomycin mixture or gentamycin (Gibco BRL, 1 mL of stock solution to 100 mL medium) may be helpful.

2. Both DMEM and IMDM can be used for efficient cardiac and myogenic differentiation of ES cells as D3 and R1. If IMDM is used, in additives I MTG (final concentration 450 μM) instead of β-ME is used for differentiation.
3. DEPC is a suspected carcinogen and should be handled with care.
4. If possible, the solutions should be treated with 0.1% DEPC at 37°C for 1 h, and then heated to 100°C for 15 min or autoclaved for 15 min. DEPC reacts rapidly with amines and cannot be used to treat solutions containing buffers, such as Tris.
5. Whereas EC cell lines are cultivated without feeder cells, ES and EG cell lines need feeder cells for growth in the undifferentiated state. Some ES or EG cell lines (e.g., EG-1 cells) need both feeder cells and LIF to keep growth in the undifferentiated state. LIF is commercially available or may be prepared from LIF expression vectors.
6. Good quality FCS is critical for long-term culture of ES cells, and failure to acquire good quality serum may be one reason why ES cells fail to differentiate appropriately. Extensive serum testing is necessary, therefore, to achieve good results. The most sensitive tests for sera include: (i) comparative plating efficiencies at 10, 15, and 30% serum concentrations; (ii) alkaline phosphatase activity in undifferentiated ES cells; and (iii) test of in vitro differentiation capacity after 3 to 5 passages in selected serum.
7. For preparation of EBs, three different protocols may be used: the hanging drop method, the mass culture, or the methylcellulose technique. The hanging drop method generates EBs of a defined cell number (and size). Therefore, this technique is used for developmental studies, because the differentiation pattern is dependent on the number of ES cells that differentiate within the EBs. For mass culture, plate 5×10^5 to 2×10^6 cells (depending on ES cell lines used) into 60-mm bacteriological Petri dishes containing 5 mL differentiation medium. After 2 d, let the aggregates settle in a centrifuge tube, remove medium, and carefully transfer the aggregates with 5 mL fresh differentiation medium into a new bacteriological dish. Change the medium every second day. Mass cultures of EBs may be used for differentiation of a large number of cells. For hematopoietic differentiation, the methylcellulose method is used. Methylcellulose is added to the differentiation medium at a final concentration of 0.9%.
8. Use gloves and filtertips throughout the whole procedure.

9. Do not leave RNA lysis buffer in culture dishes longer than 5 min, as polystyrene is not resistant to lysis buffer.
10. mRNA isolation from small samples of cells or tissues (i.e., 1 EB) can be performed using the Dynabeads mRNA DIRECT Micro kit.
11. Never mix and disturb the organic and the aqueous phases.
12. For RT-PCR, rTth DNA polymerase can also be used as both reverse transcriptase and DNA polymerase. In this case, the components of both RT- and PCR-mastermixture are different from using MuLV reverse transcriptase and *Taq* DNA polymerase.
13. Semiquantitative RT-PCR is used to detect the relative levels of mRNA expression and includes at least two sets of primer pairs in separate or co-amplification reactions. A "*housekeeping*" gene, i.e., β-tubulin or HPRT, is used as an internal standard to control variations in product abundance caused by differences in individual RT reaction and PCR efficiencies. The co-amplification reactions of two primer sets can be performed by using the primer-dropping method. In this method, the final yield of more efficiently amplified templates is reduced below saturating levels by using fewer PCR amplification cycles than for less efficiently amplified templates. This difference in amplification, between the templates with variable starting amounts, is achieved by dropping primer sets into the reactions at distinct cycle numbers. The first primer sets for less efficiently amplified templates are added at the beginning of reactions, the equal aliquots (4 μL) of second primer sets for more efficiently amplified templates are added at the appropriate cycle number (preliminary titration experiments for optimal cycle numbers have to be performed before). The relative signal strengths of PCR products of the target gene and the internal standard gene can be controlled by using varying numbers of cycles in multiplex reactions.
14. CLSM analysis can be used to study EBs. For immunofluorescence analysis of EBs cultivated in suspension, it is necessary to use a CLSM, because EBs are three-dimensional aggregates, which require an extended depth of focus. Some EBs are up to 400 μm in diameter. Therefore, EBs are scanned in thin sections (0.5–10 μm) using the appropriate filter combinations depending on the fluorescent dyes used.
15. The use of random insertion events or gene targeting technologies to ES cells in cell culture enables studies of gain- or loss-of-

function. Targeting events are extremely powerful, because individual genes can be altered or modified, and the consequences on an individual gene product can be studied both in vitro and/or, if desired, in vivo. The difficulty for a traditional "*knock-out*" ES cell line is the need to target alleles on both chromosomes of a diploid cell. Increased concentrations of selective agent are required for appropriate selection or the use of widespread DNA screening. For selection of a double-targeting event (i.e., targeting of both alleles), use approximately 2.5 μg/mL puromycin and 1 mg/mL G418. In cases in which gene targeting results in embryonic lethality in mice, such a technique in ES cells affords the opportunity of examining loss-of-function on individual cell types.

16. LoxP sites are 34-bp sequences containing an 8-bp core region and flanking 13-bp inverted sequences. The bacteriophage P1 Cre-recombinase recognizes specific loxP recognition sequences, and in an ATP independent manner, specifically catalyzes recombination events (deletion, insertion, inversion, and translocation). Insertion events require a single loxP site; whereas the other recombinantion events require 2 sites. The orientation–direction of the central 8-bp cores determine the type of recombination event between 2 sites. Authentic loxP sites are not found in mammalian genomes, but when inserted randomly or targeted into mammalian systems are recognized by Cre-recombinase. All four recombination events can be catalyzed in either dividing or nondividing mammalian cells, and, when used appropriately, can be used to make gene chimeras between the endogenous mouse gene and cDNAs targeted to the loxP sites.
17. When DNA is introduced into ES cells by electroporation, transfection of a single DNA molecule containing selection-cassettes such as neo^R or pur^R or cotransfection with one plasmid containing the gain-of-function construct and a second with a selection cassette can be performed.
18. In cases in which the sterility of the DNA is of concern, filter-sterilize the DNA by passing a solution containing the DNA through a pre-wet 0.22-μm filter. In the absence of filter-sterilization, it is recommended that the plasmid be ethanol-precipitated 2 times to minimize potential contamination.
19. To obtain the optimal PCR conditions, not only optimal Mg^{2+} concentration, but also optimal annealing temperatures, cycle numbers, and elongation temperatures have to be tested. To test

for optimal temperatures, dilute the template to 1 pg/μL in TE buffer. Set up the PCRs as given in the text, but with the Mg^{2+} concentration that gave the best results. Amplify with different annealing temperatures: 58°, 61°, 63°C, or, if no fragment was seen in the initial amplifications, reduce the temperatures to 48°, 51°, and 53°C. Determine which conditions give the maximum amount of amplified product on an agarose gel. To obtain optimal cycle number, different numbers of cycles of 24, 29, 34, and 39 have to be tested. Afterwards, using the annealing temperature and cycle number that give the best amplification (higher temperature and lower cycle number, if several reactions gave similar results), repeat the amplifications with different elongation temperatures: 65°, 68°, 71°, and 75°C. Test the products on a gel to determine what temperature gives the best amplification result.

20. Capillary transfers can be set up as follows: place a support inside a large baking dish filled with transfer buffer (10X SSPE). Put a piece of Whatman 3MM paper on the support, and let the two ends of the paper fall into the transfer buffer. Invert the gel, and place it on the Whatman paper. Place a piece of Hybond-N$^+$ Nylon membrane cut to the same size on the gel. Then, place several pieces (5–7) of 3MM paper and a stack of paper towels (5–8 cm high), just smaller than the membrane, on top. Put a glass plate on top of the entire stack and weigh it down with a 500-g weight. Transfer overnight.
21. The DNA used in the probe preparation must be taken from sequences outside (up- or down-stream) of the targeting construct. Otherwise, it cannot differentiate between a homologous recombination event and a random integration event.

9

Lymphopoiesis

B lymphocytes, T lymphocytes, and *natural killer* (NK) cells (cells of the lymphoid lineages) and erythrocytes, megakaryocytes, platelets, granulocytes, monocytes, macrophages, osteoclasts, and dendritic cells (cells of the erythroid/myeloid lineages) are all descendants of a *pluripotent hematopoietic stem cell* (pHSC). In fact, transplantation of a single pHSC can reconstitute a lethally irradiated host with all these lineages of cells. The differentiated cells of the erythroid/myeloid and lymphoid lineages, with the exception of memory B and T cells, turn over rapidly, with half-lives between days and weeks. Therefore, in order to maintain the pools of hematopoietic cells in an individual, these cells have to be continuously generated throughout life by cell division and differentiation from stem cells.

The cells of the adaptive immune system, B and T lymphocytes, rearrange V, D, and J segments during their development to form functional IgH and L-chain genes in B cells, and TCR α, β, γ, and δ genes in T cells. Through positive and negative selection of the original antigen-recognizing repertoires in the primary lymphoid organs, for B cells in the bone marrow, for T cells in the thymus, the repertoires are shaped by selective processes. The selected repertoires of lymphocytes become available for recognition of foreign antigens in the secondary lymphoid organs of the peripheral immune system. These processes of repertoire selection must continue to operate throughout life, as lymphocytes continue to be generated from pHSC and from progenitors through continuous V(D)J rearrangements of, respectively, the Ig and TCR gene loci, and by cellular differentiation. Hematopoietic stem cells can be defined by at least four properties: Self-renewal, multipotency, long-term reconstitution capacity, and secondary transplantability. These properties are discussed below.

Self-Renewal

Upon division of a stem cell, at least one daughter cell retains the state of this stem cell. This capacity is called self-renewal. Symmetric divisions expand the number of HSCs, whereas asymmetric divisions retain HSC potential in one daughter cell, generating further differentiated progeny in the other daughter cell. Divisions that generate two differentiated progeny daughter cells delete the HSC potential. At different times of ontogeny and in different environments, the probability of an HSC dividing either symmetrically, asymmetrically, or fully differentiating may well vary. It has been a matter of much debate whether the environment, recognized by the HSC through cell contacts and via cytokines and chemokines and their receptors, does influence the type of division and the potential of differentiation of the HSC.

Clonal succession models have proposed that HSCs, once generated during embryogenesis, remain resting cells throughout life until they are called upon to enter hematopoiesis. When the progeny of the original HSC has been used up by turnover, a new HSC from the remaining pool is recruited. This model predicts that HSC potential is limited and can be used up by complete recruitment. On the other hand, models of random symmetric HSC division have been proposed that would not only provide HSCs, but also preserve them. In these models, HSCs could never be completely used up by recruitment into hematopoietic differentiation. In favor of this type of model, it has been observed that 10% of all HSCs divide symmetrically every day, so that within 1–3 months, all HSCs are called into the cell cycle at least once. However, within the total population of HSCs, only the resting cells provide long-term repopulation potential. Therefore, the cycling HSCs could be those that are about to be used up for hematopoietic differentiation. In any case, bone marrow transplantation of aged donors into young or aged recipients shows that HSC potential remains intact for most, if not all, of the life of a mouse.

Multipotency

In asymmetric or fully differentiating divisions of a single HSC, the differentiated daughter cell(s) can develop among different lineage pathways in distinct cellular steps. Their choices include erythrocytes, megakaryocytes and platelets, granulocytes, monocytes and macrophages, osteoclasts and dendritic cells, NK cells, thymocytes, T cells, and B cells. When a single HSC can give rise to all these lineages of blood cells, it is called a *pluripotent* HSC (pHSC). Stem cells with self-renewal capacity but with more limited potencies for hematopoietic differentiation are usually called *progenitors*, e.g., *common lymphoid progenitor* (CLP) for stem cells with B, T, and NK-lymphoid differentiation capacities; *common myeloid progenitor* (CMP) for stem cells with granulocyte, monocyte, macrophage, osteoclast, and dendritic cell potential; *megakaryocyte and erythrocyte progenitor* (MEP); and *granulocyte and macrophage progenitor* (GMP). Even more differentiated stages of these different lineages may have stem cell properties.

Reconstitution Potential

Upon transplantation into a suitably receptive host, stem cells can home to their proper sites in the body—unless already implanted at the proper site—and establish differentiation along all or parts of the hematopoietic lineages. Reconstitution can be short term, generating differentiated cells in a wave of developmental steps in which the progenitors vanish as the more differentiated cells develop. The differentiated cells generated in such a wave will usually disappear by normal turnover from the host. Whenever the host compartments are properly filled, turnover will be of the order of days or weeks. In lymphoid development this may be different, especially when empty compartments (such as in the RAG-or SCID severe combined immunodeficient mice) are to be filled with donor cells. This may be a consequence of a different homeostasis, or because these donor lymphocytes become long-lived memory cells due to exposure to foreign antigens. The cycling pHSCs and most of the progenitors with more limited hematopoietic potentials, i.e., the CMP, CLP, MEP, GMP, proT, and preBI cells, all exhibit only this short-term reconstitution potential.

The exception is the progeny of preB-I cells, i.e., some of the mature B cells that develop from them in vivo. PreB-I cells from wild-type mice have short-term reconstitution potential, because they appear not to be able to home to their proper sites in the bone marrow, and because they develop in a wave of differentiation first to preB-II, then to immature, and finally to mature B cells. However, some of the mature B cells generated in this developmental wave acquire long-term reconstitution potential as individually V(D)J-rearranged clones of $sIgM^+$ B cells, particularly when these B cells are selected into the B1 lineage. They not only become long-lived, but also gain long-term reconstitution potential, which is apparent when such B1 cells are transplanted into a secondary host, where they again fill the B1 compartment. These B1 cells can be considered to be V(D)J-rearranged, B-lymphocyte-committed, and B-lineage-restricted "*stem*" cells that, upon symmetric divisions, generate daughter cells with the same properties. It is suspected that their longevity and continuously activated cell cycle status are a consequence of chronic exposure to autoantigens and/or to cross-reactive foreign antigens from infectious agents.

Clones of IgG-producing B cells, i.e., derived by T-cell-dependent stimulation, which may be from conventional B cells, have also been seen to have long-term reconstitution potential. Again, this potential may be dependent on the presence of a stimulating antigen.

The only known long-term reconstituting early hematopoietic cell is the resting pHSC. The resting population of pHSCs home to their proper sites, e.g., in the bone marrow, stably populate them in normal numbers, and continuously generate cell lineages of hematopoietic cell development. Successful migration and seeding into the proper niches of pHSCs depends on an intact environment of reticular cells, stromal cells, and vascular endothelium. It is likely that this environment develops only in interactions with the hematopoietic progenitors, as it has been seen in the development of the thymus. Further development of the transplanted pHSCs to cells that fill the compartments of the different committed lineages of hematopoiesis can also be aided when these further differentiated compartments are missing as a consequence of mutation arresting that development in the transplanted host. As one example, defects in the V(D)J rearrangement machinery, such as the $RAG^{-/-}$ and SCID severe combined immunodeficiencies, arrest B- and T-lymphocyte development at the pre-T and preB-cell stages. Drug treatment or high-dose irradiation destroys the preT and preB cells and their progenitors, allowing the seeding of donor pHSCs and their progeny so that the T- and B-lymphoid compartments are filled with donor lymphocytes. Similarly, hosts that are defective in the gene encoding the common γ chain of the IL2/IL7-receptor lack NK cells and therefore support some in vivo cell development from donor pHSCs and their progeny. Furthermore, c-fos-deficient mice, lacking osteoclast development, are suitable recipients to develop these osteoclasts from transplanted donor cells.

pHSCs and their progenitors may also possess different properties to reconstitute different hematopoietic lineages, dependent on a genetic makeup of the donor that is currently being analyzed. This strength is measured by producing mixed chimeras from graded numbers of progenitors from two genetically different hosts, which are transplanted together. Transgenic expression of bcl-2 increases both the numbers of pHSCs and their repopulation potential, whereas mice homozygous for an Ikaros-transcription factor null-mutation have a 30-fold reduced number of LT-pHSC units. The generalized defect in all hematopoietic lineages appears to be, in part, compensated by unknown genetic factors in the erythroid/megakaryocytic cell lineage. A reduced expression of the tyrosine kinases c-kit and flt-3 on early progenitors may account for some of the hematopoietic defects in this Ikaros null-mutant mouse strain.

The capacity of pHSCs to home to the right environment is one of the essential prerequisites for successful reconstitution of the transplanted host. This migratory capacity of pHSC is first played out during embryogenesis. The earliest site of pHSC generation is the splanchnopleura of the *aorta-gonad-mesonephros* (AGM) region, and its anterior part in the mouse embryo between day 8 and 9 of gestation, when embryonic blood circulation is established. pHSCs are expected to migrate from the AGM region first to sites where primitive hematopoiesis takes place, e.g., in the yolk sac. Later, they must arrive at sites of definite hematopoiesis, and of myeloid and lymphoid cell development, e.g., in the fetal liver, omentum, bone marrow, and, subsequently, in the thymus. Molecules involved in adhesion, e.g., β1-integrin, or chemoattraction, e.g., the chemokine SDF-1 and its receptor, might be involved in the direction of pHSC traffic. It might well be that pHSCs of adults still use the same molecular modes to migrate to their proper niches, e.g., in bone marrow, and it is certain that most of these molecular modes of pHSC traffic still need to be discovered.

Secondary and Subsequent Reconstitution Potential

When a pHSC has populated the proper sites in the transplanted host, it should, in principle, be transplantable again in a secondary host, and thereafter even in subsequent hosts. In practice, however, the efficiency of secondary pHSC transplantation has been found to drop substantially.

Two factors may contribute to such losses of reconstitution potential: loss of environmental contact and/or loss of telomeres. When removed from their normal environment of stromal cells, epithelial cells and vascular endothelium pHSC could well enter differentiation and thereby lose reconstitution potential, because they are no longer exposed to the proper cell-to-cell contacts and no longer receive stimuli from the right cytokines and chemokines to keep them in the state of a pHSC. This might easily happen while pHSC are isolated from the tissue, are transplanted, and have to migrate until they reach the proper site. In addition, a larger number of cell divisions (*symmetric* or *asymmetric*) may result in telomere shortening, whenever telomerase activity is decreased or absent.

In fact, pHSCs can be separated into two subpopulations, one with high (LT-pHSC), the other with low (ST-pHSC) long-term reconstitution potential. LT-pHSCs have been found to have as much telomerase activity as cancer cells do, while ST-pHSCs have less, suggesting that telomeres in pHSCs are kept at a constant, unshortened length when they enter cell cycle only occasionally and are usually resting. They lose their telomeres when they continue to proliferate. Human pHSCs shorten their telomere lengths with age, and telomere length in mouse pHSCs still needs to be measured. It is conceivable that parts of the signals given to pHSCs by their proper environment keep telomerase activity up-regulated.

Models of Hematopoiesis

Models of the hierarchies of hematopoeitic differentiation pathways have been proposed either on the basis of cellular development from progenitors to differentiated, mature cells observed in in vitro cultures and colony assays, and in in vivo transplantation experiments, or on the basis of developmental arrests in hematopoiesis introduced by targeted disruption of a series of genes mostly encoding transcription factors.

With the help of characteristic marker genes expressed in specific combinations in and on pHSCs, progenitors, and mature cells of the different lineages, the branched development. In principle, this hierarchy has been deduced from in vitro cultures and colony assays, and from in vivo transplantation experiments, in which a given progenitor was isolated by cell sorting with the help of fluorescent-labeled monoclonal antibodies detecting a characteristic combination of surface markers. Either the purified cell populations were then mass-cultured, or colonies were grown from single cells transplanted into appropriately receptive hosts as genetically distinct donor cells. The differentiated progeny were thereafter analyzed and detectable in the transplanted host with their genetically determined markers.

Targeted disruption of genes encoding the transcription factors GATA-2, Spi-1 encoding PU.1, Ikaros, E2A encoding E12 and E47, EBF, SOX-4, PAX-5, and the IL-7 receptor α chain were found to arrest hematopoiesis, resulting in a complete arrest of all lineages of hematopoiesis (GATA-2), allowing only erythropoiesis and megakaryocyte/platelet formation (PU.1), allowing erythroid and myeloid, but not lymphoid development (Ikaros), blocking B-cell development before D_HJ_H-rearrangements and expression of sterile transcripts of the μH-chain gene (E2A, EBF), blocking B-cell development at the $CD19^-$ partially D_HJ_H-rearranged stage (IL-7R), arresting B-cell development before the preB-I cell stage (SOX-4), or blocking the development of $V_HD_HJ_H$-rearranged large preB-II cells from preB-I cells (PAX-5). These experiments have suggested a linear developmental model of hematopoiesis.

One major difference in the two models is the way by which B and T lymphocytes are generated from pHSCs. This becomes an issue if a hematopoietic cell of a given lineage could not only proceed along a given pathway in one direction of differentiation, but could also reverse its track and dedifferentiate from one type of committed cell, e.g., a precursor B cell, back to an earlier progenitor, and then assume redifferentiation along another lineage of hematopoiesis. If the same pathways were used in both directions, a precursor B cell would have to dedifferentiate as far back as a pHSC in model 1 before it could redifferentiate to a myeloid cell. In contrast, in model 2, dedifferentiation to only the common myeloid/lymphoid progenitor would suffice.

Both models of hematopoiesis are supported by limited experimental evidence. Against model 1 it can always be argued that the proper physiological conditions of in vivo hematopoietic differentiation of progenitors and precursors are not met in vitro because certain cytokines, chemokines, or cell contacts and their proper dosage and application with time of differentiation are missing or wrong for a given lineage to develop. Even in vivo differentiation after progenitor transplantation may be limited by the inability of a given progenitor to find its right environment, either because it cannot home to that site, or because the site has not developed properly.

Different mutant forms of the same transcription factor can elicit quite different phenotypes of hematopoietic defects. Examples are the two types of mutants that have been generated for the PU.1 encoding SPI-1 gene, and the Ikaros gene. In both cases either the whole gene, or only the DNA-binding domain of it, have been deleted in the germ line of mice. Deleting only the DNA-binding domain generates in both cases dominant-negative forms that block hematopoietic differentiation. With age, some thymocytes develop in both the PU.1 and the Ikaros mutated mice. This suggests that very rare successes to overcome these developmental blocks are most strongly used for T-cell development.

In contrast, in PU.1$^{-/-}$ mice only the formation of macrophages and osteoclasts is inhibited. Ikaros$^{-/-}$ mice are defective in fetal B- and T-cell development, and in adult B cell development. They also lack a large proportion of the NK cells, some subsets of γ/δ TCR T cells, and thymic dendritic cells. However, postnatally CD4^{+} T cells appear to expand in almost normal numbers. These results suggest that it is easier to complement a given transcription factor in a complex of factors which regulates the expression of key target genes involved in the commitment to different hematopoietic lineages when that factor is absent, than in a complex that still contains the factor, capable of interactions within the complex, but not with DNA. If the target genes of the mutated transcription factor include those that control homing and/or attachment to the proper environment, then the observed mutant phenotypes are no more informative for the hierarchies of hematopoiesis than the in vitro and in vivo assays with progenitors.

Differentiation, Dedifferentiation, and Redifferentiation of Hematopoietic Cells: Plasticity versus Dierctioanl Commitment

In normal hematopoiesis, cells proceed most probably in unidirectional steps through sequential stages that commit these cells to a given differentiated type of lineage. At the same time, they become more and more restricted to that chosen lineage, with fewer and fewer options to enter other lineages. These pathways of increasing commitments are experimentally observed either by in vitro or in vivo assays of cell development without detectable dedifferentiation or redifferentiation into other pathways. As cells become committed, even the exposure to different environmental influences, such as cell-to-cell contacts, cytokines and chemokines, at best only stops further development but usually does not lead to redifferentiation. One example of such commitment, studied in detail in our laboratory, is the commitment of early progenitors to B-lineage lymphocytes.

The earliest cell with B-lineage potential is a B220^{+}, CD19^{-}, AA4.1^{+}, ckit^{low}, flt-3^{+}, *surrogate light* (SL) chain progenitor that develops with high clonal proliferative capacity in vitro on stromal cells in the presence of interleukin-7 (IL-7) to preB-I cells. Both the ligands for c-kit and for flt-3 are involved in the stimulation of this development as they, together with IL-7, are provided by the stromal cells of this hematopoietic environment.

These progenitors develop via a SL-chain$^+$ CD19$^-$ intermediate to B220$^+$, CD19$^+$, AA4.1$^+$, c-kit^{low}, flt-3$^-$ SL chain$^+$, CD43$^+$ preB-I cell. During this cellular development both Ig heavy (H)-chain alleles become D_HJ_H-rearranged, so that every preB-I cell has a characteristic set of individually D_HJ_H-rearranged H-chain alleles. As preB-I cells, they retain the capacity to proliferate extensively on stromal cells in the presence of IL-7 so that the individual D_HJ_H-rearranged alleles become clonal markers of the progeny of this proliferative expansion. As long as IL-7 is present, preB-I cells will not differentiate further along the B-lineage pathway. Removal of IL-7 induces V_H to D_HJ_H rearrangements on the H-chain locus, and V_L to J_L rearrangements. Whenever both IgH and IgL chain loci have been rearranged productively, sIgM$^+$ immature B cells are formed. They are then selected negatively as well as positively to generate the peripheral, mature B-cell pools reactive to foreign antigens.

The cellular stages of this stepwise repertoire development of B cells can be distinguished by the expression of characteristic surface markers. PreB-cell-receptor-expressing, i.e., $V_HD_HJ_H$-rearranged μH chain$^+$/SL chain$^+$ proliferating preB-II cells, are c-kit–CD25$^+$AA4.1$^+$, while immature sIgM$^+$ B cells lose CD25 expression, and AA4.1 expression is lost as immature B cells become sIgM$^+$/sIgD$^+$ mature B cells.

Progenitor B and preB-I cells from wild-type mice have a strong unidirectional tendency to develop along this pathway of differentiation to B cells. Thus, either induced in vitro, first on stromal cells and IL-7, subsequently by the removal of IL-7, or by transplantation into lymphocyte-deficient RAG$^{-/-}$ or SCID hosts in vivo, only B-lineage cells, but no T cells nor myeloid cells, can be seen to develop from these pro and preB-I cells. Culturing preB-I cells in the absence of IL-7, and in the presence of cytokines capable of inducing myeloid cell differentiation, does not induce such myeloid differentiation.

Commitment to B Lymphopoiesis

Transcription factors appear to control the commitment of progenitors along different lineages. Three such factors have been identified for the commitment to the B-lymphocyte lineage (E2A, EBF, PAX-5). Targeted disruption of either the E2A or the EBF gene results in an arrest of B lymphopoiesis at the stage. E2A$^{-/-}$ and EBF$^{-/-}$ mutant mice show an arrest before any rearrangements of either IgH or IgL chain gene segments are initiated. Target genes controlled by E2A and EBF are RAG1 and RAG2, Igα and Igβ, VpreB and λ5, as well as the intron enhancer of the μH chain locus (Eμ) controlling sterile transcription of the μH chain locus. Transgenic expression of E2A-encoded E47 activates some of these target genes even in fibroblasts. E2A and EBF also appear to control the expression of PAX-5.

PAX-5$^{-/-}$ mutant mice are arrested at a later stage of B-lineage development, i.e., at a preB-I-like state, in which the cells have D_HJ_H rearranged both IgH chain alleles, but continue to express earlier markers, such as the flt-3 receptor tyrosine kinase. Hence, it may well be that their cellular arrest is at the earlier, proB-cell stage but that D_H to J_H rearrangements continue in these PAX-5$^{-/-}$ cells . PAX-5$^{-/-}$ preB-I cells, like wild-type pro- and preB-I cells, proliferate for long periods of time on stromal cells in the presence of IL-7, and can be cloned as individually D_HJ_H-rearranged cell clones. PAX-5$^{-/-}$ mice are blocked in B-cell development at the transition of D_HJ_H-rearranged preB-I to $V_HD_HJ_H$-rearranged large preB-II cells. Hence, whereas withdrawal of IL-7 from wild-type preB-I cell cultures results in the development of sIgM$^+$ immature and mature B cells, this development is blocked for PAX-5$^{-/-}$ preB-I cells. PAX-5 is expressed monoallelically from the earliest proB cell to the stage of an immature B cell, where it changes to biallelic expression. PAX-5$^{-/-}$ preB-I cells express almost all of the proB-cell-specific genes (with the exception of CD19). Several genes expressed in other myeloid or lymphoid lineage cells are also expressed in PAX-5$^{-/-}$ preB-I cells, although some at very low levels, such as preTα (lymphoid-related), perforin (NK-cell-related), myeloperoxidase and M-GSF receptor (myeloid-related). It has been suggested that PAX-5 not only positively controls B-lineage development but, at the same time, represses myeloid and T-lymphoid lineage development. PAX-5 could, in fact, act like a traffic sign. When it is on, traffic of hematopoiesis will proceed along the B-lineage way; when it is off, traffic is routed into the myeloid and T-lymphoid directions.

Plasticity of PAX-5$^{-/-}$ PERB-I Cells

It had previously been observed that preB-I cells from the fetal liver of wild-type mice could be grown on stromal cells in the presence of IL-7 for more than 60 divisions without losing their capacity to develop to sIgM$^+$ B cells in vitro and in vivo. In contrast, wild-type preB-I cells from bone marrow tend to cease to proliferate after 4–6 weeks; i.e., after 30–50 divisions. Bone-marrow-derived preB-I cells of PAX-5$^{-/-}$ mice appear to proliferate as well as wild-type cells from fetal liver. Clones of these cells have now been expanded for at least 100 divisions; hence, one PAX-5$^{-/-}$ preB-I cell can, in principle, be expanded to 10^{30} cells.

It was a surprise to discover that PAX-$5^{-/-}$ preB-I cells, in fact, possess stem cell-like properties. Not only are they self-renewing (for up to 100 divisions), they are also pluripotent (with the exception that B-cell development is blocked, and that erythroid and megakaryocyte development has so far not been observed), and they home back to the bone marrow, from where they can be reisolated as preB-I cells. Therefore, they have long-term reconstituting capacity.

The pluripotency of PAX-$5^{-/-}$ preB-I cells is documented in the following experiments. Clones of these cells develop into macrophages when IL-7 is removed and M-CSF is added to in vitro cultures. Removal of IL-7, followed by exposure to IL-3/IL-6 and SCF first and G-CSF later, develops granulocytes. Exposure of the IL-7-deprived cells to M-CSF and GM-CSF generates antigen-representing dendritic cells, exposure to IL-2 generates NK cells, and cultures on stromal cells expressing TRANCE lead to osteoclast development. The last two lineages of hematopoiesis can also be induced in vivo by transplanting PAX-$5^{-/-}$ preB-I cells into mice that are deficient in the common γ chain of the IL-2/IL-7-receptor ($\gamma c^{-/-}$) and into c-fos-deficient mice, respectively. In the γc–/–-deficient mice, endogenous NK-cell development is defective. It is replenished by NK cells derived from the PAX-$5^{-/-}$ preB-I cells, and the same PAX-$5^{-/-}$ preB-I cells become the source of osteoclast generation in the osteoclast-deficient c-$fos^{-/-}$ mice.

Equally impressive is the complete reconstitution of T-cell development in lymphocyte-deficient $RAG^{-/-}$ hosts by the transplantation of as little as 5×10^4 ex vivo isolated, or 5×10^6 in vitro grown, PAX-$5^{-/-}$ preB-I cells. All stages of thymocyte development in the recipient thymus, i.e., *double-negative* (DN) stage 1–4, double-positive and single-positive thymocytes, as well as peripheral T cells, all develop in normal numbers and retain this homeostatic potential for many months. T cells expressing α/β TCR as well as γ/δ TCR develop normally. Positive and negative selection on MHC class I and class II is normal, and the peripheral, mature T-cell repertoire reacts normally in cytolytic and helper responses to alloantigens and foreign antigens.

These experiments show that the apparently pluripotent PAX-5–/– preB-I cells can fill vacant hematopoietic compartments in the transplanted, partially deficient host. They also show that a single PAX-$5^{-/-}$ preB-I cell genetically marked by a characteristic set of D_HJ_H-rearranged IgH chain alleles generates all these different hematopoietic lineages, thereby endowing every progeny cell of these lineages with the same set of D_HJ_H rearrangements. Furthermore, they show that neither the expression of surrogate L chain, nor of Igβ, nor the D_HJ_H rearrangements irreversibly commit a hematopoietic cell to the B-lineage pathway. Finally, repeated isolation of the transplanted, clonal PAX-5$^{-/-}$ preB-I cells from the bone marrow of the recipient followed each time by in vitro and in vivo induction to the different hematopoietic lineages demonstrates, for at least 100 divisions, a surprisingly strong long-term reconstitution capacity.

At present, we do not know to what extent the earliest progenitor of the B-lineage pathway (the B220$^+$, CD19$^-$, AA4.1$^+$, c-kit^{low}, flt-3$^+$, pro-/ pre-B cell) bears similarities in cellular behavior to PAX-5$^{-/-}$ preB-I-like cells. In particular, do they respond similarly to inductive signals by cytokines, and are they inducible into different directions? The rearrangement status of the IgH chain locus may not be a reliable marker for B cells of mutant mice, as they have been found in macrophages and T cells. Thus, the plasticity of earlier stages of hematopoiesis may be greater, as previously thought, and influenced by the environment, rather than only autonomously determined by the hematopoietic cell.

It is clear that targeted deletion of the PAX-5 gene can hardly be called a normal state. However, if PAX-5 acts as a traffic signal, so that myeloid and T cells are made when PAX-5 is "off," and B cells are made when PAX-5 is "on," then one could imagine that the regulation of PAX-5 expression by end cells and/or their products could, for instance, influence which of the two routes of traffic is generated. Myeloid and T cells and/or their secreted products would feed back on the earlier progenitor to stop further production of cells of this lineage by up-regulating PAX-5 expression, whereas B cells and/or their secreted products would do the opposite; namely, turn off PAX-5 expression to stop further B-cell production. Depending on whether hematopoiesis occurs, the target cell of this homeostatic regulation of myeloid and lymphoid cell production would be pHSC or the CMLP, respectively.

Practical Consequences

It is important to realize that all hematopoietic lineages developed from the PAX-5$^{-/-}$ preB-I cells generate cells that have been found to function normally. This could be expected, since only early progenitors and B-lineage cells, but no other hematopoietic cells, express PAX-5. Therefore, PAX-5$^{-/-}$ preB-I cell clones offer the exciting possibility to study, and genetically influence, the development of all these normal hematopoietic pathways. The PAX-5$^{-/-}$ preB-I cells can be transgenically modified, since they can be infected with high efficiencies by a retrovirus.

Transfection of genes, or homologous recombination following such transfections, has so far not been possible. Retroviral infection allows us to mark the cells; e.g., the expression of a fluorescent protein gene, a procedure that facilitates the identification and FACS purification of progeny cells in vitro and in the transplanted host. Retroviral infection of PAX-5$^{-/-}$ preB-I cells by promoter- or polyA-tail trap vectors opens the possibilities for a genetic and functional screening of mutations; i.e., of genes that are active and functional at a given stage of development of T cells, NK cells, or myeloid cells. By analogy, retroviral infection of wild-type preB-I cells can be used for such a mutational analysis of genes active in B-cell development.

By crossing the PAX-5$^{-/-}$ mutation onto other hematopoietically defective mutant mice, it should be possible to use preB-I cells of double mutant mice for the study of the detailed structure–function relationships of such a defective gene.

It is predictable that the genetic programs controlling the development and the interactions of cells of different hematopoietic lineages with their environment, self, or non-self, will involve many genes. If homologous recombination of such genes into PAX-5$^{-/-}$ preB-I cells becomes an experimental possibility, this will obviate the need to construct "*knock-out*" and "*knock-in*" strains of mice for these genes. Still more exciting are the prospects that a human preB-I cell could even proliferate in tissue culture like its counterpart from mice, and that homologous recombination with an exogenously inducible, normally repressed form of the human PAX-5 gene on both alleles could create a human PAX-5$^{-/-}$ preB-I cell.

10

Nervous Stem Cells

Neural differentiation is an early embryonic event that occurs soon after germ layer specification. The newly formed ectoderm undergoes further patterning to separate into two identifiable components, the presumptive neural ectoderm and the presumptive epidermis. Neural tissue segregates as a clearly demarcated epithelium termed the *neuroepithelium* (or neuroectoderm). The neuroepithelium generates the *central nervous system* (CNS) whereas cells at the margins of the neuroepithelium will generate the *peripheral nervous system* (PNS). Two groups of cells contribute to the PNS. Neural crest stem cells differentiate at the neuroectodermal/epithelial junction, and placodal precursors differentiate from cranial ectoderm that lies more laterally than the zone that generates neural crest. Precursors that generate the PNS also contribute to nonneural structures, including pigment cells of the skin and craniofacial mesenchyme.

Undifferentiated neural precursor cells, whether in the CNS, neural crest, or the placodes, proliferate, differentiate, and migrate to appropriate locations. Cells undergo further maturation and become postmitotic. Neuronal cells go on to project to appropriate targets, make synapses, and acquire the correct rostrocaudal and dorsoventral identity. Seminal work by a number of laboratories has led to rapid advances in our understanding of phenotypic specification. An accumulating body of evidence suggests that neurogenesis follows a pattern of development similar to developmental patterns described in the liver, skin, and the hematopoietic system. In each of these systems, tissue-specific stem cells are generated and undergo a series of developmental restrictions to generate more restricted progeny that ultimately give rise to fully differentiated cells.

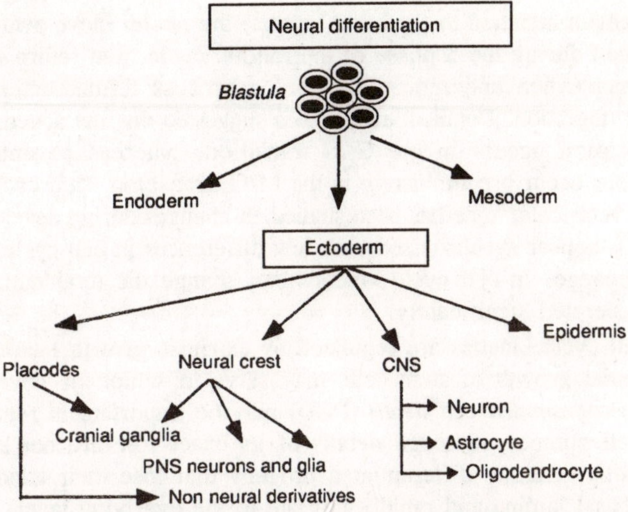

Fig. 10.1. Neural differentiation during development.

In this chapter, we discuss recent advances in our understanding of stem cells in the developing nervous system, their isolation, characterization, and role in normal development. We also discuss how some of the cell fate transitions may be achieved in vitro and the potential fates of these cells in vivo. More recently, increased interest and study of embryonic stem cells has allowed the generation of specific cell types including neurally restricted precursors and differentiated neural precursor cells. These cells have great potential both in understanding basic developmental programs and in therapeutic applications.

Multipotent Stem Cells in Normal Development

The early neuroepithelium that will generate the CNS undergoes further growth to form a closed neural tube with a central canal either by folding or by aggregation and subsequent cavitation. Initially, the neural tube consists of proliferating, morphologically homogeneous cells termed *neuroepithelial* (NEP) stem cells. NEP cells or *neural stem cells* (NSCs) are initially present in a single layer of pseudostratified epithelium spanning the entire distance from the central canal to the external limiting membrane. NEP cells continue to proliferate, and are patterned over several days in vivo to generate mature neurons, oligodendrocytes, and astrocytes in a characteristic spatial and temporal pattern. Mitotic activity in the proliferating ventricular zone is accompanied by nuclear translocation within the cytoplasm. While the cells remain attached to the basal lamina, the nuclei move away from the lumen during the S phase of the mitotic cycle, and return near to the lumen when undergoing division, a process termed interkinetic nuclear migration. Detailed analysis has suggested that the abventricular translocation occurs in the G_1/S transition, whereas adventricular transitions occur predominantly in the M/G_1 transition. Cell cycle time for the ventricular zone has been studied. It changes during development and cells appear synchronized with few differences in cell cycle times. Small changes in cell cycle kinetics can change the total number of cells generated significantly.

Cell cycle kinetics are regulated by extrinsic growth factors, and differential growth of stem cells may serve to sculpt the developing brain. *Programmed cell death* (PCD) may be important in regulating stem cell number, although details of its exact role are not known. Stem cells generate differentiated progeny that lose their attachment to the basal lamina and rapidly migrate to the overlying layers of the cortex. Progenitor cells begin to express differentiation markers while migrating and can be distinguished from stem cells by such expression. Multipotent cells in the ventricular zone undergo both symmetrical and asymmetrical divisions. Cai et al. (1997) used retroviral labeling to show that approximately 48% of labeled cells formed clusters located entirely within the ventricular zone, suggesting self-renewal via symmetrical divisions. Approximately 20% of cells, however, appeared to generate cells in both the ventricular zone and the mantle, suggesting at least some asymmetrical divisions. Whether a cell will undergo symmetric or asymmetric divisions depends on whether molecules asymmetrically distributed within the cytoplasm are differentially partitioned into the two daughter cells.

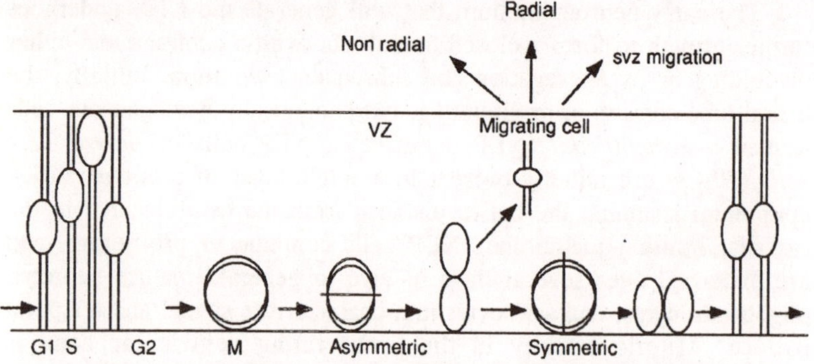

Fig. 10.2. Nuclear translocation and symmetric and asymmetric differentiation.

It has been suggested that notch, numb, and possibly other molecules are distributed apically or basally. Divisions in the horizontal plane will result in asymmetrical localization of notch and numb to the two daughter cells with consequent differences in cell fate. Divisions in an oblique or perpendicular plane will result in a more equal partitioning with a greater likelihood of symmetric divisions. How the axis of cell division is determined is a subject of study, and possible mechanisms are discussed by Jan and Jan (1998). The relationship between nuclear migration and symmetric and asymmetric cell division has been studied primarily in the telencephalic neuroepithelium, but it is likely that similar relationships are also important in the spinal cord.

As development proceeds, the ventricle becomes significantly diminished in size and the characteristics of the cells present in the ventricular zone change. It is thought that the ependymal cells represent the remnants of the ventricular zone stem cells, although some cells persist as undifferentiated cells in or near the subependymal layer. Proliferating cells are seen in an additional zone of precursor cells termed the ***subventricular zone*** (svz), which is derived from the ventricular zone during embryonic development and persists throughout life. The svz contains a mixed population of precursor cells that can generate neurons and glia. This zone may contribute to ongoing neurogenesis that continues throughout life. A svz is not present in the spinal cord, hindbrain, or cerebellum, and little postnatal neurogenesis is seen in these brain regions. Within the cerebellum there exists a separate zone of proliferating precursors termed the external *granule cell layer* (egl). The egl is a transient layer of proliferating precursors derived from the ventricular zone of the midbrain and hindbrain that disappears in the early postnatal period and generates the granule cells of the cerebellum.

Characteristics of cells present in each of these developmental zones have been described. Much of the focus has been on the cortical ventricular zone, although pluripotent stem cells that self-renew and generate neurons, astrocytes, and oligodendrocytes have been identified from the entire neural axis. Evidence of pluripotentiality includes retroviral labeling, transplantation of clonal cell populations, and characterizing stem cells in clonal and mass culture. Fetal precursor cells appear to express certain characteristic features. They require *fibroblast growth factor* (FGF) for their proliferation in vivo and in vitro and express nestin and additional markers. An emerging consensus is that although cells present in all of these different regions are multipotent and display common features, they are nevertheless distinguishable from each other in growth factor dependence, cell cycle time, markers, and ability to differentiate. Stem cells isolated from the spinal cord and early stages of embryogenesis appear to require FGF for their survival, whereas stem cells isolated from more rostral regions at later developmental stages seem equally responsive to FGF and/or *epidermal growth factor* (EGF). The cell cycle time progressively lengthens as the embryo matures and, in the adult, stem cells can be detected in multiple regions, including the cortex. Cortical stem cells express polysialated N-CAM whereas cells that remain in the adult svz express *glial fibrillary acid protein* (GFAP) immunoreactivity. Stem cells in vivo also express rostrocaudal positional markers, and these markers are maintained in culture. Response to neurotransmitters is also different. Ventricular zone stem cells proliferate in response to glutamate whereas subventricular zone stem cells respond to glutamate with a reduction in proliferation. Transplant studies have also suggested a bias in differentiation potential. For example, cells present in the *lateral ganglionic eminence* (LGE) differ in their migration and differentiation properties from cells present in the *medial ganglionic eminence* (MGE).

Table 10.1. Multiple types of multipotent cells are present in the nervous system

Classification	*Characteristics*
Embryonic stem cells	FGF-dependent, nestin-immunoreactive cells present in the ventricular zone along the entire rostrocaudal axis including the cerebellum, midbrain, and hindbrain
Adult stem cells	EGF-dependent, nestin-immunoreactive cells present in the subventricular zone, spinal cord, and cortex
Retinal stem cells	nestin-immunoreactive, Chx10-expressing, dividing cells that are biased toward retinal differentiation
Hypothalamic-pituitary stem cells	nestin-immunoreactive, dividing cells that require NOS for activity
Neural crest stem cells, stem cells	nestin, p75, HNK-1-immunoreactive
Placodal stem cells	nestin-immunoreactive

The available data do not allow us to draw any strong conclusions about how these differences arise. Either a common multipotent stem cell undergoes specification as development proceeds or multiple classes of multipotent cells arise at an early developmental age. Distinguishing between these possibilities, except in select instances, has been difficult. In the case of embryonic FGF-dependent neuroepithelial stem cells and later arising EGF-dependent neurospheres, a direct lineage relationship has been established. Early rat multipotent stem cells present from E10 onward are FGF-dependent. FGF-dependent cells undergo maturation and begin to respond to EGF by proliferation. The EGF-responsive cells can be maintained solely in EGF and appear to be present from E14 onward. At early developmental ages, both EGF- and FGF-dependent stem cells are present and likely continue to coexist in the adult. Proof that FGF-dependent stem cells generate EGF-dependent cells came from culture experiments and the use of chimeric mice. In these experiments, initially only an FGF-dependent population of stem cells was present, but later in development FGF-dependent and EGF-dependent cells (which arise from FGF-dependent cells) are both present.

Stem cells have now been isolated from proliferating ventricular and subventricular zones of the developing forebrain as well as from the spinal cord, cortex, midbrain, and hindbrain. In addition, stem cells have also been identified in another area of the mammalian CNS, the neurohypophyseal system of the endocrine hypothalamus. Stem cells emerge from the subependymal regions of the third cerebral ventricle, migrate, generate neurons, and repopulate the region after hypophysectomy. Up-regulation of *nitric oxide synthase* (NOS) seems to be essential for this process. Another specialized region of the brain from which stem cells have been identified is the neural retina. Two laboratories have independently isolated and characterized retinal stem cells. Both groups showed that embryonic and adult retinas contain a multipotent, self-renewing stem cell. This cell is preferentially localized to the ciliary margin and requires FGF, and perhaps EGF, for survival. Individual retinal stem cells are multipotent and can generate retinal derivatives in mass and clonal culture. The retinal stem cell differed from CNS stem cells in the expression of retinal-specific genes such as Chx-10 and appears to preferentially differentiate into retinal cells.

Overall, the available evidence shows that stem cells exist in multiple regions of the developing and adult brain, and specialized stem cells exist in specific regions. These multipotent cells are present at the appropriate time and place to play a role in normal development and can be distinguished on the basis of growth factor dependence, time of isolation, cell surface receptors, transcription factors expressed, and their ability to differentiate into specific subtypes of cells.

RESTRICTED PRECURSORS

Multipotent stem cells do not appear to migrate extensively and are largely localized to the ventricular zones, but their derivatives do migrate. During development, differentiating cells lose their attachment to the basal lamina and migrate from the ventricular zone to appropriate target sites. Migrating neuronal precursor cells express markers of differentiation and follow radial glial and nonradial glial pathways of migration to terminal sites. Glial precursors do not appear to require radial glia for migration, and their exact mode of migration remains to be determined. In the spinal cord the first postmitotic cells to differentiate are ventral motoneurons, followed by more dorsal neuronal phenotypes. In the cortex, differentiation is layer-specific, with early-born neurons being present in the lowest layers. Neuronal precursors migrate past formed layers to form successively more superficial layers. Neurogenesis, except in certain specific regions, persists from E11 to E17 in rats and precedes (but overlaps) differentiation of oligodendrocytes and astrocytes. A variety of evidence suggests that multipotent stem cells generate differentiated mature neurons via the generation of more restricted precursors.

Bromodeoxyuridine (BrdU) labeling studies have shown that some cell division occurs in migrating cells as well as in white matter, suggesting that cells may continue to divide after leaving the ventricular zone. These migrating precursors have been isolated and shown to possess at least limited self-renewal potential and retain the ability to generate multiple derivatives. We have termed these cells restricted precursors and classified them on the basis of their differentiation ability. Thus, neuronal restricted precursors can generate multiple classes of neurons, but not astrocytes or oligodendrocytes, under conditions in which multipotent stem cells generate such derivatives.

By analogy to the hematopoietic system, these cells would be termed blast cells or restricted progenitor cells. Compared to multipotent stem cells, they have a more limited self-renewing ability and are more restricted in their ability to differentiate. In the hematopoietic system, where prolonged self-renewal is a necessary requisite for therapy, blast cells are of limited therapeutic value. In the nervous system, however, cellular replacement is not an ongoing requirement, and thus replacement by cells that have a more limited self-renewal ability may be preferred. Transplanting cells with a restricted differentiation potential would reduce the chance of tumor formation or heterotopias. Furthermore, it is more likely that cues for normal differentiation would be present, and more differentiated cells would be far more likely to respond to cues present in the adult. Finally, the availability of more restricted precursor cells may provide the ability to replace selected populations of cells. In the following section, we describe some of the lineage-restricted precursors that have been characterized and discuss their therapeutic potential.

Neuron-restricted Precursors

A number of in vivo and in vitro experiments over the years have provided evidence that NRPs are present and play an important role in normal development. Strong evidence for the existence of NRPs in vivo came from retroviral-labeling experiments. Labeled single cells and their progeny gave rise to clones that comprised solely neurons. Clones that contained a single class of neurons as well as clones that contained multiple kinds of neurons were identified, suggesting precursors exist that could undergo at least limited self-renewal and generate solely neurons. Progeny arising from these clones were present in clusters or were widely dispersed, suggesting that they could undertake radial and nonradial migration. Since in the same experiments other clones could be identified that generated solely glial cells, it appeared likely that the restriction seen was intrinsic to the cells and that the environment at this stage could support both neuronal and glial differentiation. The generation of restricted precursors appears to be a late developmental event as analysis in vivo and in vitro at early embryonic ages generated largely multipotent clones, whereas at later stages, primarily unipotent clones were seen. If cells were followed at later stages of development, most precursor cells gave rise to either neurons or oligodendrocytes or astrocytes (Luskin et al. 1988). Few mixed clones were detected, suggesting that initially multipotent stem cells generate differentiated cells via the generation of more restricted precursors.

The existence of NRPs has also been corroborated by in vitro culture assays. Early work by Gensburger et al. (1987) had shown the presence of neurofilament-positive cells from 13-day-old rat embryos that proliferate in culture in response to bFGF. Other investigators have maintained dividing neuronal precursors from a number of regions of the mammalian brain although, except in selected cases, the cultures were not pure NRP populations. Other experiments suggested that embryonal N-CAM was expressed by neuronal and astrocytic precursors and could be used to isolate these cells. Mayer-Proschel and colleagues used antibody panning to isolate a pure population of NRP cells that could differentiate into multiple classes of neurons. They did not differentiate into astrocytes and oligodendrocytes, even in conditions where multipotent stem cells or glial precursors did. The authors have termed these cells NRP cells and have shown that, when transplanted in vivo, they remain restricted to differentiate into neurons. Although restricted to neuronal differentiation, these early NRPs could differentiate into excitatory, inhibitory, and cholinergic neurons. Direct isolation of neuronal precursors, which do not generate glial cells, has been achieved by fluorescence sorting after transfection with GFP constructs under the control of the Tα1 tubulin early promoter.

Equally important are recent results that clearly establish a lineage relationship between multipotent stem cells and more restricted precursor cells. *Neuroepithelial precursors* (NEP) from the spinal cord have been shown to give rise to NRPs. NEP cells generate E-NCAM-immunoreactive cells that can further differentiate into neurons of multiple phenotypes in culture, including some that are not normally observed in the spinal cord. They do not give rise to oligodendrocytes or astrocytes. NRP cells isolated from NEP cells seem identical to those generated directly from the developing spinal cord. Vescovi et al. (1993) have also showed that transient exposure to bFGF can cause the differentiation of EGF-responsive precursors into secondary progenitors, some of which give rise to neurons alone. It is important to emphasize here that work with cultured precursor cells from the embryonic CNS has been consistent with much of the in vivo observations. In particular, the retroviral labeling of cells in vivo or following single cells in culture has allowed the unambiguous delineation of lineage and potential of proliferating precursors.

Neuronal precursors have also been isolated from the developing human brain by fluorescence sorting after transfection with GFP constructs under the control of the Tα1 tubulin promoter. Goldman and colleagues have recently reported the successful use of neuronal promoter targeted screening techniques to obtain neuronal precursors from adult human hippocampus. These precursors divide in vitro and mature to form physiologically active neurons. Using the same technique, neuronal precursors have been isolated from the adult ventricular zone. Not surprisingly, a human immortalized and neuron-restricted spinal precursor cell line (HSP-1) has been identified. This cell line may represent the equivalent of NRPs in humans, an idea strongly supported by the fact that it does not differentiate into astrocytes or oligodendrocytes under conditions where they are usually generated.

Are there Subclasses of NRPs?

When marked SVZa cells from postnatal mice are transplanted in utero into the ventricles of E15 mice, they integrate at multiple levels throughout the neuraxis and form neurons. Interestingly, the migration is not uniform into all areas of the brain. Transplanted cells were not seen in the cortex or the hippocampus of the recipient. Stem cells from the adult forebrain of rats also could be directed to increased generation of neurons in the presence of *insulin-like growth factor* 1 (IGF-1) and heparin. External cues as well as intrinsic potential, therefore, seem to play a role. Similarly, Luskin and coworkers have shown that the origin of the precursor cells seems to be important in determining the efficiency with which NRPs integrate in a homotypic or heterotypic site. Embryonic neurons from the medial ganglionic eminence seem to be more widely dispersed and differentiate in multiple areas of the adult brain in comparison to cells from the lateral ganglionic eminence.

Progenitor cells, isolated from postnatal mouse hippocampus, cultured in the presence of EGF, divide, and on addition of brain-derived neurotrophic factor, give rise to pyramidal-like neurons. When transplanted into the cerebellum, however, these progenitors do not give rise to neurons, and when introduced into the rostral olfactory tract, only a few cells migrated into the olfactory bulb to give *tyrosine hydroxylase* (TH)-positive neurons. Thus, multipotent precursors isolated from different regions of the developing brain are likely to differ depending on the developmental stage and region from which they were isolated.

Glial-restricted Precursors

Radial glia are the first identified glial population to develop, followed by glial precursors, followed by astrocytes and oligodendrocytes. Unlike most neurons, which cease to divide postnatally, astrocytes continue to be generated in adults. Oligodendrocytes do not divide, dedifferentiate, or reenter the cell cycle. However, precursors to oligodendrocytes exist and their division persists, albeit at a slow rate, throughout life. Extrinsic signals can modulate cell division, and the rate of oligodendrocyte differentiation is increased after demyelinating lesions. Current evidence suggests that glial cells differentiate from multipotential stem cells by sequential stages of restriction in developmental potential, and several types of glial precursors have been identified.

Glial precursors are first seen at E13 in mice and E14.5 in rats in multiple restricted foci. In the developing caudal neural tube, glial precursors arise in ventral regions and migrate both dorsally and ventrally, and differentiate into oligodendrocytes. Astrocyte differentiation is first observed in the dorsal region although the location of astrocyte precursors within the neural tube remains to be determined.

Rao and colleagues have suggested that the early precursor identified on the basis of A2B5 immunoreactivity, the expression of PLP-DM20 and Nkx2.2, and the absence of *platelet-derived growth factor receptor-α* (PDGFR-α) immunoreactivity, is the precursor for both astrocytes and oligodendrocytes. These investigators used immunopanning to isolate A2B5 immunoreactive cells and showed that these cells differed from multipotent stem cells and NRPs and that these glial precursors could generate astrocytes and oligodendrocytes both in vitro and in vivo. The investigators termed these cells *glial restricted precursors* (GRPs) and showed that they differentiated from multipotent stem cells, thereby establishing a direct relationship between these two precursor populations.

Mayer-Proschel and colleagues have further characterized the process of differentiation. They have shown that GRP cells acquire PDGFR-α immunoreactivity and PDGF dependence upon prolonged culture. GRP cells can generate more differentiated precursors that do not generate astrocytes in vivo but can generate oligodendrocytes and thus resemble the well-characterized O2A progenitor cell. This transition is also likely to occur in vivo. Mengsheng Qiu and colleagues have used Nkx2.2 expression to follow glial precursors and were able to show that glial precursors do not initially express PDGFR-α immunoreactivity but acquire it within 2 days of developmental time. Spassky and colleagues have likewise shown that when the expression of LacZ is driven by the PLP promoter, they see multiple foci of expression that match regions of A2B5 immunoreactivity and do not overlap regions of PDGFR-α immunoreactivity. Cells isolated at this early stage appear capable of differentiating into both astrocytes and oligodendrocytes, providing independent confirmation of the existence of an oligodendrocyte–astrocyte precursor.

Another class of glial precursors found in a structure termed the oligosphere has been described by Avellana-Adalid et al. (1996).

Oligospheres are floating aggregates, generated from neonatal rat brains, containing presumably homogeneous populations of A2B5 immunoreactive cells. These aggregates can be propagated as undifferentiated cells for long periods of time and can differentiate into oligodendrocytes and astrocytes in culture or after transplantation. Similar cultures of oligospheres have been obtained from the subependymal striata of adult rats and canines.

Apart from precursors that can generate both astrocytes and oligodendrocytes, more restricted precursors have been isolated. The best characterized of all such precursors is the O2A or oligodendrocyte type-2 astrocyte precursor cell. This cell was initially isolated from the optic nerve and has been subsequently shown to exist in multiple brain regions. Unlike the GRP cell, it does not differentiate into astrocytes when transplanted in vivo and thus appears more restricted in its differentiation potential. It is likely that at least some O2A cells arise from GRP cells, but it is not clear whether additional pathways of glial differentiation exist. A variety of astrocyte precursors have also been identified. One such population is the *astrocyte precursor cell* (APC) that gives rise to astrocytes alone. These have been isolated from embryonic and P1 rat optic nerves by immunopanning with an astrocyte-lineage-specific anti-neuroepithelial antibody (C5). In culture, APCs differentiate into type-1 astrocytes alone. APCs have an antigenic phenotype that is $GFAP^-$, $A2B5^+$, and $Pax2^+$, whereas type-1 astrocytes are $GFAP^+$, $A2B5^-$, and $Pax2^+$. APCs are different from O2A cells, since no oligodendrocytes are obtained. Levels of the A2B5 antigen are found to be lower in APCs than in other precursors. Astrocytes have been generated from the dorsal spinal cord, which suggests the existence of an astrocyte precursor. $A2B5^+$ cells have also been identified in the dorsal spinal cord. This raises the possibility that cells similar to APCs exist in the spinal cord. Seidman et al. (1997) have also described an immortalized astrocyte-restricted precursor cell line. This cell line, isolated from E16 mouse cerebellum, proliferates in response to EGF and, upon EGF withdrawal and the addition of bFGF, develops the morphology and expression pattern of a fibrous astrocyte.

Specialized glia such as Bergmann glia or Muller glia provide a different view of glial cell production and differentiation. Recently, Gudino-Cabrera and Nieto-Sampedro have suggested that Muller glia and olfactory ensheathing cells belong to a third glial population, distinguished on the basis of antigenic and functional criteria. This population of glial cells, termed aldynoglia, includes olfactory ensheathing cells, tanycytes, pituocytes, pineal glial, Muller glial, and Bergman glial cells. These cells are distinguished from astrocytes and oligodendrocytes in that they concomitantly express O4, GFAP, P75, and estrogen receptor α. Differentiating these cells from other glial populations is primarily their potential functional importance. They appear to have dramatic growth- and regeneration-promoting properties. In particular, transplanted olfactory ensheathing cells have shown remarkable ability to promote axonal regeneration and functional recovery. How these cells are lineally related to the more prevalent glial populations present in the brain is not known. Muller glia, a specialized retinal glial cell, is one of the last cell types to mature in the retina. It does not seem to arise from a glial-restricted precursor, but rather from a multipotential precursor that also gives rise to neurons. It has been suggested that Muller glia can dedifferentiate to give rise to retinal stem cells in amphibian and avian retina.

More recently, stem cells have been identified in the mammalian retina that give rise to Muller glia in culture. Another specialized glial cell type is the intrafascicular ensheathing cells in the olfactory bulb, which ensheathe olfactory axons. They are unique in that they seem to have a mixture of astrocytic and oligodendrocytic properties. These cells may arise from the olfactory placode and express the homeoprotein Otx-2. Thus, these specialized glia, although functionally similar, may have distinct embryological origins.

Of interest, however, is a recent observation by Blakemore and colleagues on the differentiation of CNS glial precursors. They isolated CNS glial precursors and showed that they generated Schwann-like cells in addition to astrocytes and oligodendrocytes. This result raises the possibility that some aldynoglia share a common lineage with other glial precursors. In conclusion, the recent identification of GRPs and the relationship between different glial cell types has led to a clearer understanding of gliogenesis. The molecular mechanisms involved await further studies.

Neural Crest Stem Cells

The neural crest is a late development in the evolution of vertebrates and is of importance because of the variety of derivatives that arise from this specialized population of cells. Crest cells segregate from the developing neural tube at or around the time of neural tube closure. The exact process of delamination is species dependent. In mice, cells begin to delaminate before the tube has completely closed, whereas in the chick, a premigratory form of crest exists and can be reconstituted from surrounding tissue.

Crest stem cells differ from CNS stem cells discussed earlier because of their extensive migration ability, their epithelial to mesenchymal transition, and the phenotypes that they can differentiate into. The ***neural crest stem cell*** (NCSC) has been isolated and characterized in multiple species and, although it is a transient population, it nevertheless fulfills all the criteria of a stem cell. The NCSC cell is multipotent, undergoes self-renewal, and can generate multiple phenotypes that include craniofacial mesoderm, melanocytes, and the neurons and glia of the PNS. It is important to note that NCSC, like mesodermal stem cells, can generate mesodermal derivatives including bone, cartilage, smooth muscle, etc. The potential of neural crest to generate craniofacial mesenchyme was initially thought to be limited to the cranial crest, but several groups have shown that crest from more caudal regions is equally capable of differentiating into nonneural derivatives. What restricts this developmental potential in more caudal regions of the embryo remains undefined.

Evidence from chick embryo experiments at early developmental stages suggests that NCSC and differentiated CNS cells share a common progenitor. Additional evidence that the CNS and the PNS share a common precursor comes from chick neural fold ablation experiments which demonstrate that cells of the remaining neural tube have the regulative capacity to compensate for the ablated neural crest cells. These results suggest that neuroepithelial cells normally destined to form the CNS have the ability to regulate their prospective fates to generate PNS derivatives. Recent work from two different laboratories has shown that FGF-dependent multipotent CNS stem cells likely represent the common CNS-PNS precursor in rats. Mujtaba and colleagues (1998) have shown that NEP cells can generate p75/nestin immunoreactive cells that are morphologically and antigenically similar to previously characterized NCSCs. NEP-derived p75 immunoreactive cells differentiate into peripheral neurons, smooth muscle, and Schwann cells in both mass and clonal culture.

More importantly, clonal analysis indicates that individual NEP cells could generate both CNS and PNS derivatives, providing evidence for the first time of a direct lineage relationship between these two distinct cell types. In addition, McKay and colleagues have shown that cortical stem cells which were isolated at a stage well after neural crest migration has taken place still retain their ability to differentiate into PNS derivatives. No data regarding the ability of EGF-dependent neurosphere cells to generate crest or PNS derivatives exist. Although the NEP cell can generate NCSCs, it is unclear whether NCSCs can differentiate into CNS derivatives. Transplants of neural crest into the spinal cord result in Schwann cell differentiation but not into CNS derivatives, suggesting that the NCSC is a restricted precursor cell.

NCSCs isolated during crest migration have a characteristic phenotype but may not represent the sole kind of multipotent crest precursor. Multiple classes of stem cells are likely to be present, and in this respect, PNS differentiation may be similar to CNS differentiation. Initial results have suggested that there is some rostrocaudal patterning that distinguishes crest cells from each other

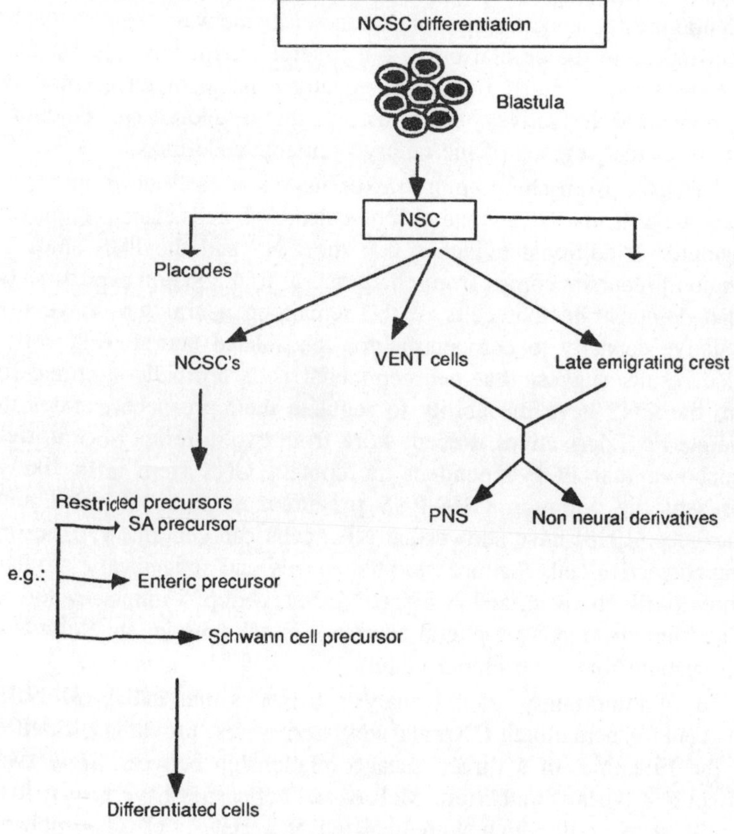

Fig. 10.3. Neural crest stem cells.

and thus biases their differentiation potential. For example, the ability to contribute to heart development is normally restricted to a specific rostrocaudal level. Clonal assays have suggested, however, that even caudal crest can differentiate into smooth muscle and when transplanted will migrate to the heart. Sommer and colleagues have identified a subpopulation of crest precursors that, unlike most NCSCs, express PMP-22. These cells, nevertheless, are multipotent by the same criteria that have been used to establish the stem cell potential of neural crest.

Like the CNS stem cell, it is likely that the NCSC also generates differentiated progeny via the generation of more restricted precursors. Although a detailed description of the multiple classes of more restricted precursors identified is beyond the scope of this chapter, several restricted precursors have been identified. Of interest are Schwann cell precursors that appear to be the equivalent of PNS glial restricted precursors. Schwann cells perform the functions of both astrocytes and oligodendrocytes and are known to be important in promoting axonal regeneration and recovery in peripheral nerves. A melanocyte precursor may also exist, and Weston and colleagues have suggested the existence of such restricted precursors.

Anderson and colleagues have immortalized a sympathoadrenal precursor that makes sympathetic neurons and adrenal chromaffin cells. The enteric nervous system may also be populated by more restricted precursors. Thus, NCSCs represent a lineage-restricted precursor cell that can generate PNS but not CNS derivatives. As in the CNS, differentiation appears to progress through sequential stages of more differentiated cells. Although such classes of cells have been identified, these cells have not been characterized in as much detail due to the limited numbers of cells available for analysis and the small number of stage-specific markers available.

Two other populations of cells that contribute to the developing PNS have been described. Both appear to differentiate later than NCSCs and migrate along characteristic pathways. One population, termed the late emigrating crest cell population, emigrates along dorsal nerve roots along the rostrocaudal axis and can contribute neurons and glia to the sensory ganglia. Frank and colleagues have shown that this population may also contribute to pigment cell development and perhaps to sensory neuron derivatives. Sohal and colleagues have described yet another population of cells that arise from the neural tube and migrate through the ventral roots of cranial nerves and contribute to a variety of tissues. These cells have been termed ventrally emigrating neural cells or VENT cells. VENT cells differ from neural crest in that they arise later from ventral neural tube and do not express HNK-1 immunoreactivity (a marker for emigrating neural crest), and the repertoire of their phenotypes is distinct. Sohal and colleagues used replication-deficient retrovirus to label VENT cells and follow their differentiation. VENT cells followed nerve roots of the cranial ganglia and could be localized to a variety of tissue that included the trigeminal ganglion, heart, cartilage, vascular smooth muscle, and liver. These results suggest that neural precursor cells are more pluripotent than previously supposed and may be capable of generating ectodermal and mesodermal cells. The developmental significance of these populations of late emigrating cells is unclear. No specific ablation experiments have been performed, and the importance of discrete multipotent populations that contribute to the same developing structure remains to be determined. The existence of these populations, however, needs to be taken into account as a potential atypical source of cells for cellular replacement strategies.

Adult Precursor Cells

Recent data have suggested that both neurogenesis and gliogenesis persist in the adult and can be modulated by external environmental influences. It is not clear from the available data whether this represents activation of a quiescent population of multipotent stem cells, activation of more restricted neuronal or glial precursors, or dedifferentiation of postmitotic cells. Evidence demonstrating the presence of multipotent cells in the adult suggests that multipotent stem cells may contribute to some of the ongoing neuro-gliogenesis. Controversy exists as to the location of adult stem cells, and multipotent cells may be present in the ependymal, subventricular, and cortical regions. The structure of the subependymal region has been characterized and at least five different classes of cells have been described. Which of these cells represents the putative stem cell remains unclear. Johansson et al. (1999) have argued that the ependymal cells represent the adult stem cell population. These authors showed that ependymal cells are a quiescent population that generates a more rapidly dividing cell (termed *transit amplifying cell*), which then generates neurons and astrocytes.

Conversely, van der Kooy and colleagues have argued that in their in vitro culture system only subependymal cells (and not ependymal cells) can self-renew and generate multipotent cells. Which subependymal population represents the true stem cell is still under investigation. Doetsch and others have suggested that type C cells represent a stem cell, as they express nestin but no other differentiation markers and are rapidly proliferating. More recently, Alvarez Buylla and colleagues have postulated that type B1 cells may represent the adult stem cell population. Using a retroviral label, they have shown that a GFAP immunoreactive cell is present in the subventricular zone and that this cell fulfills the criteria of a stem cell. Marmur and colleagues have suggested that in the adult, stem cells may be present not only in the remnants of ventricular or subventricular zones, but also as a quiescent population throughout the cortex. These investigators have shown that polysialated NCAM immunoreactivity can be used to isolate multipotent cortical stem cell populations. Similarly, Gage and coworkers have isolated multipotent neural progenitors from adult hippocampus.

Table 10.2. Localization of multipotent stem cells in the adult SVZ

Classification	*Characteristics*
Type A	neuronal markers, rapidly dividing, limited self-renewal
Type B1	GFAP-immunoreactive, rapidly dividing cells
Type B2	GFAP-immunoreactive, slow dividing cells
Type C	nestin-immunoreactive, dividing cells
Type D	tanycytes, possess microvilli and are GFAP-immunoreactive
Type E	ependymal cell, villi, nestin, and vimentin-immunoreactive

Localizing stem cells in the adult is important for detailed analysis in vivo as well as for future therapeutic uses. There has been a tremendous excitement in the field because the presence of significant numbers of stem cells in the adult was not expected. The ability to maintain these cells in culture for prolonged periods using genetic or epigenetic means has provided the potential for virtually unlimited numbers of cells for therapy.

The development of techniques to obtain stem cells from small pieces of human tissue has further heightened expectations that stem cells will be clinically useful. Although the excitement is warranted, caution should be maintained, as several questions need to be resolved before therapeutic transplants can be attempted. An important issue that still remains is whether the environment contains the requisite signals to direct differentiation of pluripotent cells to the desired cell type.

In most reported transplant experiments, a majority of the transplanted multipotent stem cells fail to differentiate, suggesting the absence of instructive cues. Indeed, the best examples of differentiation and migration are usually in regions of ongoing neurogenesis such as the hippocampus and olfactory bulb. Little differentiation is seen in the spinal cord where there is no ongoing neurogenesis. Likewise, the ability to migrate in response to endogenous signals appears limited. Neuronal precursors (see below) readily migrate but multipotent stem cells when transplanted to the same region fail to do so. Indeed, several investigators have suggested that transplanting predifferentiating stem cells may be a better option.

Adult NRPs

It had been widely believed that neurogenesis in the mammalian brain ceases after early postnatal life. In fact, postnatal neurogenesis was believed to be an anomaly occurring in exceptional areas of the avian and rodent brain. Considerable evidence has now accumulated that neurogenesis in the adult primate (including human) brain is possible. Therefore, there must be stem cells that divide and generate neurons in the adult mammalian brain. What is not clear is whether these are multipotent stem cells biased toward neuronal differentiation or NRPs. Even dedifferentiation of mature neurons may occur, as has been dramatically demonstrated in the hippocampus. It is likely that NRPs exist in at least two regions where there is ongoing adult neurogenesis: the ependymal/subependymal zone and the dentate gyrus of the hippocampus.

Neuronal progenitors from this region, in particular the anterior part often termed SVZ(a), give rise to neurons exclusively. They migrate along the rostral olfactory tract and terminally differentiate to form interneurons of the olfactory bulb. Labeling of the dentate gyrus cells by BrdU in adult humans and in rats in vivo has been observed with dramatic increases seen postseizure.

Using the early neuronal Tα1 tubulin promoter to drive the expression of GFP, Goldman and coworkers have isolated cells from 3-month-old rat forebrain that were morphologically neuronal and expressed MAP-2 and b-III tubulin after a week in culture. This promoter is active in early neuronal precursors and, as expected, these precursors differentiate into neurons in vitro but do not generate any glial cells. Fewer than 5% of the cells sorted exhibited astrocytic markers. The cells also incorporated BrdU in culture, indicating that they were capable of replication. All the recent data, therefore, suggest that in the hippocampus, neuronal precursors either already exist or can be induced, and these precursor cells can repopulate some regions of the adult or fetal nervous system.

It is increasingly evident that neurogenesis can take place in the adult brain and that new neurons are generated and can integrate into the existing network and establish synapses. This has raised the possibility that, rather than transplanting cells for therapy, it may be possible to enhance endogenous stem cell proliferation or bias their differentiation by providing extrinsic cues. Several investigators have reported on the ability to alter neuronal generation. These include providing an enriched environment in juvenile or adult mice or infusion of growth factors. Increased neurogenesis may not be beneficial and has been noted in seizures both in vivo and in vitro.

In response to ischemia, there is a 12-fold increase in neurogenesis in the dentate gyrus of gerbils. The endogenous proliferation may lead to heterotopias and increased chances of seizures. Not all regions exhibit ongoing or regulated neurogenesis; the spinal cord is a notable example. Dividing glial cells have been observed. Transplants of embryonic neurons, on the other hand, survive in the adult spinal cord, and new neurons are generated from newborn and adult rat spinal cords in culture, as determined by BrdU incorporation. This de novo generation of neurons suggests that a quiescent population of multipotent stem cells may exist in the spinal cord.

ES Cells and Neural Precursors

The neural precursor cells that have been described above are derived from tissues of the advanced embryo and adult. In normal development, these neural precursor cells must arise from earlier, more undifferentiated cells, and it should be possible to isolate neural precursors from such undifferentiated cells in vitro. Indeed, neural precursors have been isolated from ES cells and primordial germ cells.

The idea that ES cells can be used as a reservoir of cells from which differentiated cells can be harvested has been validated in rodent models. Several groups have shown that cell surface markers, manipulation of culture conditions, or utilizing tissue-specific promoters can be used to isolate neural stem cells or more restricted precursors. Recently transplanted oligodendrocyte precursor cells isolated from ES cells have been shown to generate new myelin in a rat model of demyelination. McKay and colleagues have manipulated culture conditions to isolate multipotent stem cells from ES cell cultures, and Li et al. (1998) have used an elegant selection strategy to isolate neural precursor cells.

Using cell-type-specific markers, we have isolated neuronal and glial restricted precursors from mouse ES cells and have shown that these cells are antigenically and phenotypically identical to restricted precursors isolated from fetal tissue. Data on isolation of specific neural precursor types from human ES cell cultures are not available. However, preliminary results suggest that markers utilized to isolate neuron and glial precursor populations from rodent ES cells are also expressed by differentiating human ES cell cultures. Thus, it is likely that human ES cells could also serve a reservoir function.

One potential therapeutic advantage of neurons, glia, or precursor cells derived from ES cells is that they may not be regionally specified and thus may be more capable of site-specific integration. Furthermore, since it is possible to obtain regionally specific phenotypes by manipulating ES cell culture conditions, it may be possible to obtain appropriately specified phenotypes. Thus, ES-cell-derived precursors may be preferable to fetal or adult precursor cells. McDonald and colleagues have presented suggestive evidence that ES-cell-derived cells may be useful in spinal cord injury. When ES-cell-derived neural cells were transplanted in a spinal cord injury model, significant functional improvement was seen.

Suggestive evidence that neural cells derived from human ES cells will prove useful in spinal cord injury is provided by experiments using a NRP cell derived from a human embryonic carcinoma cell line. Trojanowski and colleagues isolated a subclone (HNT-2) from a human embryonic carcinoma that showed exclusive differentiation into neurons. Neurons derived from this subclone were used in rat models of spinal cord injury. Transplanted cells differentiate into neurons that send long axonal projections, which even project into the nerve roots. No tumors have been seen in any of the transplants done so far. These results suggest that it will be possible to obtain NRPs from human ES cell lines and that these are likely to be functionally useful. Several other classes of precursors have been isolated. The recent demonstration that melanocytes and potential neural crest precursors can be directly isolated from ES cells is of great interest. Melanocytes were generated from ES cells by co-culturing them with a bone marrow-derived stromal cell line, dexamethasone, and the steel factor. All the above results taken together suggest that it should be possible to obtain both CNS and PNS derivatives from differentiating ES cells.

Are Tissue-specific Stem Cells Irreversibly Determined?

Although we have discussed the ability of neural stem cells to generate CNS derivatives, data from several groups have suggested that these cells are more plastic than previously proposed and perhaps it is premature to classify stem cells based on the tissue from which they were isolated. Recent studies have shown that neural stem cells, when transplanted in the bone marrow of irradiated mice, will generate hematopoietic derivatives. Similarly, in an impressive demonstration of pluripotentiality, Frisen and colleagues injected adult neural stem cells into chick embryos and mouse blastocysts and showed that neural stem cells contributed to ectodermal, endodermal, and mesodermal tissue. Contributions to tissue were large and occasionally comprised as much as 30% of the entire organ. These data suggest that neural stem cells are capable of contributing to multiple tissues, and under appropriate conditions this contribution may be very large. Thus, the term neural may be too restrictive. It should be noted, however, that this ability to contribute to other tissues appears restricted to multipotent neural stem cells. Available data suggest that more restricted neural precursor cells do not trans-differentiate in a similar fashion.

Likewise, mesenchymal stem cells and bone marrow cells will generate astrocytes and possibly neurons when infused into the brain. Since these precursor cells usually generate mesodermal derivatives, their ability to generate neuroectodermal derivatives is somewhat surprising. These and other trans-tissue differentiation results suggest that tissue-specific stem cells may be more pluripotent than previously thought or still may retain the ability to dedifferentiate. Whether this ability is simply a by-product of the plasticity of these early cells or whether this ability has some developmental significance remains to be determined.

Ectoderm to mesoderm transformation is normally seen in neural crest differentiation. Indeed, neural crest cells have been shown to generate muscle, bone, cartilage melanocytes, fibroblasts, and smooth muscle as well as neural components of the PNS. Thus, a potential pathway for dedifferentiation from mesoderm to ectoderm or vice versa may exist. Quiescent neural stem cells present in ectopic neural tissue transplants may generate neural crest that subsequently can generate bone, cartilage, and smooth and striated muscle. Alternatively, an additional neural derivative, VENT cells (*ventrally emigrating neural cells*), which have also been shown to contribute to cartilage, bone, heart muscle, and hepatocytes during normal development, may be present in neural tissue. These cells are distinct from crest cells and appear to have a truly broad spectrum of differentiation. Thus, the ability of neuroectoderm to differentiate into mesodermal derivatives may reflect an activation of this VENT cell differentiation pathway or a neural crest differentiation pathway. Finally, it is possible that rare totipotent cells are present in all tissues, and it is these cells rather than tissue-specific stem cells that generate nonneural derivatives.

It is interesting to note that mesodermal to ectodermal transformation is a normal aspect of some organ development. In kidney, for example, mesodermal cells undergo an epithelial transformation to generate kidney tubules. Mesodermal stem cells or *mesenchymal stem cells* (MSC) have also been identified. It is possible that these MSCs can generate all mesodermal derivatives (including hematopoietic lineages) as well as undergo an ectodermal transformation to generate ectodermal derivatives. Alternatively, quiescent neural crest stem cells (or VENT cells) may persist in peripheral tissue and be activated when transplanted into the CNS. Most transplants have not utilized a homogeneous population of cells. Irrespective of the source of cells, it should be noted that the effectiveness of neuronal differentiation has been relatively modest, and it is unclear whether these results merely illustrate the plasticity of stem cells or whether they provide a novel therapeutic source of cells. Additional experiments are clearly required to refine our understanding of the extent and limitation of the differentiation potential of stem cells isolated from different tissues. It is unlikely, in our opinion, that cells derived from the adult or late fetal stages can contribute to all germ layers and also contribute to germ-line cells (ES cell equivalents). However, in the absence of direct evidence to the contrary, this remains a formal possibility that should be tested.

11

Stem Cells and Parkinson's Disease

In vertebrate embryogenesis, the primordia of the nervous systems arise from uncommitted ectoderm during gastrulation. Spemann and Mangold demonstrated that the dorsal lip of the amphibian blastopore, which gives rise mainly to axial mesoderm, emanates inductive factors that direct neural differentiation in the ectoderm. During the last decade, much progress has been made in the molecular understanding of early neural differentiation in *Xenopus*. Neural inducer molecules, such as chordin, noggin and follistatin, were identified, and several intracellular mediators of neural differentiation have been characterized. These neural inducers do not have their own receptors on target cells, instead they act by binding to and inactivating *bone morphogenic protein* 4 (BMP4), which suppresses neural differentiation and promotes epidermogenesis. Noggin and chordin bind to BMP with dissociation constants comparable to that of BMP receptors to BMP. Thus, neural induction in *Xenopus* is controlled by a morphogenetic signaling involving a BMP activity gradient.

By contrast, relatively little is known about the regulatory factors involved in mammalian neural induction. One main reason for this is that good experimental systems for in vitro neural differentiation are still lacking in mice, that is, something comparable to the animal cap assay commonly used in *Xenopus* studies. Mammalian *embryonic stem* (ES) cells can differentiate into all embryonic cell types when injected into blastocyst-stage embryos. This pluripotency of ES cells can be partially recapitulated in vitro by a floating culture of ES cell aggregates or *embryoid bodies* (EBs). After a few weel of culture without *leukemia inhibitory factor* (LIF), EBs frequently contain ectodermal, mesodermal, and endodermal derivatives. However, there had not been any good methods that can induce "*selective*" differentiation to a particular cell type such as neurons. Therefore, we attempted to develop such an experimental system by using colony assays.

Stromal Cell-Derived Inducing Activity Promotes Neural and Neuronal Differentiation of ES Cells

We first asked whether attenuation of BMP signaling is sufficient to induce neural differentiation of mouse ES cells by administering BMP antagonist molecules. Neither transfection of *cytomegalovirus* (CMV)-chordin plasmid into ES cells nor the addition of follistatin protein or neutralizing BMPR-Fc to culture medium induced significant neural differentiation of ES cells. On the other hand, the administration of BMP4 protein efficiently suppressed in vitro neural differentiation of mouse ES cells (e.g., EB + *retinoic acid* [RA] method) or of isolated mouse epiblasts even at a low concentration (0.5 nM). These results indicate that a blockade of BMP4 signaling is required but not sufficient for neural differentiation of undifferentiated mouse cells. This suggests that mouse ES cells, unlike *Xenopus* cells, require some unknown signals for initiating neural differentiation of ES cells, in addition to the attenuation of BMP signals.

By using a co-culture system, we screened various primary culture cells and cell lines for activities promoting neural differentiation of ES cells under serum-free conditions. Most of the cell types screened, including *mouse embryonic fibroblasts* (MEF), MDCK, and COS cells, did not significantly induce neural markers such as NCAM (pan-neural) in the overlying ES cells. However, some stromal (or *mesenchymal*) cells promoted neural differentiation when used as feeders. A low yet

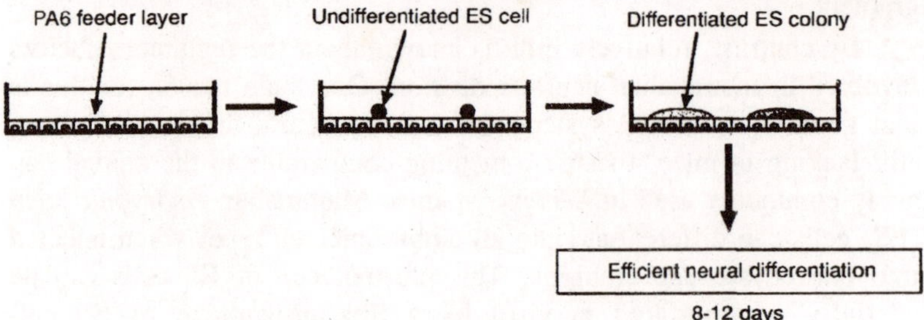

Fig. 11.1. Overview of the SDIA method.

significant number of NCAM-positive colonies were observed in the presence of OP9 (stromal line derived from mouse calvaria) and NIH3T3 (an embryonic fibroblast line) cells. PA6 cells (stromal cells derived from skull bone marrow) induced remarkably efficient neural differentiation when co-cultured with ES cells, resulting in 92% of colonies becoming NCAM-positive by d 12.

In these colonies, the majority of cells were stained with either the neuronal marker TuJ (class III β-tubulin) or the neural precursor marker nestin. Very few colonies contained *glial fibrillary acidic protein* (GFAP)-positive cells (2%). The TuJ-positive neurons also expressed other neuronal markers such as MAP2 and neurofilament, and the presynaptic marker synaptophysin was detected on the induced neurons. To confirm that nestin-positive cells were neural precursors, cells were double-stained with nestin and RC2, a marker for neuroepithelium and radial glia. All of the nestin-positive cells were co-stained with the RC2 antibody. We next tested whether neural differentiation induced by PA6 was accompanied by mesodermal induction. In contrast to the high NCAM-positive percentage, very few colonies expressed mesodermal markers such as *platelet-derived growth factor receptor*-α (PDGFR α), Flk1, and MF20 (all <2% colonies). This is consistent with a previous report showing that PDGFR α and Flk-1 are induced in ES cells co-cultured with OP9 cells, but not in those co-cultured with PA6 cells. Thus, PA6 can promote neural differentiation of co-cultured ES cells without inducing mesodermal markers. We named the neural-inducing activity on the stromal cells "*stromal cell-derived inducing activity*" or "*SDIA*".

Efficient Induction of Dopaminergic Neurons by the SDIA Method

Immunohistochemical analyses of the characteristics of SDIA-induced neurons revealed that 92% of colonies contained *tyrosine hydroxylase* (TH)-positive neurons. This value was much higher than those for GABAergic, cholinergic, and serotonergic neuron markers (43, 28, and 7%, respectively). TuJ-positive neurons and nestin-positive cells represented 52 ± 9% and 47 ± 10% of the total cells, respectively. TH neurons occupied 30 ± 4% of TuJ-positive neurons (i.e., approx 16% of the total cells). This value is again significantly higher than percentages of GABAergic, cholinergic and serotonergic neurons in TuJ-positive neurons (18 ± 8%, 9 ± 5%, and 2 ± 1%, respectively) at the cell level.

A time course study showed that TH-positive neurons appeared between d 6–8 of the induction period, following the appearance of nestin and tubulin markers. The cells remain negative for dopamine-β-hydroxylase (DBH), a marker for norepinephrine and epinephrine neurons, even after 13 d of induction. The mesencephalic dopaminergic neuron markers Nurr1 and Ptx3 were induced in SDIA-treated ES cells. *High-pressure liquid chromatography* (HPLC) analyses showed that ES cell-derived neurons released a significant amount of dopamine into the medium (7.7 pmol/10^6 cells) in response to a depolarizing stimulus. These data show that functional neurons producing dopamine were generated with this method.

Epidermal Differentiation by Treating ES Cells with SDIA and BMP4

In *Xenopus*, BMP4 inhibits neural differentiation of animal cap ectoderm and promotes epidermogenesis. We therefore tested whether BMP4 has a similar effect on the neural differentiation of mouse ES cells induced by SDIA. The addition of 0.5 n*M* BMP4 suppressed neural differentiation and significantly increased the number of colonies positive for the non-neural ectoderm marker E-cadherin, without inducing mesodermal markers. Furthermore, after 11 d of culture, BMP4 induced keratin 14-positive colonies (0% without BMP4, 34% with BMP4). A time-course study showed that neural differentiation in ES cells was most sensitive to BMP4 during d 3–5 of the induction period.

Thus, with SDIA, neural and epidermal differentiation of ES cells are selectively induced in the absence and the presence of BMP signals, respectively.

Advantages of the SDIA Methods

A superphysiological dose (up to 1 μM) of RA promotes neural differentiation in EBs. Although this method can produce a good proportion of neural cells, it has two apparent problems. First, it is difficult to analyze and control each regulatory step of differentiation in this method, because EBs contain many different kinds of cells, including mesodermal and endodermal cells. Second, RA, a strong teratogen, is supposed to perturb neural patterning and neuronal identities in EBs, as it does in vivo. It is, therefore, preferable to avoid RA treatment unless RA induces the particular type of neurons of one's interest. In fact, we have noticed that RA treatment efficiently suppresses dopaminergic neuron differentiation in EB and SDIA methods.

The SDIA method is technically simple, and the induction is efficient and speedy. Studies with various differentiation markers have demonstrated that time course features of neural differentiation by SDIA in vitro mimic well those observed in the developing embryo.

The SDIA method does not involve EB formation or RA treatment, and each differentiating colony grows from a single ES cell in two dimensions under serum-free conditions. Because of these features, the SDIA method has advantages over the EB/RA methods when used for detailed analyses of differentiation, such as effects of exogenous growth factors.

Mesencephalic dopaminergic neurons can be efficiently produced by the SDIA method. Recently, it was reported that dopaminergic neurons could be generated from neural precursor cells amplified from EBs. In contrast to our single-step method, they used the following 4 steps for the production of dopaminergic neurons. Since only small numbers of neural precursors are present in RA-untreated EBs, one needs to select and amplify nestin-positive cells for a long time (14 d) in this method before inducing differentiation (>24 d in total). This 4-step method produced TH-positive neurons at an efficiency of approx 7% of TuJ-positive neurons in conventional medium. When Shh, FGF8, and ascorbate were added, the production increased to about 30%, which is comparable to the efficiency of our method.

We infer that the quickness of dopaminergic neuron production could be crucial once this method is applied to primate–human ES cells, as primate–human ES cells seem to grow and differentiate much slower than mouse cells. Given that human ES cells take a 3 to 4 longer time period to become dopaminergic neurons than mouse cells in vitro, it seems that 8 d for mouse cells is a maximal tolerable length of time for the production of human cells by culturing cells in a practical sense.

Possible Mechanisms of Neural Induction by SDIA

At present, the molecular nature of SDIA remains to be identified. One important question that arises is whether PA6-derived factors act by antagonizing BMP4 in a similar manner as chordin and noggin. We actively blocked BMP signaling by administering neutralizing BMPR-Fc and found that the presence of BMPR-Fc did not affect the extent of neural differentiation in the ES cells under any conditions tested. Thus, BMP antagonism is unlikely to explain the neuralizing activity of SDIA.

One possible explanation is that some additional factors such as SDIA are required for neural differentiation before mouse cells make the neural–epidermal binary decision in a BMP-dependent manner. Consistent with this idea, the SDIA-treated cells acquire their highest sensitivity to BMP4 subsequent to the onset of nestin expression (d 3).

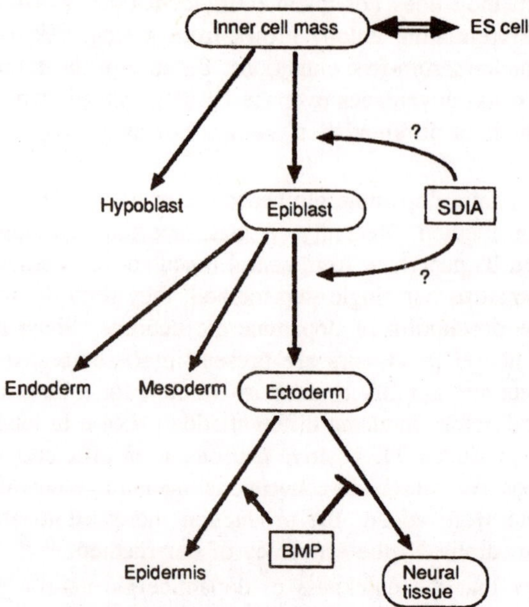

Fig. 11.2. A working model for the role of SDIA.

This indicates that SDIA has already exerted some effects (nestin induction) before the cells react to BMP signals. The following scenario might be applicable to the mechanism of neuralization occurring in SDIA-treated ES cells. First, ES cells cultured on PA6 move in an ectodermal direction under the influence of SDIA. SDIA-treated ES cells then adopt a default neural status unless they receive a sufficient level of BMP4 signals. However, as the molecular nature of SDIA remains to be elucidated, we must await further study to judge this proposition and to understand the relevant roles of SDIA in the embryo.

Application to Cell Transplantation Therapies and its Prospective

We tested whether SDIA-treated ES cells could be integrated into the mouse striatum after implantation. ES cell colonies were cultured on PA6 cells for 12 d and detached en bloc (approx 50 μm) from the feeders by a mild protease treatment without EDTA. In order to enrich for postmitotic neurons by eliminating mitotic cells, the SDIA-treated ES cells were treated with mitomycin C (MMC) before grafting. The isolated ES cell colonies were then implanted into the mouse striatum, which had been treated with 6-hydroxydopamine (6-OHDA). Ipsilateral implantation of SDIA-induced neurons significantly restored TH-positive areas in and around the DiI-positive graft. Two weeks after implantation of 4 × 10^5 SDIA-treated ES cells, 3.9 ± 0.6 × 10^4 grafted cells were found in the brain, and 74% of them were TuJ-positive neurons. The estimated survival rate of TH-positive neurons was approx 22% after these procedures. No teratoma formation was observed in the grafted tissue by histological analysis.

As the SDIA method produces a high yield of mesencephalic dopaminergic neurons, and once this method is successfully applied to human ES cells, SDIA-induced neurons may provide a noninvasive alternative to embryonic brain tissues and neural stem cells for neuronal replacement therapy of Parkinson's disease. Then, long-term survival of SDIA-induced neurons and its functional consequences, such as motor recovery, should be examined in Parkinsonian model monkeys before applied to patients. One advantage of using ES cells over neural stem cells is that genetic manipulations, such as modifying histocompatibility, are theoretically feasible by homologous recombination. In addition, we have to develop an even more efficient way to enrich for dopaminergic neurons from the total population of differentiated ES cells. Sorting by flow cytometry or separating with magnetic beads should be feasible once appropriate surface antigens for early dopaminergic neurons become available. We also need to test the safety of ES cell therapy in long-term implantation studies. It may be beneficial to select only postmitotic neurons by eliminating dividing cells with MMC (and/or cytosine β-D-arabinofuranoside [Ara-C]) before grafting to prevent tumor formation.

To explore further application possibilities, it is important to study how to regulate the differentiation of SDIA-treated ES cells into many specific types of *central nervous system* (CNS) neurons. For instance, as the generation of motor neurons from mouse ES cells has been previously demonstrated, the SDIA method may provide a protocol for more efficient production of dopaminergic neurons.

Materials

Cell Lines

1. PA6 cells: PA6 is a stromal cell line derived from newborn mouse calvaria. Among various cell lines that we screened, PA6 cells have a marked activity to induce neural differentiation of ES cells.
2. ES cell lines: the ES cell line used in this study is EB5. EB5 cells carry the blasticidin S-resistant selection marker gene driven by the Oct3/4 promoter (active under the undifferentiated status) and are maintained in medium containing 20 μg/mL blasticidin S

to eliminate spontaneously differentiated cells. EB5 is a subline derived from E14tg2a ES cells and was generated by the targeted integration of the Oct-3/4-IRES-BSD-pA vector into the Oct-3/4 allele. EB5 and CCE behaved similarly using our procedures.

Reagents

1. PA6 culture medium: we use α minimum essential medium (αMEM) supplemented with 50 U/mL penicillin and 50 μg/mL streptomycin. *Fetal calf serum* (FCS) is added at a final concentration of 10%.
2. ES differentiation medium: *Glasgow minimum essential medium* (GMEM), *Knockout serum replacement* (KSR); 10% final concentration, nonessential amino acids; 0.1 mM final concentration, sodium pyruvate; 1 m*M* final concentration, 2-mercaptoethanol (WAKO); 0.1 m*M* final concentration.
3. N-2 medium (optional): G-MEM, N-2 supplement (100X); 1X final concentration, nonessential amino acids; 0.1 m*M* final concentration, sodium pyruvate; 1 m*M* final concentration, 2-mercaptoethanol (WAKO); 0.1 m*M* final concentration.

Methods

Maintenance of PA6 Cells

1. Remove PA6 culture medium and rinse with Ca^{2+} and Mg^{2+}-free *phosphate-buffered saline* (PBS).
2. Add 0.05% trypsin-EDTA and incubate for 5 min at 37°C.
3. Add PA6 culture medium and break cell aggregates by gentle pipetting.
4. Spin down the cells and resuspend the pellet in PA6 culture medium and plate one-fifth of the cells onto tissue culture dishes.
5. Incubate in a 37°C incubator with 5% CO_2 environment and passage cells every third day when the density reaches near the confluency.

Preparation of PA6 Feeder Layers

1. Into each well of a collagen type I-coated 8-well Biocoat culture slide, dispense 500 μL of PA6 cell suspension in PA6 culture medium. It is important to plate PA6 cells to ensure that a confluent uniform monolayer is produced.
2. Allow PA6 cells to attach overnight.

Induction of Neural Differentiation of ES Cells

1. Rinse undifferentiated ES cells with PBS and collect the cells by trypsinization.

2. Add ES differentiation medium and disperse the cells into a single-cell suspension by pipetting.
3. Collect the cells by centrifugation at 250*g* for 5 min.
4. Resuspend the pellet with ES differentiation medium, count the cells, and dilute them into a final cell density of 2×10^3/mL.
5. Remove PA6 culture medium from the culture slide covered with a PA6 feeder monolayer and rinse with ES differentiation medium.
6. Plate 500 μL of ES cell suspension into each well and incubate at 37°C under 5% CO_2.
7. Change the medium with fresh ES differentiation medium on d 4 and 6, and afterwards, everyday.
8. (Optional) Change the medium to N-2 medium on d 8. This medium promotes the differentiation of neural precursor cells to postmitotic neurons.
9. Fix the cells with an appropriate fixative and stain the cells with appropriate antibodies. Neural precursor cells, postmitotic neurons, and dopaminergic neurons first appear in culture after 3, 5, and 7 d, respectively, and 2 d later, the frequencies of these cells increase rapidly. In general, more than 90% of colonies become NCAM-positive. In these colonies, the majority of the cells are neural precursor cells or postmitotic neurons. Dopaminergic neurons (tyrosine hydroxylase-positive and DBH-negative neurons) occupy about 30% of postmitotic neurons. Very few colonies contain GFAP-, PDGFRα-, Flk1-, and MF20-positive cells.

Mitomycin C Treatment of Differentiated ES Cells

1. Prepare differentiated ES colonies in a 10-cm dish.
2. Remove the medium and add 6 mL of fresh ES differentiation medium containing 10 μg/mL MMC onto the dish and then incubate at 37°C for 2 to 3 h.
3. Rinse the cells twice with PBS and use them for further experiments such as transplantation.

Isolation of Differentiated Colonies from PA6 Feeder Layer

We routinely use the Papain Dissociation System.

1. Prepare differentiated ES colonies in a 10-cm dish and, according to the manufacturer's protocol, reconstitute ovomucoid inhibitor, papain, and DNase.
2. Wash differentiated ES cells with PBS containing Ca^{2+} and Mg^{2+}. Discard PBS and pour 2 ml of the reconstituted papain solution.

3. Incubate at 37°C for 5 min.
4. Rock the dish gently. The colonies are detached from the feeder layer en bloc.
5. Collect the isolated colonies into a 15-mL tube. Spin down the cells briefly and resuspend the pellet in the reconstituted ovomucoid inhibitor solution.
6. Spin down briefly and replace the supernatant with appropriate medium.

Notes

1. We screened various primary culture cells and cell lines for activities promoting the neural differentiation of ES cells. While certain stromal cells such as OP9 cells and NIH3T3 cells generated neural cells to some degree, PA6 cells resulted in the highest yield and are therefore routinely used.
2. All tissue culture procedures described must be carried out under sterile conditions using sterile plasticware and detergent-free glassware. Water quality is very important. Medium or solutions should be warmed to 37°C before use. We recommend using medium less than 4 wk old to obtain a high yield of neuronal cells.
3. We noticed that the quality of KSR varies from lot to lot. We recommend lot checks to find appropriate KSR batches. It is essential to determine the optimal concentration of KSR to induce efficient neural differentiation. Optimal concentration of KSR may range from 5–15%. Inappropriate batches and concentration of KSR result in excessive E-cadherin-positive cells instead of neural cells.
4. Inappropriate concentration of 2-mercaptoethanol leads to low yields of neural cells.
5. Precoating of tissue culture dishes is not required. PA6 cells should be maintained carefully, so as not to lose their supportive activity. We recommend to passage the cells frequently (every 3 days) with a minimum dilution of the cells (1:5) using fresh medium.
6. Culture slides must be coated with collagen type I. Otherwise, PA6 cells dissociate during the differentiation period. Plasticware pretreated with a 0.1% solution of gelatin for 1 h is also useful. In the case that large numbers of the colonies are needed, we use a 10-cm tissue culture dish coated with 0.1% gelatin and usually place a coverglass coated with 0.1% gelatin onto the dish. Plate 5×10^4 undifferentiated ES cells into a 10-cm dish covered with

PA6 cells. After the differentiation period, the coverglass is used for immunocytochemistry to examine the efficiency of neuronal differentiation.

7. As long as PA6 cells are confluent, small differences in the cell density of PA6 cells do not significantly affect the efficiency of differentiation. MMC-treated PA6 cells can be used, although the treatment occasionally results in less efficient induction of neuronal cells. PA6 feeder slides may be used for up to 2 d after they become confluent.
8. ES cells are cultured to form a colony from a single cell. Within 8 d of differentiation, the cell number will increase more than 100-fold in a typical case.
9. Because serum strongly inhibits the neural differentiation of ES cells, it is essential to remove serum completely by washing cells with ES differentiation medium.
10. The addition of 0.5 n*M* BMP4 into ES differentiation medium markedly inhibits neural differentiation and promotes epidermogenesis.
11. In the case that PA6 cells dissociate during the differentiation period, the reasons may be as follows: (i) the cells and/or the media are old; (ii) the plating density of PA6 cells in culture slides is extremely high; and (iii) PA6 cells are incubated more than 3 d in culture slides before the differentiation.
12. In general, 8 d differentiation is enough to obtain high yields of postmitotic neurons, and neurites become visible by using a phase-contrast microscope. However, differentiation speed can vary depending on cell lines and other conditions. In the case that neuronal differentiation is not enough, 10–12 d differentiation should be tried. To obtain specific types of neurons that appear at later stage in vivo development, a longer incubation period should be considered.
13. Neural precursor cells, postmitotic neurons and dopaminergic neurons can be identified using anti-nestin antibody, anti-class III β-tubulin antibody, and anti-tyrosine hydroxylase antibody, respectively.
14. If yields of neural cells are low, the parameters that need to be considered are: (i) the cells and/or the media are old; (ii) serum remains in the differentiation medium; and (iii) the quality or concentration of KSR is inappropriate.

15. RA, which is commonly used in the EB method to induce the neural differentiation of ES cells, suppresses the induction of dopaminergic neurons without reducing the number of postmitotic neurons. Thus, the absence of RA is required for the successful induction of dopaminergic neurons.
16. When we transplant differentiated ES cells into mouse brains, we treat the cells with MMC in advance to prevent teratoma formation.
17. If necessary, add Ara-C at a final concentration of 5 μg/mL for the last several days of the differentiation period.
18. The isolated colonies are used for transplantation, RNA analysis, and protein analysis. Although this method is useful to obtain large numbers of colonies, it is possible that a small number of PA6 cells remain on the isolated colonies. It should be noted that neurons, especially neurons whose neurites go out of colonies, might be damaged by the isolation.
19. Among several proteases that we have analyzed for their ability to isolate the colonies, papain resulted in the highest viability of the cells.
20. Removing Ca^{2+} from the medium weakens the binding between differentiated ES cells in the colonies and prevents the isolation of the colonies en bloc. Therefore, PBS containing Ca^{2+} and Mg^{2+} without EDTA is recommended.
21. After the papain treatment, attachment of ES and PA6 cell is weakened, while the integrities of a PA6 monolayer and ES colonies are not much affected. Do not incubate for more than 10 min, or a PA6 monolayer and ES colonies will be broken into pieces.

12

Multipotent Stem Cells

The crowns of human teeth consist of enamel, dentin, and dental pulp tissue. During tooth growth and development, ameloblasts form enamel and odontoblasts generate primary dentin. After tooth eruption, ameloblasts disappear from the surface of the enamel; consequently, enamel formation ceases to occur naturally in vivo. In contrast, odontoblasts, along the inner surface of the dentin inside the pulp chamber, continue to deposit dentin matrix to form secondary dentin throughout life. In addition to secondary dentin, odontoblasts can form tertiary (reactionary/reparative) dentin in response to several stimuli, such as mechanical, chemical, and/or bacterial stimulation. Even when odontoblasts have been damaged, the reparative dentin can be formed in the dental pulp to protect against further disruption of the pulp tissue. This reparative dentinogenesis has been thought to be mediated by newly generated odontoblasts that seem to arise from dental pulp tissue. These findings have led to the speculation that odontogenic progenitor cells or stem cells may exist in dental pulp tissue. Several reports have demonstrated that pulp tissue contains proliferating odontoblast-like cells and that these cells are capable of forming mineralized nodules in vitro. However, in these studies, the cells isolated from dental pulp appear to have a limited capacity to differentiate into odontoblast-like cells and an inability to differentiate into other cell types such as adipocytes or neurons. More recently, Tecles and colleagues confirmed that proliferating odontogenic precursor cells appear to be mobilized from blood vessels to sites of damaged pulp or dentin tissue. Collectively, these studies describe the presence of preodontoblast cells present in dental pulp tissue, leading to speculation of the existence of putative dental stem cell populations.

Mesenchymal stem cells were first isolated from bone marrow (*bone marrow mesenchymal stem cells*; BM-MSCs); they are a population of multipotent postnatal stem cells. One of the most important characteristics of BM-MSCs is their capacity to form single-cell-derived colony clusters called *colony forming unit-fibroblast* (CFU-F) in vitro. Accumulated knowledge regarding the phenotypic characteristics of BMMSCs has permitted us to isolate putative stem cell populations from the dental pulp of human third molars (*dental pulp stem cells*; DPSCs) and deciduous teeth, which exhibit properties similar to those of BM-MSCs. Stem cells in dental pulp were found to reside in a specific perivascular microenvironment, where they are quiescent and maintain their basic stem cell characteristics, including a self-renewal capacity and undifferentiated status. This specific microenvironment is called the "*stem cell niche*". Thus any isolation of mesenchymal stem cells must give consideration to their niche microenvironment in order to identify the factors that maintain the "*stemness*" of cultured mesenchymal stem cells, which gradually lose their stem cell-like properties after ex vivo expansion. In this chapter, we provide detailed procedures for the isolation, purified preparation, expansion, and tissue regeneration potential of DPSC and SHED.

Preparation of Media and Reagents

Culture Media

All media should be sterilized by filtration through a 0.22-μm membrane filter and stored at 4°C.

Mesenchymal stem cell medium (MSC Medium)

Prepare α-modified minimal essential medium (α-MEM) with 2 mM glutamine and supplemented with 15% *fetal bovine serum* (FBS), 0.1 mM l-ascorbic acid phosphate, 100 U/ml penicillin, and 100 μg/ml streptomycin. Selection of a suitable lot of FBS is critical for successful MSC culture. We select FBS on the basis of its colony-forming efficiency. Briefly, primary MSCs are seeded at the same density with several kinds of FBS, and then colonies are counted. Usually, a higher colony number is associated with better proliferation of MSCs.

Odontogenic differentiation medium

Add 0.01 μM dexamethasone sodium phosphate and 1.8 mM monopotassium phosphate (KH_2PO_4) to MSC medium.

Adipogenic differentiation medium

Add 0.5 μM isobutylmethylxanthine, 60 μM indomethacin, 0.5 μM hydrocortisone, and 10 μg/ml insulin to MSC medium.

Neural differentiation medium

Prepare Neurobasal A with B27 supplement, 20 ng/ml *epidermal growth factor* (EGF), 40 ng/ml basic fibroblast growth factor (bFGF), 100 U/ml penicillin, and 100 μg/ml streptomycin.

Solutions for Tissue Digestion

1. *Collagenase solution*. Dissolve 4 mg/ml collagenase (type I) in Dulbecco's phosphate-buffered saline without Ca^{2+} and Mg^{2+} (PBSA), and sterilize by filtration. Store at –20°C.
2. *Dispase solution*. Dissolve 2 mg/ml dispase in PBSA, and sterilize by filtration. Store at –20°C.
3. *Freezing medium*:
   (i) *Freezing medium I*. FBS, 100%. Store at 4°C.
   (ii) *Freezing medium II*. FBS, 80%, and dimethyl sulfoxide (DMSO), 20%. Sterilize by filtration. Store at 4°C.

Detailed Protocols for Culture

Safety Precautions

The health status of donors may not always be known. Therefore, great care should be taken to avoid transmission of diseases in handling tissue and possibly contaminated instruments and materials.

Tissue Handling

The extracted tooth should be immersed in a sterile, normal saline solution or PBSA immediately after extraction. Addition of double-strength antibiotics (e.g., penicillin and streptomycin) to these solutions is optional but advised. When a baby tooth is shed unexpectedly, it can be kept in sterile solution as described above. In either case, it should be stored at 4°C to maintain cell vitality. Although it is better to isolate cells from a tooth as soon as possible, cells can be isolated as much as 24 h later if the tooth is stored under appropriate conditions.

Isolation of Dental Pulp Tissue

To maintain cell vitality, it is important to keep the pulp tissue at a low temperature and in a wet condition even during the cutting of the tooth.

The tooth can be contaminated because it is exposed to a huge amount of bacteria in the oral cavity, although dental pulp surrounded by dental hard tissues (enamel, dentin, and cementum) is not exposed to the oral cavity. However, in SHED isolation, dental pulp of the baby tooth is sometimes exposed as a result of resorption of the tooth root. Therefore, it is critical for successful cell isolation to disinfect the teeth on the outside with a disinfectant reagent before isolation of pulp tissue.

Protocol—1. Isolation of dental pulp tissue

Reagents and materials

- *Sterile*

  PBS without Ca^{2+} and Mg^{2+} (PBSA)

  MSC medium

  Petri dishes

  Fine forceps (small and large)

  Dental carbide burs

  Dental excavator

  High-speed dental hand piece

- *Nonsterile*

  Equipment to use high-speed dental hand piece (e.g., air compressor)

  Disinfectant reagent (e.g., povidone-iodine)

Procedure

1. Clean tooth surface well by washing three times with PBSA.
2. Disinfect with disinfectant reagent and again wash well with PBSA.
3. Cut the tooth around the cementum–enamel junction with a dental carbide bur equipped with a high-speed dental hand piece, and thereby reveal the pulp chamber. Soak the tooth in ice-cold PBSA intermittently to avoid heating while cutting the tooth. In SHED isolation, this process is sometimes not required because the pulp is already exposed by root resorption.
4. Separate the pulp tissue gently from the pulp chamber with small fine forceps or a dental excavator.
5. Put the pulp tissue into a small amount of MSC medium on a Petri dish. Be careful not to allow the tissue to become dry.

Primary Culture of DPSCs/SHED

This isolation protocol is based on the ability of stem cells to adhere to culture dishes and form discrete colony clusters. To obtain colonies each derived from a single cell (i.e., clones), it is very important to release DPSCs and SHED from their perivascular niche in the pulp tissue. Use a sterile scalpel blade and chop up the pulp into small segments. Enzyme digestion is then required for harvesting putative stem cells from pulp tissue. However, it may be harmful for the stem cells if enzyme digestion lasts for more than 1 h. Each researcher should find optimal conditions for the digestion. Usually, it takes 30–60 min to fully digest well-minced pulp tissue.

It is recommended that the medium be changed 1 day after cell isolation; this minimizes the incidence of contamination. However, when the risk of contamination is high, the medium can be changed 5 h after the isolation.

Cells isolated from dental pulp are usually seeded at $1–10 \times 10^3$ cells/cm^2 for culture. To evaluate their colony formation rate (CFU-F assay), a lower cell density ($0.1–1.0 \times 10^3$ cells/cm^2) is recommended in order to distinguish each single colony cluster.

Purified preparations of dental pulp stem cells can be obtained by immunoselection prior to culture. Single-cell suspensions of enzyme-digested dental pulp tissue are obtained by passing the cells through a 70-μm cell strainer. The cells are then incubated with primary antibodies reactive with DPSC, using STRO-1 (mouse anti-human MSC; IgM), CC9 (mouse anti-human CD146/MUC-18; IgG_{2a}), or 3G5 (mouse anti-human pericyte; IgM). After this, the cells are washed with PBSA containing 1% *bovine serum albumin* (BSA) and then incubated with either sheep anti-mouse IgG-conjugated or rat anti-mouse IgM-conjugated magnetic Dynabeads. Cells bound to beads are removed with the Dynal MPC-1 magnetic particle concentrator. The STRO-1-, CD146-, or 3G5-positive cells are then seeded at $1–10 \times 10^3$/cm^2 in growth medium as described below.

Protocol—2. Primary culture of DPSCs or SHED

Reagents and materials

- *Sterile*

  PBSA

  MSC medium

  Collagenase

  Dispase

  Reagents for immunomagnetic separation, if used:

  (a) PBSA containing 1% BSA

  (b) Primary antibodies reactive with DPSC using STRO-1 (mouse anti-human MSC; IgM), CC9 (mouse anti-human CD146/MUC-18; IgG_{2a}), or 3G5 (mouse anti-human pericyte; IgM)

  (c) Sheep anti-mouse IgG-conjugated or rat anti-mouse IgM-conjugated magnetic Dynabeads

  Culture flasks or dishes

Surgical blades and their folders

Cell strainer, 70 μm

- *Nonsterile equipment for immunomagnetic separation, if used*:

  Rotary mixer

  Dynal MPC-1 magnetic particle concentrator

Procedure

1. Mince the dental pulp tissue into tiny pieces with a surgical blade.
2. Immerse the minced tissue into a mixed collagenase/dispase solution (1:1).
3. Incubate at 37°C for up to 30–60 min and mix well intermittently.
4. After the digestion, inactivate the enzyme by dilution in sufficient MSC medium.
5. Pass the cells through a 70-μm cell strainer to remove tissue debris to obtain a single-cell suspension.
6. Centrifuge at 500 *g* for 6 min.
7. Remove supernate and resuspend pellet with MSC medium. Immunomagnetic bead selection can be performed at this stage:
   (a) Incubate with primary antibodies reactive with DPSC using STRO-1 (mouse anti-human MSC; IgM), CC9 (mouse anti-human CD146/MUC-18; IgG_{2a}), or 3G5 (mouse anti-human pericyte; IgM) for 1 h on ice at a concentration of 20 μg/ml.
   (b) Wash twice with PBSA containing 1% BSA, spinning at 600 *g* for 6 min.
   (c) Incubate with either sheep anti-mouse IgG-conjugated or rat anti-mouse IgM-conjugated magnetic Dynabeads (4 beads per cell) for 40 min on a rotary mixer at 4°C.
   (d) Remove cells bound to beads with the Dynal MPC-1 magnetic particle concentrator according to the manufacturer's recommended protocol.
8. Count the cells and seed them into culture flasks or dishes at 1–$10 \times 10^3/cm^2$.
9. Culture cells in MSC medium at 37°C and 5% CO_2 in the incubator.
10. Seven days after the cell isolation, wash the culture vessels with PBSA and change the medium. After that, the medium can be changed twice a week until cell confluence is reached.

Subculture

Usually around 1 week after the cell isolation, colonies are easily identified in the culture vessels, where the cells have a typical fibroblast-like spindle shape. Before the cells become 100% confluent (usually after about 2–3 weeks), they should be subcultured.

Each colony is theoretically derived from a single CFU-F and can be isolated by use of a cloning cylinder. In addition, single colonies can be collected by serial dilution following subculture or by *fluorescence activated cell sorting* (FACS) using STRO-1 and CD146 antibodies.

If cultures are established with unselected preparations, occasionally colonies of cells with a morphology resembling epithelial cells or endothelial cells can be observed. Usually, these contaminating cells disappear in the course of successive cell passages. If the contamination is extensive, the following three procedures can be performed on subculture. The first procedure involves trypsinizing the culture for a shorter time so that only stromal cells are detached, because epithelial- or endothelial-like cells are more strongly attached to the culture flask or dish. The second is changing the medium 4–6 h after subculture, because stromal cells attach to the culture surface earlier than the contaminating cells. The third and most reliable approach to separation of DPSCs from epithelial cells is to use FACS, in which STRO-1 or CD146 can be used to select DPSCs as previously described.

Protocol—3. Subculture of DPSCs

Reagents and materials

- *Sterile*

  PBSA

  MSC medium

  Trypsin-EDTA solution: trypsin, 0.05%, EDTA, 0.54 mM (0.2%), in Hanks' balanced salt solution without Ca^{2+} and Mg^{2+}

Procedure

1. Wash the flasks/dishes with PBSA three times.
2. Add a sufficient volume of trypsin-EDTA solution and incubate at 37°C.
3. Confirm that 80% of the attached cells have become detached from the culture surface, then add MSC medium.
4. Centrifuge at 500 *g* for 6 min.
5. Remove supernate and resuspend with MSC medium.
6. Count the cells and seed them at the desired density.
7. Culture them in MSC medium at 37°C and 5% CO_2 in air.

Cryopreservation and Recovery

Dental pulp stem cells can be cryopreserved and recovered by the usual procedure. The more important points in the procedure are as follows:

1. Cells harvested near the end of log-phase growth (approximately 80–90% confluent) are best for cryopreservation.
2. The number of cells should be around 1–2 × 10^6/vial containing 1.5 ml of freezing medium. Too low or too high a cell number may decrease the recovery rate.
3. The cyroprotective agent (DMSO) should be added gradually up to 10% at a low temperature (e.g., on ice).
4. A high serum concentration (90% FBS and 10% DMSO at final concentration) should be used to assist in cell survival.

Protocol—4. Cryopreservation of DPSCs

Reagents and materials

- Sterile

  PBSA

  Ice-cold Freezing Medium I

  Ice-cold Freezing Medium II

  Trypsin-EDTA solution: trypsin, 0.05%, EDTA, 0.54 mM (0.2%), in Hanks' balanced salt solution without Ca^{2+} and Mg^{2+}

  Cryovials, 1.8 mL

Procedure

1. Wash the flasks/dishes with PBSA three times.
2. Add a sufficient volume of trypsin-EDTA solution and incubate at 37°C.
3. Confirm that 80% of the attached cells have become detached from the surface, then add MSC medium.
4. Centrifuge at 500 *g* for 6 min.
5. Remove supernate and resuspend cells in ice-cold Freezing Medium I.
6. Count the cells, then dilute or concentrate them to twice the desired final concentration with Freezing Medium I. Keep the tubes containing the cells on ice.
7. Add an equal volume of Freezing Medium II little by little on ice while rotating the tube.
8. Add aliquots to cryopreservation vials.

9. Place the vials in a programmed freezer. A controlled freezing container and -80°C freezer may also be used to achieve a slow cooling rate. Place the vials in a liquid nitrogen freezer for long-term storage.

Protocol—5. Recovery of DPSCs from frozen storage

Reagents and materials

- *Sterile or aseptically prepared*

  Cryovial taken directly from nitrogen freezer

  MSC medium

Procedure

1. Place the vials in a warm water bath (37°C). Quick thawing (1–2 min) is important for the best recovery.
2. Add sufficient volume of MSC medium and mix well.
3. Centrifuge at 500 *g* for 6 min.
4. Remove supernate and resuspend cells in MSC medium.
5. Count the viable cells with Trypan Blue solution and then seed them at the desired viable cell density.
6. Culture them in MSC medium at 37°C and 5% CO_2 in air.

Characterization of Undifferentiated DPSCs and SHED

STRO-1 is one of the early cell surface markers for mesenchymal stem cells that can be used to evaluate the undifferentiated status of DPSCs and SHED. Monoclonal antibody for STRO-1 was first described as a potential reagent that reacted with a cell surface molecule highly expressed on human bone marrow CFU-F. STRO-1-positive cells from adult bone marrow contain a CFU-F population and exhibit an ability to differentiate into multiple cell lineages including myelo-supportive stromal cells, osteoblasts, adipocytes, and chondrocytes. DPSCs and SHED have been found to contain a STRO-1-positive fraction at around 10–20%. CD146 (MUC18), known as a possible marker for BM-MSCs, is also expressed in DPSCs and SHED. These markers can be used in immunocytochemistry or FACS analysis.

Dentin sialophosphoprotein (DSPP), alkaline phosphatase (AP), bone sialoprotein (BSP), and *osteocalcin* (OSC) are representative lineage markers of odontoblasts and osteoblasts. Core binding factor 1 (Cbfa1)/ Runt-related gene factor 2 (Runx2) and osterix are also known as osteoblast-specific transcription factors and are thought to play a pivotal role in tooth development. The genes for these molecules are not expressed, or are markedly less expressed, in undifferentiated DPSCs and SHED as analyzed by the reverse transcription-polymerase chain reaction (RT-PCR) or Western blot analysis. Lineage markers of other cell types, including proliferator-activated receptor-γ 2 (PPARγ 2) and *lipoprotein lipase* (LPL) for adipocytes or *glial fibrillary acidic protein* (GFAP) and nestin for neural cells, are also not expressed, or markedly less expressed, by RT-PCR or Western blot analysis before stimulation for cell-specific differentiation.

It is important to note that DPSCs and SHED are heterogeneous populations after expansion ex vivo and begin gradually to lose their stemness (loss of STRO-1 and 3G5 expression) with successive culture passages. For example, DPSCs seem to partially differentiate into an odontoblast lineage after continuous passage without any specific inductive stimulation. The genes for DSPP and OSC become are expressed after passage 4 or 5 (cumulative population doubling is approximately 20) as detected by RT-PCR. Therefore, it is very important to use DPSCs and SHED at an early passage if they are to be used as "stem" cells. Further studies are required to establish adequate and ideal culture conditions to maintain the stemness of DPSCs and SHED.

Differentiation Capacity and Assay

Odontogenic/Osteogenic Differentiation

DPSCs and SHED can differentiate into odontoblasts and osteoblasts in vitro when they are cultured with a medium containing dexamethasone, inorganic phosphate, and l-ascorbic acid. This differentiation is confirmed by up-regulation of odontoblast- and osteoblastrelated markers, for example, DSPP, BSP, AP, cbfa1, and osterix, within a few weeks after induction. In the meantime, a down-regulation of STRO1 should be observed. After culture for several weeks, mineralized nodules are observed under the microscope as a result of calcium accumulation. This calcium accumulation can be analyzed by Alizarin Red staining or quantified with commercially available kits.

Importantly, DPSCs and SHED can regenerate a dentin–pulp complex when transplanted subcutaneously into immunocompromised mice with a hydroxyapatite/tricalcium phosphate (HA/TCP) carrier. The regenerated mineralized matrix has a typical tubular structure along with odontoblastic cells, which is a characteristic of the natural dentin–pulp complex. The origin and characteristics of this matrix are confirmed by immunostaining with antibodies against human-specific mitochondria, DSPP, and BSP. Cells that maintain odontoblastic capacity can be recovered from transplanted tissue, and this suggests a self-renewal potential of DPSCs. Interestingly, SHED can be induced to form a bone-like matrix with a lamellar structure by recruiting host cells. This capacity for bone generation might be correlated with the nature of the baby tooth, whose root resorption is followed by bone and permanent tooth eruption. Another interesting feature of the bone-like matrix generated by SHED transplants is the lack of bone marrow components. In contrast, the bone-like matrix generated by BM-MSC transplants contains material compatible with bone marrow.

Protocol—6. Odontogenic and osteogenic differentiation of DPSCs in vitro

Reagents and materials

- *Sterile*

  MSC medium

  Odontogenic differentiation medium

Procedure

1. Culture the cells with regular MSC medium until they reach complete confluence.
2. Switch the medium to odontogenic differentiation medium, and change 2–3 times a week for up to 6 weeks. (After confluence, cells may easily be detached from the culture surface. Therefore, great care should be taken to avoid detachment on changing the medium.)

Protocol—7. Odontogenic/osteogenic differentiation of DSPCs in vivo

Reagents and materials

- *Sterile*

  PBSA

  MSC medium

  Trypsin-EDTA solution: trypsin, 0.05%, EDTA, 0.54 mM (0.2%), in Hanks' balanced salt solution without Ca^{2+} and Mg^{2+}

  Cryovials, 1.8 mL

  Appropriate carrier (hydroxyapatite/tricalcium phosphate carrier)

Procedure

1. Culture the cells with MSC medium until 90% confluent.
2. Wash the dishes three times with PBSA.
3. Add sufficient volume of Trypsin-EDTA solution and incubate at 37°C.

4. Confirm that 80% of cells are detached from the culture surface, then add MSC medium.
5. Centrifuge at 500 *g* for 6 min.
6. Remove supernate and resuspend cells in MSC medium.
7. Count the cells.
8. Mix 2–4 × 10^6 cells suspended in MSC medium with carrier in a 1.8-mL cryovial.
9. Incubate them at 37°C with rotation for 1 h.
10. Transplant them aseptically and subcutaneously into an immunocompromised mouse. We usually use NIH *bg-nu/nu-xid* mice for transplantation.
11. Harvest them at an appropriate time, and examine histologically. (In the system with HA/TCP carrier and NIH *bg-nu/nu-xid* mice, a duration of 8 weeks is adequate for the generation of sufficient mineralized matrix by DPSCs and SHED. The time point for harvest depends on the system.)

Adipogenic Differentiation

Although adipocytes have not been observed in dental pulp, DPSCs and SHED can, under appropriate inductive conditions, differentiate into adipocytes along with accumulation of lipid clusters in their cytoplasm. These lipid clusters are easily identified under the microscope and can be examined with Oil Red O staining after several weeks of adipogenic induction. Also, up-regulation of adipocyte-related genes (PPARγ 2 and LPL) are is by RT-PCR.

It is no surprise to find that DPSCs and SHED have a lower capacity to differentiate into adipocytes than BM-MSCs. This might be caused by a difference in the nature of their original tissue (adipose tissue occurs in the bone marrow cavity of long bones in aged humans).

Protocol—8. Adipogenic differentiation of DPSCs in vitro

Reagents and materials

- *Sterile*

  MSC medium

  Adipogenic differentiation medium

Procedure

1. Culture the cells with MSC medium until they reach complete confluence.
2. Switch the medium to adipogenic differentiation medium, and change 2–3 times a week for up to 6 weeks. (Differentiated

adipocytes easily detach from the culture surface. Therefore, great care should be taken to avoid detachment on changing the medium.)

Neural Differentiation

Development of dental pulp is closely associated with neural crest cells. Therefore, it is reasonable to assume that DPSCs and SHED might have the capacity to differentiate into neural cells. Indeed, under specified inductive conditions in vitro, DPSCs and SHED can differentiate into neural cells with protruding elongated cytoplasmic processes. Neural differentiation is confirmed by morphological appearance and expression of neural-specific molecules, including GFAP, nestin, neurofilament M, neuronal nuclear marker (NeuN), 2', 3'-cyclic nucleotide-3'-phosphodiesterase (CNPase), βIII tubulin, and *glutamic acid decarboxylase* (GAD). Some cells are capable of forming spherelike structures, which are observed in neural stem cells. Transplanted SHED, precultured in neural differentiation medium, can survive with neural phenotypes for at least 1 week in mouse brain.

Protocol—9. Neural differentiation of DPSCs in vitro

Reagents and materials

- *Sterile*

  MSC medium

  Neural differentiation medium

  Gelatin, 0.1%

Procedure

1. Coat culture plates with gelatin for 30 min and dry.
2. Seed cells at relatively low density 0.1–1.0 × 10^3 cells/cm^2 with neural differentiation medium.
3. Change medium every day, usually for 4 weeks.

Concluding Remark

DPSCs and SHED are derived from an easily accessible tissue resource and can be expanded to reach a sufficient number of cells for therapy as a result of their extremely high population-doubling ability. They may be an important source of stem cells for autologous stem cell transplantation. Although DPSCs and SHED are capable of regenerating dentin/pulp and bone tissues in vivo, considerable work is still required to maintain their stemness in vitro, and to achieve optimal tissue regeneration in vivo.

13

Anchorage-dependent Animal Cells

In this chapter, scale-up is described in a laboratory context (10-20 L), but the principles and techniques employed have been successfully adapted so that cells are now grown industrially in unit volumes of up to 10,000 L for vaccine, interferon, and monoclonal antibody production. The need to scale-up cell cultures has been expanded from the historical requirement for vaccine manufacture to include not only interferon and antibodies, but many important medical products such as tissue plasminogen activator, erythropoietin, and a range of hormones and blood factors. The low productivity of animal cells, resulting from their slow growth rate and low expression of product, plus the complexity of the growth conditions and media, led to attempts to use recombinant bacteria to express mammalian cell and virus proteins. However, this has proven unsuitable for many products, mainly because of incomplete expression and contamination with bacterial toxins, and more importance is now being put on expression of recombinant proteins from mammalian cells. This has allowed the use of faster growing and less fastidious cell lines, such as CHO, and amplification of product expression by multiple copies of the gene.

Principles of Scale-Up

Animals cells are grown in two completely separate systems. For those cells that grow individually in suspension, the range of fermentation equipment developed for bacteria can be readily modified. This is a great advantage, since these culture vessels are economic in terms of space, the environment is homogeneous and can be critically controlled, and scale-up is relatively straightforward. Many cell types, however, will only grow when attached to a substrate or, in some cases, will only produce significant levels of a product when grown in this mode. Scale-up of substrate attached cells is far more difficult to achieve and has given rise to a wide range of alternative culture systems.

Two approaches to scale-up can be taken. The first is volumetric-a simple increase in volume while retaining the same cell density or process intensity. The second method is to increase the cell density/unit vol 10-100-fold by means of medium perfusion techniques. Cell densities of over 10^8/mL can be achieved in a variety of systems, but they are difficult to volumetrically scale-up because of the effects of nutrient and waste metabolite concentration gradients, however, the development of cultured systems based on the immobilization of cells in porous carriers has overcome this limitation. Compromise is possible with large-scale (100-500 L) cultures operating at just 10-20 times above the conventional cell densities (1-3 $\times$ 10^7/mL) by means of special perfusion devices, such as the spin filter.

The environmental factors that can most readily be controlled are pH (and redox) and oxygen. The limiting factor in scale-up, particularly in cell density, is usually oxygen. Surface aeration used in small cultures soon becomes inadequate, since the volume (and therefore depth) of medium increases. Bubbling of air/oxygen mixtures into cultures, with turbulent stirring/agitation, is the most efficient means of oxygenation. Unfortunately, cells are fragile, compared to bacteria and only slow stirring and bubbling rates can be used, which are often inadequate for maintaining a sufficient oxygen supply. To overcome this problem, most cultures rely on several oxygenation methods, and many ingenious methods have been developed for this purpose.

Two further points should be taken into account during scale-up. The first of these is the increased risk of contamination and the proportionally higher costs of culture failure. The second is a question of logistics in the preparation of medium and particularly cell macula. It is a small matter to harvest 10^8 cells and maculate them in a good physiological state into a new culture. An inoculum of 10^{10} cells takes a long time to prepare, cells can be left for long periods in damaging conditions, and media can lose its temperature and pH can change while these handling procedures are carried out. The objective is to keep both the process and the culture system as simple as possible, having everything well prepared and ready, and not to be overambitious with regard to scale. This will ensure that cultures are initiated with cells in good physiological condition and reduce the risks of microbial contamination.

METHODS

Suspension Culture

Culture vessels

The simplest means of growing cells in agitated suspension is the spinner culture vessel. The culture pot has a magnetic bar, usually placed a few millimeters from the bottom of the vessel, and is placed on a magnetic stirrer. As long as the bar is able to rotate freely and the stirrer is of sufficient quality to maintain constant stirring speeds, and not overheat, this methodology works extremely well for growing most cells up to densities of 1-2 × 10^6/mL. These glass spinner vessels are available from a wide range of suppliers in sizes from 0.2-20 L.

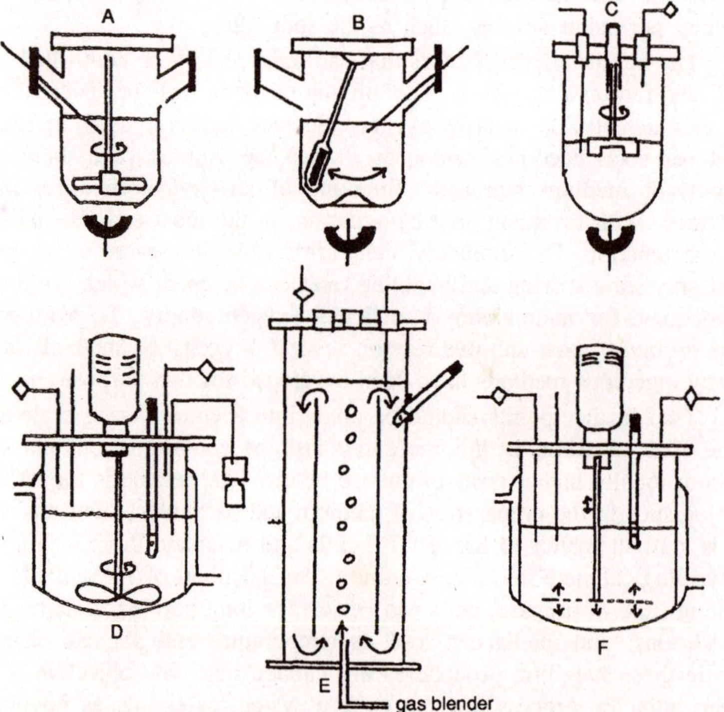

Fig. 13.1. Range of culture apparatus for suspension cells (A) magnetic bar spinner culture, (B) Techne MCS stirrer, (C) surface stirrer, (D) small scale fermenter with marine impeller, (E) airlift fermenter, (F) vibro-fermenter.

A modification of this principal for shear-sensitive cells is the spinner vessel using surface agitation as exemplified by the BR-06 Bioreactor. Spinner vessels are only satisfactory up to a certain size between 2 and 10 L depending on the cell line and its required use. Above 10 L, glass vessels become inconvenient to handle and the progression to in situ stainless steel vessels should be considered. The other reason for change is that, with scale-up, the need to control the culture environment and carry out specialized manipulations (e.g., perfusion, media changes, and so on) increases. For this level of sophistication, a fermentation system needs to be used. The main differences are that starring is by a direct-drive mechanism with a motor, and the vessel has a complex top that allows the inclusion of a range of sensors, probes, feed supply lanes, and sampling devices for contamination-free monitoring and control. Fermenter kits (laboratory scale) are available in the range from 1-40 L and cost in the region of $4500-37,500 (£3,000-25,000) (with control of stirring speed, temperature, pH, and oxygen).

Culture vessels for animal cells should have the following features:

1. Curved or domed bottom to increase mixing efficiency at the low stirring speeds that are used (100-350 rpm).
2. Water Jacket temperature control to avoid the use of immersion heaters that give localized high temperatures Electrically operated silicone pads are also suitable at volumes up to 5 L—the only disadvantage is the reduction in visibility into the vessel.
3. Absence of baffles and other sharp protrusions that cause turbulence. The interior is finished to a high grade of smoothness to minimize mechanical damage and for cleanliness.
4. An aspect ratio (height to diameter) of 2 1 maximum, and preferably no more than 1.5.1.
5. Suitable impeller to achieve nondamaging bulk flow patterns (e.g., modified marine or pitched blade impellers) with top drive, so that there are no combinations of moving parts that would grind up the cells.

Some animal cells, including many hybridoma lines, are sensitive to the mechanical effects of stirring. For such cells, there are two alternative means of mixing besides stirring.

1. Vibromixer: This is a nonrotary device using a plate that vibrates in the vertical plane a distance of 0 1-3 mm. Conical perforations in the plate affect the mixing.
2. Airlift medium is circulated in a low velocity bulkflow pattern by being lifted up a central draft tube by rising air bubbles, and

recirculated downward in the outer ring formed by the draft tube. This system forms a near ideal mixing pattern and allows near-linear scale-up to at least 1000 L. Unfortunately, the apparatus, with a 12:1 aspect ratio, is very high and a 30-L fermenter needs 3-m ceiling height. Airlift fermenters are available commercially either as complete systems or as disposable 570 mL unit.

Culture procedure for suspension cells

1. Inoculum should be prepared from a growing suspension of cells (i.e., in mid-to-late logarithmic phase). Stationary phase cells are either slow to start in a fresh culture or do not grow.
2. Prewarm to 37°C and equilibrate the pH of the culture medium with the CO_2/air gas mix, before inoculating the cells.
3. Inoculate cells at over 1×10^5/mL. Recommended level is $2\text{-}3 \times 10^5$/mL for many cell (hybridoma) lines.
4. Stir the culture within the range 100-300 rpm. This speed depends on the individual type of culture reactor. Star at a speed sufficient to keep the cells in homogeneous suspension. Do not use speeds that allow cells to settle out at the bottom of the culture.
5. Monitor cell growth at least daily by taking a small sample, either through a special sampling device or removing the vessel to a laminar flow cabinet and using a pipet, and carry out a viable cell count.
6. pH. If the culture is closed (i.e., all ports stoppered with no filters), then the pH will fall. You should remove the culture to a cabinet and gas the head space with an If the culture is very acid, sterile sodium bicarbonate (5.5%) can be added or, when the cells have settled out, remove 50% of the medium and replace with fresh (prewarmed) medium Return culture to stirrer. It is preferable to have inlet and outlet filters, so that there is continuous head-space gassing, initially (i.e., first 24 h) with 5% CO_2 in air, followed by air only. Suitable filters are nonwettable with a 0 22-μm rating.
7. After 3-4 d, the saturation density of $1\text{-}2 \times 10^6$ cells/ml should be attained. Most suspension cells will then die at a rapid rate unless harvested or maintained with medium changes

Special procedures

1. *Airlift*. Follow the protocol described previously except, instead of stirring, a gas flow rate of approximately 5-20 cc air/L/min is used for mixing.

2. *Increase cell density by perfusion.* To perfuse a culture (i e., the continuous or semicontinuous addition of fresh medium and withdrawal of an equal volume of spent medium) means that methods of separating the cells from the medium are needed. There are a number of ways to do this, i.e., spin filters, membranes (e.g., hollow fibers), and porous carriers.

Spin filter

The problem with most filtration techniques is that the filter rapidly becomes blocked with cells. A spin filter, so called because it is attached to the stirrer shaft, reduces the problem of blockage, because as it spins it produces a boundary effect on its surface that reduces cell contact. Also, they normally have a large surface area and, thus, have a low flow rate at any one point. A porosity of about 6-10 μm is needed, and stainless steel mesh can be used. This is thin enough to be cut to form a cone in which the join can be double-folded and machine-pressed. This allows a perfusion rate in the order of 1-2 vol/d. This device will allow cell densities of over 10^7 cells/ml to be achieved.

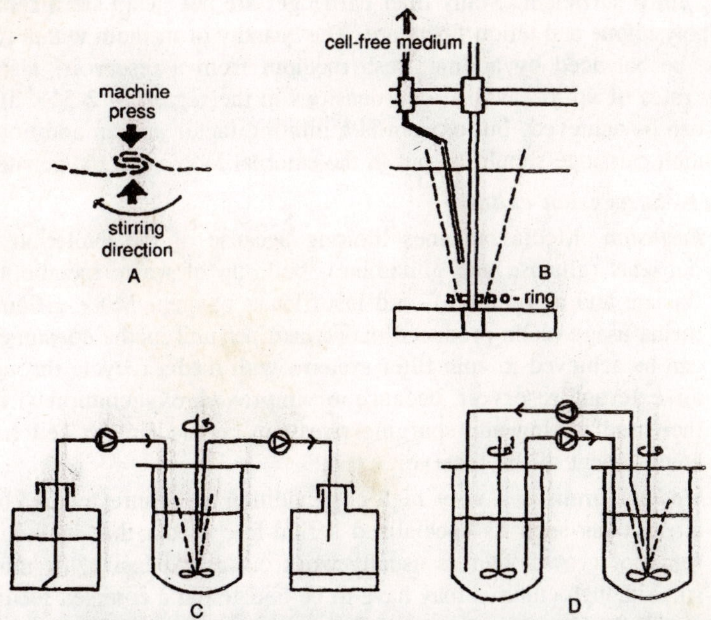

Fig. 13.2. Spin filters: (A) construction using folded interleaving edges, (B) diagrammatic representation of a spin filter, (C) open perfusion systems, (D) perfusion system with media recycle.

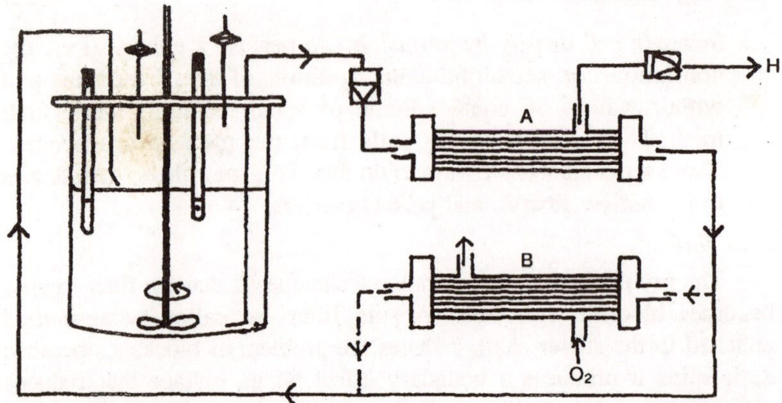

Fig. 13.3. Hollow-fiber cartridges in perfusion loop for (A) removing spent medium (H) and returning cells to the culture and (B) oxygenating the medium.

Hollow-fiber cartridges

Hollow-fiber units are available at both ultrafiltration and microfiltration membranes. For the purposes of withdrawing medium and returning the cells to the culture in a loop outside the vessel filtration 0.22 μm is sufficient. Many fiber cartridges are not steam sterilizable, but polysulfone and teflon fibers are. The quantity of medium withdrawn must be balanced by adding fresh medium from a reservoir. Using flow rates of up to 1 vol/h, cell densities in the region of 2-5 $\times$ 10^7/mL can be achieved, but oxygen is a limiting factor and an additional filtration cartridge should be put in the external loop as an oxygenator.

Notes—Suspension culture

1. *Perfusion*. Media becomes limiting because of the depletion of nutrients (glucose and glutamine), build-up of waste metabolites (lactate and ammonium), and insufficient oxygen. More efficient media usage (cells produced/maintained per unit media consumed) can be achieved in spin filter systems with media recycle through an external reservoir because of improved oxygenation (i.e., increased mixing and sparging rates can be used in the cell free environment of the reservoir).
2. *Media*. Serum is a very high cost addition to culture media, but alternatives such as specialized serum-free media that include a range of growth factors usually work out as expensive, or more so. Although cultures may have to be initiated in a complex media, as cell density increases, so cells become less dependent on serum and growth factors. Thus, at densities above 5 $\times$ 10^6-10^7/mL,

serum concentration can be drastically reduced (to 2%) or even excluded. If serum-free medium is used, cells are more susceptible to damage by stirring and sparging, but this can be offset to a certain extent by adding pluronic F68 (polyglycol) at 0 1%.

3. *Contamination/sterility*. The larger the scale, the more expensive a culture failure becomes carry out stringent quality-control procedures by testing growth media several days before it is to be used for bacterial contamination. Do not take shortcuts on the support equipment, but use specialized tubing connectors and sampling devices supplied by fermenter equipment companies Also, do not overuse the air filter (6-10 sterilization cycles maximum), and do not allow them to get wet (either during autoclaving or with media or condensation).
4. *Suspension culture*. Many cells either attach to the surfaces of the vessel or form unwanted clumps. Media for suspension culture should have a reduced calcium and magnesium-ion concentration (special formulations are commercially available) because of the role of these ions in cell attachment. Attachment to the vessel can be discouraged with a pretreatment of a proprietary silicone solution.

Anchorage-dependent Culture

Materials

Anchorage-dependent culture systems are far more difficult to scale-up than suspension cultures because of the additional requirement of providing the extra surface area in an economical (in terms of space) way and still maintain homogeneity throughout the system. For this reason, suspension culture, in which a 1-L stirred vessel is conceptually similar to a 1000-L vessel, is always the preferred culture method. The first step in scale-up usually involves the change from stationary flasks (available in sizes up to 200 cm^2) to roller bottles (sizes up to 1750 cm^2). The larger size of roller bottle will yield in the range of 2-5 $\times$ 10^8 cells, and therefore, for most purposes, a multiplicity of rollers has to be used. The next step in scale-up is to use roller bottles that have an increased surface area resulting from the inclusion of glass tubing or plastic spiral films. By this means, the surface area within a roller bottle can be increased to 8500 cm^2 (spiral film) and 15,000 cm^2 (glass tubing). An alternative to investing in specialized, and costly, roller culture equipment is to use plastic multi-tray units. Each tray has a surface area of 600 cm^2 and units of 6, 10, and 40 (24,000 cm^2) plates can be obtained. Two systems that allow a huge

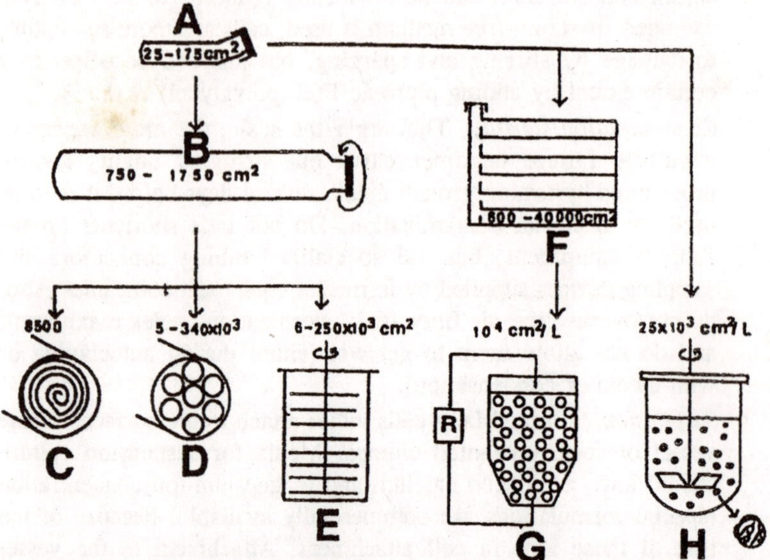

Fig. 13.4. Scale-up of anchorage-dependent cultures (a) flask, (b) roller bottle, (c) plastic spiral, (d) glass tubing, (e) stack plates, (f) multi-tray units, (g) immobilized bed; (h) microcarrier culture.

unit scale-up of substrate attached cells are immobilized beds (e.g., constructed of glass spheres) and microcarrier culture (cells growing on 200-μm spheres that are stirred in suspension culture apparatus). Mrcrocarriers can provide 5000 to 50,000 cm²/L and are currently being used in commercial production systems at the 1000-L scale (a total of 15×10^6 cm² surface area, which has the potential of supporting 15×10^{11} cells).

Roller culture

Reutilizable glass or disposable plastic roller cultures are used. The most commonly used sizes are 750-850 cm² and 1500-1700 cm². Complete modular systems holding up to 48 large bottles can be purchased either free-standing, for use in hot rooms, or within an incubator cabinet.

Procedure. The following procedure is based on a 1500 cm² (24 $\times$ 12 cm) roller bottle.

1. Add 200-300 mL of medium
2. Add 1-5 $\times 10^7$ cells (observe previously listed advice on preparation of inocula and medium).
3. Revolve the culture at 15 rpm.

4. Cell growth can be observed under an inverted microscope with a long-distance objective.
5. After 3-5 d, the cell sheet will be confluent yielding from 1.5 × 10^5 (human diploid) to 5 × 10^5 (heteroploid cells, e.g., HeLa) cells/cm².
6. Pour off the medium, wash the cell sheet with prewarmed phosphate buffered saline, and add 20 mL trypsin (0 25%). Place culture back on roller, and allow to revolve for 10-20 mm The cells will detach and can be harvested, diluted in fresh medium and serum, and passaged on.

Thus outline protocol can be considerably modified. An advantage of this method is that the medium volume:surface area ratio can be altered easily. Thus, after a growth phase, and when a product is to be harvested, the medium volume can be reduced to 100 mL in order to obtain higher product concentration.

Glass bead immobilized beds

This type of culture system apparatus is easily fabricated in the laboratory. A suitable cylindrical glass container is packed with borosilicate glass spheres (minimum diameter 3 mm, optimum diameter 4 or 5 mm). Medium is perfused by means of a peristaltic pump from a reservoir (which ideally is a stirred tank reactor), which can be monitored and controlled for pH, oxygen, and so on.

Table 13.1. Physical characteristics of glass sphere beds.

Bead diameter	*3 mm*	*5 mm*
Surface area (cm²)		
Total	7400	4600
Available (70%)	5200	3200
Void medium vol (cc)	295	250
Total vol (cc)	675	625
Cell count (× 105/cm²)	0.78	2.50
(× 108/kg)	4.0	8.0

Procedure

1. Prepare growth medium and add to reservoir (2 mL/cm² culture surface area).
2. Equilibriate the system for temperature and pH, and circulate the media through the packed bed (allow sufficient time for the solid glass spheres to reach 37°C).

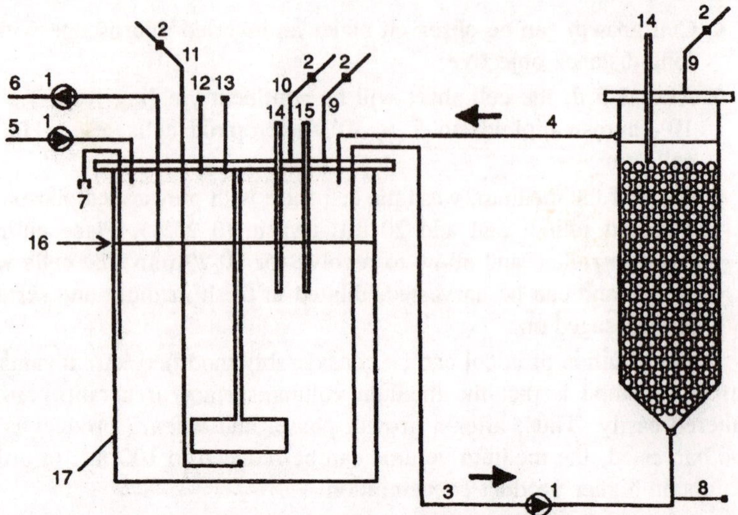

Fig. 13.5. Schematic diagram of the fixed bed culture system (1) peristaltic pump, (2) air filter, (3) media recycle out, (4) media recycle return, (5) media feed, (6) harvest line, (7) samples port, (8) inoculum line, (9) off gas, (10) head space air/CO_2, (11) air/O_2 sparge, (12 addition lines, (13) DO probe, (15) pH probe, (16) medium level in reservoir, (17) reservoir vessel.

3. Inoculate cells (1-2 × 10^4/cm^2) into a volume of medium equal to the void volume of the bed (250 mL/kg 5 mm spheres).
4. Allow the cells to attach (3-8 h depending on cell type), but the culture can safely be left overnight (16 h) at this stage.
5. Start medium recycle, initially at a rate of 0.1 linear cm/mm, but as the culture progresses, this rate is increased to a maximum of 5 cm/mm.
6. Cell growth can only be monitored by indirect measurement, and the glucose utilization rate is the most convenient. An alternative is oxygen utilization rate. Growth yields should be determined for the particular cell line and nutrient, so that an approximation of cell numbers can be made (e.g., glucose utilization is usually within the range of 2-5 × 10^8 cells produced/g glucose).
7. When the culture is estimated to be confluent (after 5-7 d), dram the medium, wash the bed with phosphate-buffered saline wash, and add trypsin/versene to harvest the cells. The efficiency of cell detachment can be increased by intermittently draining and pumping back the trypsin solution. The bed acts as a depth filter, and to recover a high percentage of the cells, the bed should be washed through several times with medium after the trypsin has

been drained off. (Note 5 mm beads allow a better drainage and cell recovery than 3 mm).

8. Wash out the culture bed immediately with a detergent, so that cell debris does not become fixed onto the glass beads.

This culture method is basically very simple, and the apparatus is cheap and reutilizable. It has a large potential scale-up and has been proven at the 100 L vol scale (30 L bed of 3 mm beads). It is more suitable for harvesting a secreted cell product over a long period of time, rather than acting as a source of cells resulting from the difficulties of removing cells from the bed.

Microcarrier culture

The advantage of this methodology is that the cells, when growing on small caners, can be treated as a suspension culture with all the advantages of large unit scale-up, homogeneity, and easily controlled environmental conditions. The range of microcarriers commercially available is extensive, and at least one type is suitable for all cell types, however demanding.

The decision of which one to use is influenced by whether a dried powder or already-prepared sterile solution is preferred, the cost/cm^2, whether a special derivitized surface is needed for a particular cell, or whether one wishes to harvest cells by dissolving the carrier (gelatin, collagen) and thus producing a higher quality cell suspension. Experience has shown that it is worthwhile to evaluate several types in small-scale cultures for each particular cell line, since significant differences in cell yield, cell-specific productivity, and longevity of culture (before cell detachment) are seen.

Culture apparatus

Modifications of suspension culture vessels are used. Spinner flasks with a magnetic bar are unsuitable, but versions are available with large paddle-type impellers (Bellco) or specially modified stirring actions. Stirring rates are much slower (20-70 rpm) than for suspension cells, and thus more efficient mixing at low speeds is required. Scale-up in laboratory fermenters can also use the large-bladed paddles, but there are several modifications of the marine impeller available that are very efficient.

Procedure

Microcarrier culture is not a difficult technique, but it does require more critical attention to experimental detail than most methods and the use of the correct culture vessels. The following procedure is

Table 13.2. Commercially available microcarriers

Name	*Type*	*cm^2/g*
Acrobead	Derivitized	5000
Bioglas	Glass/latex	350
Bioplas	Polystyrene	350
Biosilon	Polystyrene	255
Cytodex 1,2	Dextran	6000
Cytodex 3	Collagen	4600
Cytosphere	Polystyrene	250
Dormacell	Dextran	7000
Gelibead	Gelatin	3800
Mica	Polyacrylamide	—
Micarcel G	Polyacrylamide	5000
Microdex	Dextran	250
Superbeads	Dextran	6000
Ventregel	Gelatin	4000
Ventreglas	Glass/polystyrene	300

based on using Cytodex 3 or Dormacell 2.3 at 3 g/L. Prepare the microcarriers according to manufacturer's recommendations.

1. It is essential that the medium with microcarriers be prewarmed and stabilized before inoculating the cells. Cell attachment to moving spheres requires conditions to be just right. It is even more important with this method than with previously described ones to initiate the culture with growing (logarithmic cells) and not stationary-phase cells, and cells that have been rapidly prepared and are in good physiological condition (i.e., have not been standing in trypsin for extended periods).
2. Inoculate at $2 \times 10^4/cm^2$ into 30-50% of the final volume Stir at the minimum speed to maintain homogeneity (20-30 rpm) for 4-8 h.
3. When the cells have attached (expect 70-90% plating efficiency), the volume can be increased to the fall working volume. (Stirring speed may also have to be increased to give complete mixing.)
4. A great advantage of the microcarrier system is that samples can be readily removed and microscopically examined. Unstained preparations will show whether or not cells have attached, spread out, and then begun to grow. Cell counts can be made by the standard nuclei-counting procedure, which releases the stained nuclei from the attached cells.

5. As the culture progresses, the stirring rate can be increased to prevent cell-to-cell attachment, bridging microcarriers and causing clumps to form. A maximum of 75 rpm should be achieved.
6. In nonenvironmental controlled cultures, the media will become acidic after 3-4 d and a partial (50-70%) media change should be carried out. Stop stirring, allow beads to settle (10 mm), decant spent medium, add fresh (prewarmed) medium, and start stirring, gradually increasing the rate.
7. Cells can be harvested when confluent by allowing the carriers to settle out, giving a serum-free wash, allowing the carriers to settle out again, decanting off as much of the free fluid as possible, adding trypsin, and restarting stirring, but at slightly faster speed (75-125 rpm). After 20 mm, allow the beads to settle out for 2 min. Cells can either be removed by decanting, or the mixture can be filtered through a coarse sintered glass filter that allows passage of the cells, not the microcarriers. If gelatin or collagen carriers are being used, then cells can be released by treatment with trypsin/EDTA (which solubilizes gelatin) or collagenase.

Scaling-up

This can be achieved by increasing the culture volume, and increasing the microcarrier concentration from the suggested 3-15 g/L If higher concentrations are used, then it is imperative to have a perfused system with full environmental control. The easiest means of perfusing is the spin filter, but a much larger pore size can be used (60-100 μm). This allows much faster perfusion rates to be attained (1-2 vol/h). Perfusion from a reservoir that is adequately gassed is an efficient means of oxygenating the culture. Spin filter systems are commercially available.

Porous Carrier Culture

Porous carriers can be used for the immobilization of both anchorage-dependent and suspension cells to high densities (0 5-2.0 × 10^8 cells/mL carrier). They can be used in stirred tanks, fixed beds and fluidized beds. Porous carrier systems like hollow-fiber systems are high-density continuous perfusion processes, but unlike hollow-fibers they can be scaled-up volumetrically (>100-L bed volumes).

Fixed bed porous glass sphere culture system

This is the same design as the fixed bed system, except that the solid glass spheres are replaced with Siran porous glass spheres. This can lead to an increase in unit cell density of up to 50-fold. The system is very suitable for secreted cell products and can be operated as a continuous perfusion process for many months.

1. Set up the system.
2. Add 10 L of cell culture media to the reservoir.
3. Equilibrate the system for temperature, pH, and *dissolved oxygen* (DO).
4. Add the inoculum (700 mL of cell suspension containing 12×10^9 cells) to the fixed bed (1 L) of dry porous glass spheres.
5. For suspension cells start media recycle immediately at a linear flow velocity of 2 cpm increasing to 20 cpm with cell growth.
6. For anchorage-dependent cells leave the bed stationary for 3-6 h to allow the cells to attach before starting media recycle as in step 5.
7. Take a sample from the reservoir daily and carry out the following analysis:
   (a) Free cell count,
   (b) Glucose concentration,
   (c) Product concentration.
8. Start media feed and harvest when the glucose concentration drops below 1.5 mg/mL (i.e., for a feed glucose concentration of 4 mg/mL).
9. Adjust the media feed rate to give a glucose concentration in the reservior of 1.5-2 mg/mL (typically 10 L/day in steady-state culture).

Table 13.3. Characteristics of Siran porous glass spheres

Material	*Borosilicate glass*
Average diameter	3-5 mm
Pore size	60-300 μm
Pore volume	60% Open
Total surface area	75 m²/L
Biocompatible	Yes
Steam sterilizable/autoclavable	Yes
Reusable	Yes

Verax fluidized bed culture system

The Verax fluidized bed culture systems range from the System One with a 16 mL fluidized bed capable of producing about 1 L of harvest per day to the System 2000 which has a 24 L fluidized bed

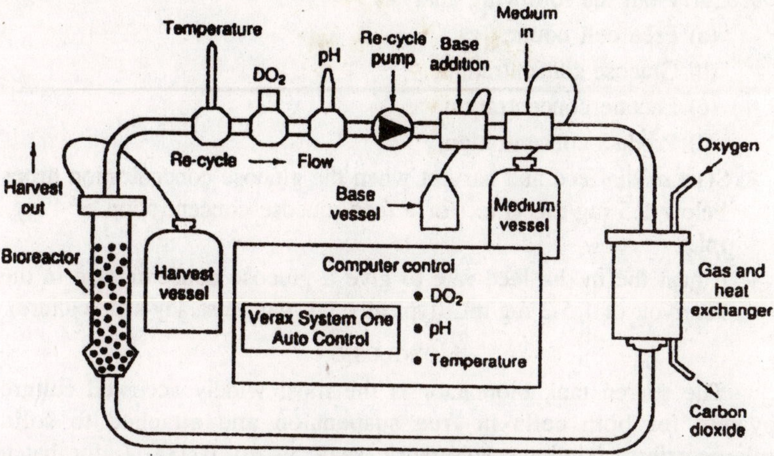

Fig. 13.6. Adjust the media feed rate to give a glucose concentration in the reservoir of 1.5-2 mg/mL (typically 1 L/day in steady-state culture).

capable of producing 1000 L of harvest per day. The process is based on the immobilization of cells (anchorage-dependent and suspension) in porous collagen carriers. The carriers are weighted (specific gravity 1.6) so that they can be used in fluidized beds with a high recycle flow rate (typically 70 linear cpm). The microspheres have a pore size of 20-40 μm, and pore volume of 85%, allowing the immobilizations of cells to high density (1-4 $\times$ 10^8 cells/mL). The culture system is based on a fluidized bed bioreactor containing the carriers, through which the culture fluid flows upward at a velocity sufficient to suspend the microspheres in the form of a slurry (i.e., approx 70% bed expansion). For oxygenation the medium is recycled through a membrane oxygenator. The system is run continuously for long culture periods (typically 100 d).

This procedure is based on the Verax System One:

1. Set up and calibrate the system as per the manufacturer's instructions.
2. Set recycle flow rate to 90 mL/mm.
3. Concentrate 5 $\times$ 10^7 viable cells into approx 2.1 in a 30-mL syringe with 25-gage needle.
4. Swab the septum 70% with alcohol.
5. Insert the needle through the septum and inject half the inoculum, watt 30 s, and inject the remaining inoculum.
6. Take a 3-mL sample from the reactor using a syringe with 25-gage needle daily.

7. Carry out the following analysts:
   (a) Free cell count.
   (b) Glucose concentration.
   (c) Lactate concentration
   (d) Product concentration.
8. Start media feed and harvest when the glucose concentration drops below 1.5 mg/mL (i.e., for a feed glucose concentration of 4 mg/mL).
9. Adjust the media feed rate to give a glucose concentration in the reservoir of 1.5-2 mg/mL (typically 1 L/day in steady-state culture).

CONCLUSION

The stirred tank bioreactor is the most widely accepted culture system for both cells in free suspension and attached to solid microcarriers. It allows volumetric scale up to 10,000 L for batch suspension culture (1-5 $\times$ 10^6/mL), 1000 liters for batch microcarrier cultures (1-5 $\times$ 10^6/mL), and density scale up (to 1-2 $\times$ 10^7/mL) in perfused spin filter cultures, However there are disadvantages with solid microcarrier-based systems that make them unsuitable for some cell lines. In particular, the effects of sparging and microcarrier collisions on the cells can lead to inefficient cell attachment, poor cell growth, and stripping of cells from confluent microcarriers (particularly in long term perfusion culture systems). These limitations are overcome by using porous instead of solid microcarriers because the cells in the porous carriers are protected from the effects of sparging and carrier collisions, and can withstand much higher sparge rates and stirrer speeds than cells attached to solid microcarriers. Porous carrier culture systems have been proven by many workers at the Laboratory scale and by Verax at an industrial scale. Although the superiority of porous carrier systems has been demonstrated, they are not yet widely used in industry.

14

Stem Cells in Physiological Analysis

Embryonic stem (ES) cells have the potential to proliferate infinitely in vitro in an undifferentiated and pluripotent state, thereby maintaining a relatively normal and stable karyotype even with continual passaging. Remarkably, in vivo, ES cells can be reincorporated into normal embryonic development by transfer into a host blastocyst or aggregation with blastomere stage embryos. They can contribute to all tissues in the resulting chimeras including gametes. When cultivated in vitro, ES cells differentiate under appropriate cell culture conditions, i.e., in the absence of *leukemia inhibitory factor* (LIF) into cell types of all three germ layers: endoderm, ectoderm and mesoderm. However, these differentiation processes occur only when ES cells are cultivated in suspension culture in which they grow to multicellular spheroidal tissues, termed *embryoid bodies* (EBs).

In recent years, a large number of different cell types have been described to differentiate within EBs. These include hematopoietic and endothelial cells, cartilage, neurons as well as smooth, skeletal and cardiac muscle cells. The capacity of ES cells to differentiate into cell types of the mesodermal cell lineage has been extensively used by us and others to investigate the molecular and physiological events occurring during the process of differentiation into cells of the cardiovascular system, i.e., cardiac and endothelial cells. We have observed cardiac development within EBs as early as 7 d after formation of the aggregates, which correlates well with the murine embryo, where the first beating is seen on day E8.5 to E9.5 (i.e., 8.5 to 9.5 d postcoitum). It was found that the earliest detectable cardiomyocytes (stage 0) were not beating, but expressed already voltage-dependent L-type Ca^{2+} channels at low density. During the further developmental stages (stage 1–4), spontaneous contracting activity occurs, and the increasing number of different ion channels causes a diversification of cardiac phenotypes finally leading to the known specialized cardiomyocytes that are found in the neonatal heart, i.e., ventricular-like, atrial-like, and sinus-nodal-like, as well as Purkinje-like cardiomyocytes. Cardiomyogenesis in EBs is paralleled by the development of vascular structures starting on d 5 of differentiation and resulting in the formation of hollow capillary-like tubules within 3 to 4 d, which improves the supply of nutrients and oxygen to EBs, as well as its export of catabolic end products.

The development of a primitive cardiovascular system giving the EB an embryolike organization has raised the idea of using the EB to screen for embryotoxic and teratogenic agents. Furthermore, EBs have been recently introduced as a novel in vitro assay system to study the effectiveness of anti-angiogenic agents, which have been proven to efficiently inhibit tumor growth in in vitro as well as in vivo anticancer trials. With the recent isolation of ES cells of human origin by Thomson and co-workers, a renewable tissue culture source of human cells capable of differentiating into a variety of different cell types is available. Their capacity to differentiate into cardiac and endothelial cells may be potentially useful for tissue transplantation and cell replacement strategies and may be exploited for the treatment of infarcted hearts and other cardiac disorders.

Research by use of the ES cell technology requires sophisticated techniques for tissue cultivation and physiological as well as biophysical cell analysis. In this chapter, we describe recent advances of our laboratory in cell culture protocols and techniques for the analysis of the function of the cardiovascular system developed in EBs. We report on the recently developed spinner flask technique, that allows mass culturing of EBs with highly synchronized and efficient differentiation of cardiac and endothelial cells. For the investigation of diffusion processes via the vascular structures differentiated in EBs, the optical probe technique was elaborated. This technique permits the quantification of fluorescent tracers in the depth of the three-dimensional tissue by an optical sectioning routine based on confocal laser scanning microscopy and a mathematical algorithm to correct for the attenuation of fluorescence light in the depth of the tissue. Furthermore, we report on electrophysiological techniques for the characterization of the differentiation of cardiac cell lineages, i.e., the patch-clamp technique and the *micro-electrode array* (MEA) technique. The patch-clamp technique has provided deep insights into the formation of electrical activity and the underlying ion channels and signaling cascades during cardiomyogenesis. Whole mount EBs can be plated onto MEAs, and the electrical signals of the field potentials are recorded over several days from a multitude of electrodes beneath the spontaneously contracting tissue. The MEA technique has been proven fruitful for the analysis of action potential propagation, the characterization of pacemaker activity, the development of intercellular communication, and the analysis of arrythmia within cardiac cell areas of spontaneously beating EBs.

Materials

Maintenance of Mouse ES Cells

1. Complete growth medium (1X stored at 4°C). To 760 mL Iscove's modified Dulbecco's medium add: 10 mL Glutamax I (100X) 10 mL penicillin–streptomycin solution 10 mL minimum essential medium (MEM) nonessential amino acid solution (100X), 100 *μM* 2-mercaptoethanol, and 200 mL fetal calf serum (FCS).
2. Phosphate-buffered saline (PBS) Ca^{2+} and Mg^{2+}-free: 137 m*M* NaCl, 2.7 m*M* KCl, 10.1 m*M* Na_2HPO_4, 2 m*M* KH_2PO_4, pH 7.4, filter-sterilized.
3. Trypsin-EDTA in PBS.

Spinner Flasks

1. Cellspin stirrer system equipped with 250-mL spinner flasks.
2. Sigmacote solution, stored at 4°C.
3. 5N NaOH solution.

Investigation of Fluorochrome Distributions and Diffusion Coefficients

1. Cell culture medium: to 860 mL Nutrient Mixture Ham's F-10 medium add: 10 mL Glutamax I (100X), 10 mL penicillin–streptomycin solution, 10 mL MEM nonessential amino acid solution (100X), 100 *μM* 2-mercaptoethanol, and 100 mL FCS.
2. PBS, Ca^{2+}- and Mg^{2+}-free: 137 m*M* NaCl, 2.7 m*M* KCl, 10.1 m*M* Na_2HPO_4, 2 m*M* KH_2PO_4, pH 7.4, filter sterilized.
3. E1-buffer: 135 m*M* NaCl, 5.4 m*M* KCl, 1.8 m*M* $CaCl_2$, 1 m*M* $MgCl_2$, 10 m*M* glucose, 10 m*M* HEPES, pH 7.4, filter-sterilized.
4. Doxorubicin.

Isolation of Cardiomyocytes from EBs

1. Dissociation buffer: 120 m*M* NaCl, 5.4 m*M* KCl, 5 m*M* $MgSO_4$, 0.03 m*M* $CaCl_2$, 5 m*M* Na pyruvate, 20 m*M* glucose, 20 m*M* taurine, 10 m*M* HEPES, pH 6.9, with NaOH.
2. KB buffer: 85 m*M* KCl, 30 m*M* K_2HPO_4, 5 m*M* $MgSO_4$, 1 m*M* EDTA, 2 m*M* Na_2ATP, 5 m*M* pyruvate, 5 m*M* creatine, 20 m*M* taurine, 20 m*M* glucose, pH 7.2, with KOH.
3. Collagenase B.

Patch-Clamp Recording

1. Patch-clamp amplifier, model Axopatch 200A.
2. Data aquisition software, Iso2.
3. Extracellular solution for the recording of I_{Ca} : 135 m*M* NaCl, 5 m*M* KCl, 10 m*M* $CaCl_2$, 2 m*M* $MgCl_2$, 5 m*M* HEPES, 10 m*M* glucose, pH 7.4, with NaOH.
4. Pipet solution for the recording of I_{Ca}: 55 m*M* CsCl, 80 m*M* Cs_2SO_4, 2 m*M* $MgCl_2$, 10 m*M* HEPES, 10 m*M* EGTA, 1 m*M* $CaCl_2$, 5 m*M* ATP (Mg), pH 7.4, with CsOH.
5. Amphotericin B solution. Dissolve amphotericin B (7.5 mg) as stock solution in 125 μL dimethyl sulfoxide (DMSO).
6. Borosilicate glass capillaries with filament.

Ca^{2+} Imaging

1. Computer-controlled monochromator.
2. Inverted microscope equipped with a 40X oil-immersion objective (Zeiss), a fura-2 filterblock (TIL Photonics), a 470/525-nm interference filter, and an intensified charge-coupled device (CCD) camera.
3. Software for data recording and analysis, Fucal fluorescence software package.
4. Fura-2,AM, fura-2-free acid.

Methods

Mass Culture Techniques for EBs: the Spinner Flask Technique

Routine screening of embryotoxic and/or anti-angiogenic agents, as well as transplantation of organ-specific cells, requires large numbers of EBs differentiated from ES cells. In most laboratories, the hanging drop method is applied, which was previously developed by Wobus and co-workers, and has been proven useful for the characterization of the developmental aspects of cardiomyogenesis, skeletal, and smooth muscle development, as well as neurogenesis. However, this technique is

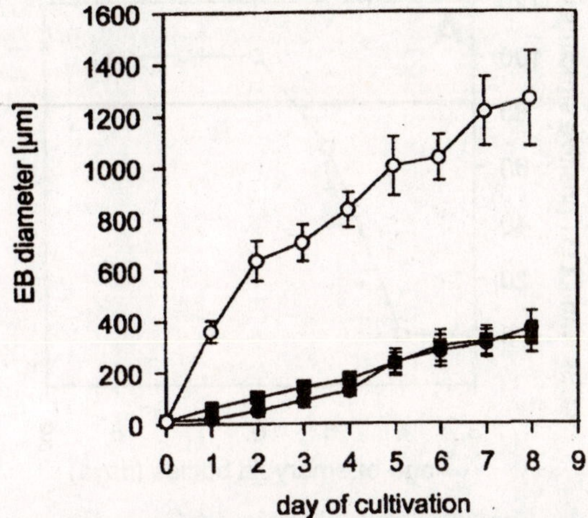

Fig. 14.1. Growth kinetics of EBs cultivated in either spinner flask culture (○) or petri dish suspension culture in normal cell culture medium (■) or methylcellulose-supplemented medium (●).

limited by the low yield of EBs, since a single hanging drop has to be prepared manually for each EB. To achieve mass cultures of EBs with a high efficiency of differentiation, a spinner flask culture technique was recently developed in our laboratory. It was observed that, by using this technique, large amounts of EBs (about 1000/spinner flask) could be cultivated. EBs cultivated in spinner flask culture grew to a significantly larger size, as compared with cultures of EBs in methylcellulosesupplemented cell culture medium or cultivation using the liquid overlay technique, thereby indicating an improved supply with oxygen and nutrients. The spinner flask technique proved efficient for the differentiation of cells of the endothelial cell and the cardiac cell lineage. For the evaluation of the differentiation of the endothelial cell lineage, EBs were fixed after different times of cultivation and immunostained using an antibody directed against *platelet endothelial cell adhesion molecule* (PECAM-1). It was demonstrated that 100% of EBs differentiated vascular structures with a more synchronized time course of differentiation as compared to conventional cell culture protocols. The spinner flask technique proved similarly efficient for the differentiation of the cardiac cell lineage. When EBs were cultivated using the spinner flask technique and were plated to cell culture dishes after 7 d, approx 90% of the EBs developed spontaneous beating areas of cardiomyocytes, which were positive for cardiac-specific contractile proteins.

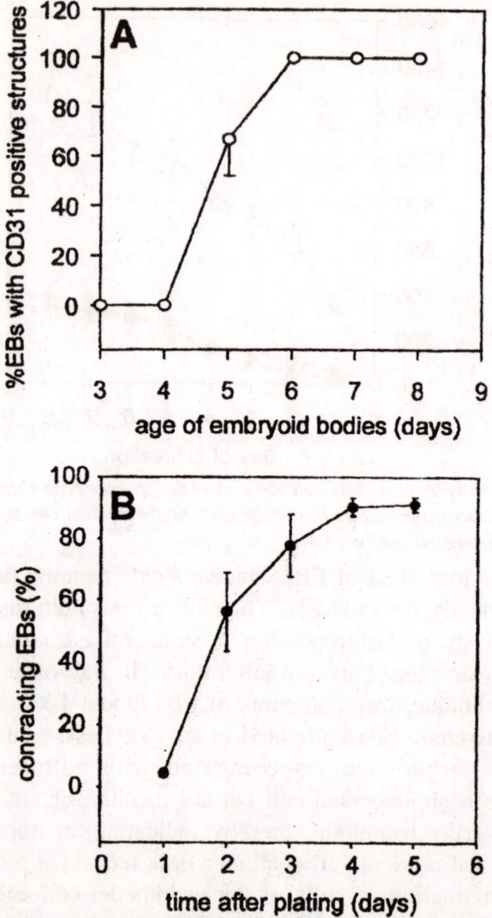

Fig. 14.2. Time-course of endothelial (A) and cardiac (B) cell differentiation in EBs.

Procedure for spinner flask preparation prior to the addition of ES cells

1. Wash clean spinner flasks with excessive Milli-Q-plus water and dry for 1 h at 60°C.
2. Siliconize spinner flasks by moistening the interior as well as the mallets with Sigmacote. Excessive Sigmacote is removed from the flasks using a 10-mL glass pipet.
3. Dry the silicon coat in an oven for 1 h at 120°C.
4. Rinse spinner flasks three times with 250 mL Milli-Q-plus water and autoclave subsequently.

5. Moisten the interior of the flasks with 20 mL of complete Iscove's medium prior to the addition of ES cells; exchange the medium for 125 mL of complete Iscove's medium.

Procedure for cleaning of spinner flask at the end of the experiment

After the end of the experiment, the spinner flasks have to be cleaned prior to the inoculation with fresh ES cells.

1. Remove the old medium with residual EBs and cells.
2. Wash the flasks with 70% ethanol and subsequently with 1 L water.
3. Remove the silicon coat by adding 250 mL 5 N NaOH to the spinner flasks for a maximum of 12 h.
4. Remove NaOH and wash the flasks with at least 5 L water. The interior of the flasks is thoroughly cleaned with a brush. Subsequently, rinse the flasks with 1 L of Milli-Q-plus water.

Inoculation of spinner flasks with ES cells

1. Wash ES cells grown in 6-cm cell culture Petri dishes once with 0.2% trypsin and 0.05% EDTA in PBS.
2. Remove the trypsin solution and incubate ES cells for 5 min with 2 mL 0.2% trypsin and 0.05% EDTA.
3. Triturate the cells with a 2-mL glass pipet until the ES cell clusters are dissociated and a single-cell suspension is achieved.
4. Prepare spinner flasks and seed ES cells at a density of 1×10^7 cells/mL in 125 mL complete Iscove's cell culture medium. Stir at a speed of 20 rotations/minute. The stirring direction is reversed every 1440°.
5. Add 125 mL Iscove's complete cell culture medium after 24 h to yield a final volume of 250 mL.
6. Exchange 125 mL of the cell culture medium every day.
7. For the differentiation of cardiomyocytes, EBs are carefully removed from the spinner flask by the use of 10-mL plastic pipets to avoid any cell injury. They are subsequently plated to 10-cm cell culture dishes filled with 20 mL complete Iscove's cell culture medium. After 24–48 h, spontaneous beating activity of the EBs indicates the differentiation of cardiac cells.

Investigation of the Distribution of Fluorescent Molecules in the 3-Dimensional Tissue of EBs: the Optical Probe Technique

Fluorescent tracers can be used in multicellular tissues to study diffusion processes. Moreover, fluorescent indicators are generally applied for the evaluation of physiological parameters in cell cultures and tissue slices, e.g., intracellular pH, Ca^{2+}, Mg^{2+}, Na^+, and Cl^- ion concentrations, nitric oxide, and reactive oxygen species, as well as the membrane potential. The optical probe technique was recently developed in our laboratory for the study of the diffusion properties of fluorochromes in thick biological specimens, the distribution of vital-lethal fluorescence dyes, and fluorescent anthracyclines, i.e., doxorubicin in the depth of the tissue as well as the fluorescence distribution of fluorescent ion indicators. The optical probe technique is based on confocal laser scanning microscopy and an optical sectioning routine and allows the determination of fluorescence distributions in the tissue down to a depth of approximately 300 μm. A mathematical algorithm was elaborated, which corrects for light absorbtion and scattering, and allows the quantification of fluorescence in the depth of a 3-dimensional tissue. By use of the optical probe technique, we recently determined the diffusion coefficients of the fluorescent anthracycline doxorubicin in the vital tissue of vascularized EBs, in EBs treated with anti-angiogenic agents, and in avascular tumor spheroids.

Dissociation of single cells from EBs and multicellular tumor spheroids and production of densely packed cell pellets

1. Wash either EBs or multicellular tumor spheroids in PBS and place the tissues in a Petri dish containing 1 mL 0.2% trypsin and 0.05% EDTA in PBS.
2. Incubate the objects for 10 min at 37°C and agitate every 30 s.
3. Triturate the tissues with a 1-mL Eppendorf pipet and stop the enzymatic reaction by addition of 4 mL fresh cell culture medium.
4. Centrifugate at 80*g* for 5 min and resuspend the cells in cell culture medium.
5. For the preparation of densely packed cell pellets, incubate the dissociated cells for 4 h in bicarbonate-buffered F10 medium supplemented with 10% FCS at 37°C.
6. Triturate the cells and incubate for 2 h with different concentrations of the respective fluorescence dye ranging from 2–60 μM.
7. After dye loading, wash the cells 3 times in serum-free HEPES-buffered F10 medium.
8. Centrifugate for 10 min at 80*g* and store the cell pellets at room temperature prior to examination.

Determination of fluorescence distribution coefficients

1. Stain either EBs or tumor spheroids for 1–2 h with the respective fluorescence dye. Wash the culture objects in E1 buffer and

transfer them in a liquid vol of 300 μL on 24 × 60 mm coverslips to the stage of an inverted confocal laser scanning microscope.

2. Adjust the motor commands of the stepper motor of the confocal setup to steps of 10 μm in z-direction. Adjust the pinhole settings of the confocal setup to achieve a full-width half-maximum of 8 μm.
3. Select a 25X oil-immersion corrected objective with a numerical aperture of 0.8.
4. Perform single x-y-scans in selected *regions of interest* (ROIs) in the center of the spherical objects and move in 10 μm steps to the periphery of the object. The area of the ROIs is selected to 600 μm^2 (40 × 40 pixel). Each ROI is scanned once in 0.064 seconds. A whole z-series of 16 ROIs is achieved in approx 7 s, the convolution time of the stepper motor thereby being the rate limiting step.
5. Determine the mean field intensity of the fluorescence signal in each ROI and plot the fluorescence intensity data as a function of the penetration depth of the laser beam in the tissue.
6. In order to correct the measured fluorescence intensity traces for fluorochrome and depth-dependent light absorption and scattering, determine attenuation coefficients C using homogenously stained densely packed cell pellets. This results in a fluorescence distribution coefficient $D = I/I_{cal}$, which is calculated as the ratio of fluorescence intensity I, at a discrete depth from the periphery of the spheroid, and the fluorescence intensity I_{cal} obtained equidistantly in the densely packed cell pellet. The distribution coefficients become D = 1, when the tissue is homogenously stained, comparable to the situation in the densely packed cell pellet. When the tissue is less stained or unstained, the distribution coefficient shows values lower than 1. In heterogenously stained samples the coefficient D differs in every optical section z.

Correction procedure for light attenuation in the depth of the tissue

The correction for light attenuation within multicellular tissues loaded with doxorubicin is performed using the following algorithm:

$$\ln (I_{corr}) = \ln (I) + C \bullet D(z) \bullet z$$

where I_{corr} is the fluorescence intensity after correction for light attenuation in a discrete depth z within the fluorochrome containing specimen, and I is the fluorescence intensity before correction. The fluorescence distribution coefficient $D(z) = I/I_{cal}$ is obtained by dividing the measured fluorescence intensity I at a discrete depth z of the laser beam in the tissue by the respective intensity value I_{cal} in the calibration function, i.e., the light attenuation function of the densely packed cell suspension. D ranges between 0 and 1. It approximates 1 in homogenously stained cell layers ($I = I_{cal}$) and approximates 0 where consecutive optical sections reach unstained cell layers.

Evaluation of dynamic processes of fluorochrome distribution: use of the optical probe technique for the study of diffusion

By use of the optical probe technique, diffusion processes of fluorescent molecules in 3-dimensional tissues can be determined, and diffusion coefficients can be calculated from these data. We have recently used this approach to determine the diffusion coefficients for doxorubicin in the vascularized tissue of EBs and EBs treated with

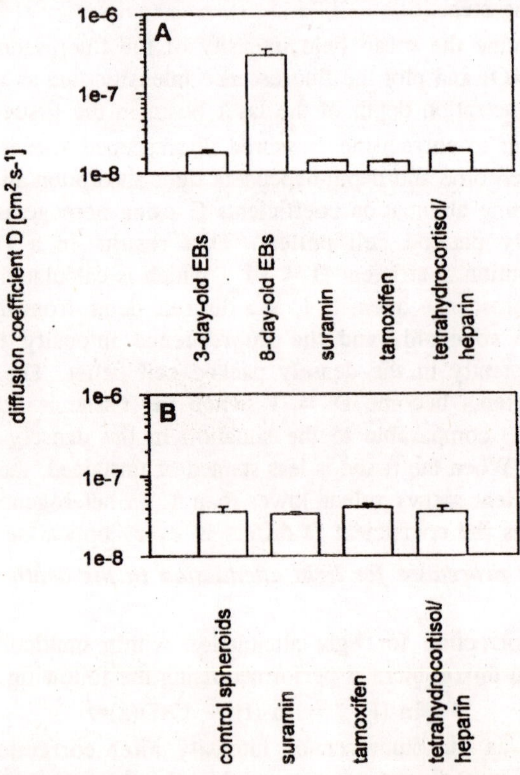

Fig. 14.3. Diffusion coefficients D for doxorubicin as evaluated by the optical probe technique.

antiangiogenic agents, and for comparison in the avascular tissue of multicellular tumor spheroids. For diffusion studies, EBs and multicellular tumor spheroids are incubated at room temperature for 10, 30, and 60 min with 10 μM doxorubicin. They are then washed, and doxorubicin fluorescence is determined by the optical probe technique as described previously. The time-dependent fluorescence increase is recorded in a depth of 80 μm from the spheroid periphery.

For the determination of diffusion coefficients, the doxorubicin distribution is evaluated by use of the optical probe technique after a 60-min incubation with doxorubicin. From the maximal diffusion distance of doxorubicin fluorescence from the tissue periphery, diffusion coefficients are calculated according to the Einstein-Smoluchovski equation: $D = x^2/2t$, where D is the diffusion coefficient, x is the maximal diffusion distance of doxorubicin from the spheroid periphery, and t is the diffusion time of 60 min.

Patch-Clamp Technique and Ca^{2+} Imaging for the Single-Cell Analysis of Cardiomyocytes

One distinct advantage of the investigation of ES cell-derived cardiomyocytes is that, unlike in mammal-derived embryonic cardiomyocytes, their functional phenotype can be investigated starting from very early stages of cardiomyogenesis. Using patch-clamp as well as single-cell Ca^{2+} imaging techniques allows, for the first time, the ability to characterize the correlation between ion channel expression and the electrical activity of early embryonic cardiomyocytes. Thus, by using these techniques not only the detailed development-dependent expression of ion channels could be described, but also their possible involvement in the determination of action potential shape and electrical activity. This also enables the functional characterization of cardiomyocytes deficient in genes (i.e., β1 integrin), thus resulting in early embryonic death.

Because ion channels can be used as sensitive tools for the analysis of signaling cascades, we have used in our studies the voltage-dependent L-type Ca^{2+} current (I_{Ca}) as well as the hyperpolarization-activated nonselective cation current (I_f) to investigate the establishment of various signaling cascades, in particular muscarinic and β-adrenergic regulation. These studies demonstrate that the inhibitory muscarinic regulation is present starting from early stages (stage 0 to 1), whereas the β-adrenergic regulation is becoming only fully established at later stages (stage 3 to 4) of development, which is similar to findings in mammals. Furthermore, we could also identify a development-dependent

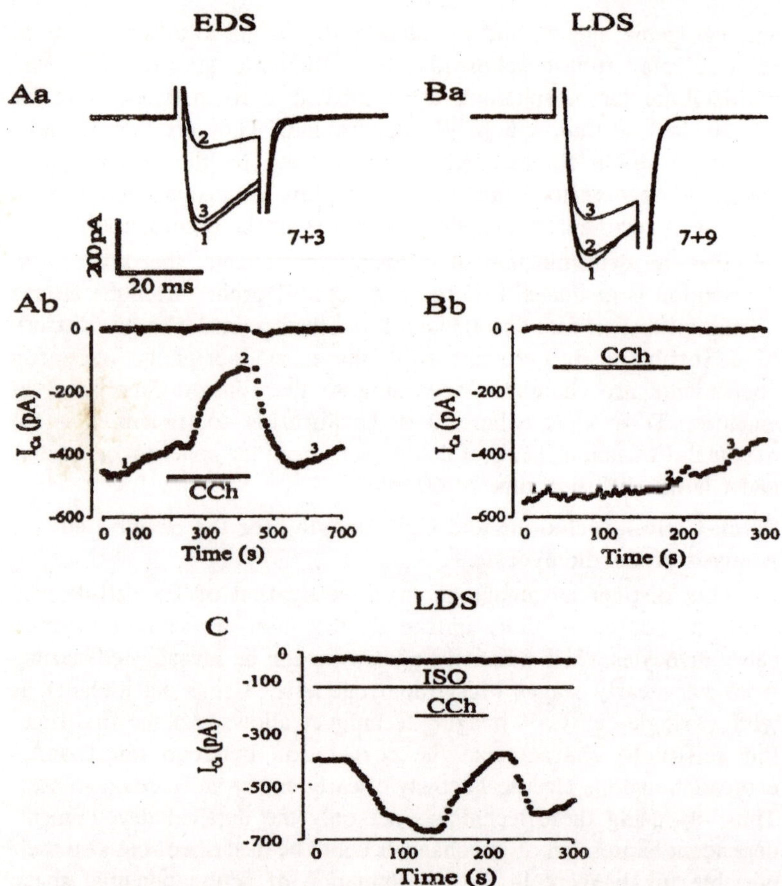

Fig. 14.4. Basal I_{Ca} was strongly depressed upon application of the muscarinic agonist carbachol (CCh, 1 μM) in early development stage cardiomyocytes (As, Ab), whereas no effect of CCh on basal I_{Ca} was detected in late stage cardiomyocytes (Ba, Bb).

switch in the muscarinic signaling cascade. These studies are particularly interesting because pathologically transformed cells can recapitulate their embryonic phenotype.

To successfully perform these physiological experiments on isolated cells, however, cell cultivation, as well as isolation procedures, need to be well established to provide appropriate cell quality. In addition, because within the EB all different cell types are generated, it is mandatory to have clear criteria to identify the cell type of interest or to alter cell culture conditions for the preferential generation of one tissue type, as shown for neuronal differentiation. In the past, spontaneously beating cells were recognized as cardiomyocytes and used for patch-clamp experiments. However, this approach excluded from the analysis cardiomyocytes prior to the initiation of beating and also differentiated ventricular cardiomyocytes known to lack spontaneous electrical activity. We have, therefore, established stably transfected ES cell lines, in which the expression of live reporter genes, such as the *enhanced green fluorescent protein* (EGFP), is under control of a tissue-specific promoter, allowing recognition of the cells of interest based on their fluorescence under excitation light.

While electrophysiological experiments have been performed on isolated ES cellderived skeletal muscle cells, smooth muscle cells, and neurons within the whole EB, we will focus on cardiomyocytes, because most data have been accumulated on this topic.

Isolation of cardiomyocytes from EBs

Cardiomyocytes are derived from pluripotent ES cell lines (i.e., D3;R1). EBs are generated by cultivating the ES cells first for 2 d in hanging drops (400 cells/20 μL), then for 5 d in suspension, and finally plated for 1–15 d on gelatin-coated glass coverslips. About 12–24 h after plating, spontaneously contracting cell clusters appear.

1. For the preparation of isolated cardiomyocytes, dissect beating areas of 20–30 EBs under the microscope using a microscalpel.
2. Disperse the tissue by enzymatic digestion using collagenase B (0.5–1 mg/mL dissolved in dissociation buffer).
3. Sediment the digested tissue by gravity and remove the supernatant.
4. Resuspend cells in 200 μL KB buffer and incubate for 1 h at 37°C.
5. Triturate the tissue until a homogenous cell suspension is achieved.
6. Plate the cell suspension onto gelatine-coated glass coverslips in 24-multiwell dishes filled with cell culture medium and store in an incubator. Within the first 12 h, cells attach to the glass surface. Ideally, the isolated cells are functionally characterized 24–48 h after dissociation.

Patch-clamp recording

ES cell-derived cardiomyocytes are well suited for electrophysiological experiments since they are small (cellular capacitance ranging between 10–25 pF) and lack the T-tubular system at least during early stages. This leads to an improvement of the quality of voltage clamp conditions and facilitates intracellular dialysis.

Moreover, upon appropriate enzymatic dissociation, gigaohm-seals are relatively easily obtained even without fire-polishing the pipet tip. In order to preserve the biophysical characteristics and the function of the ion channels of interest, it is recommended to perform the experiments at a temperature of 35–37°C.

The pipets are made on a DMZ Universal Puller from 1.5-mm borosilicate glass capillaries with filaments yielding a resistance between 2–4 MΩ. The composition of the different recording solutions varies in dependence of the ion channel of interest. For the recording of I_{Ca}, the solutions described previously are used. The series resistance reaches a steady-state level of 10–20 MΩ within 5–15 min after obtaining the gigaohm seal. For the recording of I_{Ca}, voltage-clamped cells are held at -50 mV, and trains of depolarizing pulses lasting 20 ms are applied to a test potential of 0 mV at a frequency of 0.2 Hz. Current–voltage (I/V) relationships are determined by applying 150 ms lasting depolarizing voltage steps from test potentials of –40 mV to +40 mV in 10-mV steps.

Ca^{2+} imaging

For the recording of changes in the intracellular Ca^{2+} concentration of ES cellderived cardiomyocytes, different approaches can be taken. Our group has had good experience with single-cell Ca^{2+} imaging techniques using a monochromator as the fluorescence excitation source. For the exclusion of movement artifacts, we have preferentially used the ratiometric dye fura-2.

1. Load cardiomyocytes on coverslips either with fura-2-acid (50 μM, equilibration after 3–5 min dialysis) via the patch-pipet or with the cell permeate form fura-2,AM (1 μM) for 12–15 min at 37°C in cell culture medium. In case organelle loading is observed, fura-2,AM loading at room temperature for about 30 min is recommended.
2. Transfer the coverslips to a temperature-controlled recording chamber and superfuse for 10 min by gravity at a rate of 1 mL/min with extracellular solution. A 90% volume exchange should be achieved within approx 10 s.
3. For monitoring changes of the intracellular Ca^{2+} concentration ($[Ca^{2+}]_i$) use oil immersion objectives, preferentially 40X. Monochromic excitation light (340, 380 nm) is generated by the monochromator at a frequency of 50 Hz. The emitted fluorescence is imaged through a 470-nm interference filter using an intensified CCD camera connected to the TV port of the microscope.

Fluorescence images (50–100 ms exposure time) are acquired at different rates.

4. For conversion of the ratiometric data into $(Ca^{2+})_i$, *in situ* calibration is performed according to Grynkiewicz. R_{max} is obtained in the presence of ionomycin and 10 m*M* Ca^{2+}, and R_{min} is in the presence of excess EGTA. Representative values are determined in a series of experiments and these average values are used to calculate $[Ca^{2+}]_i$. Average values used for R_{max}, R_{min}, and $F380_{max}/F380_{min}$ are 4.3, 0.44, and 3.94, respectively, for embryonic stem cells, and 2.18, 0.33, 3.57, respectively, for EBs and the dissociation constant is assumed to be 224. When changes of $[Ca^{2+}]_i$ during spontaneous contractions are monitored, fast acquisition rates (50–100 Hz) are required for temporal resolution. To avoid dye bleaching at these high acquisition rates, we have used neutrodensity filters or defocusing of the excitation light source.

Experiments using intracellular dialysis of various substances

For the investigation of the modulation of ion channels and/or the functional expression of inositoltriphosophate (IP_3)-sensitive Ca^{2+} stores, substances such as cAMP, cGMP, protein kinase inhibitor, catalytic subunit of protein kinase A, GTP-γ-S, as well as IP_3 are added to the patch pipet. After obtaining a gigaohm seal, the modulation, for example, of I_{Ca} or I_f can be monitored separately or in combined voltage protocols using appropriate solutions. In order to observe changes occurring during intracellular dialysis with the above mentioned substances, it is mandatory to initiate current recordings immediately upon obtaining the whole cell conformation.

When changes of $[Ca^{2+}]_i$ are monitored while dialyzing the cell with fura-2-acid, excitation wavelengths of 360/390 nm are used. The ensuing changes of the dye concentration are taken into account by monitoring the fluorescence emission intensity at the isobestic wavelength (360 nm), which is independent of $[Ca^{2+}]_i$.

MEAs to Monitor the Development of Excitability in Clusters of Cardiomyocytes Differentiating from ES Cells

MEAs find broad application in the study of excitable multicellular preparations. In rat hippocampus slice preparations, local field potentials and spike measurements allow the analysis of network activity with high spatiotemporal resolution. It enables a tissue-dependent description of long-term potentiation or the long-term effect of drug application.

Different techniques were applied up to now to describe the excitation spread in preparations of cardiac muscle cells. Rohr and co-workers, for example, examined the excitation spread in monolayer cultures of neonatal cardiomyocytes on a cellular level by the help of voltage-sensitive dyes and multisite optical mapping. With this approach, they analyzed the dependence of excitation spread on tissue geometry and excitability. In whole organ preparations, excitation spread can be monitored by optical mapping, thereby resolving the involvement of the cardiac structure in the occurrence of arrhythmia or reentrant phenomena.

In the approach described in this chapter, an in vitro system is used to differentiate ES cells into beating clusters of cardiomyocytes and cultivation on substrate-integrated microelectrodes allows for the long-term recordings on a single preparation. Electrophysiological studies on single cells isolated from these clusters of cardiomyocytes could demonstrate that the cells express ion channels characteristic for cardiac muscle cells. The action potentials recorded with the current clamp technique correlate with those described for sinus node, atrium, or ventricle, underlining the ability of ES cells to differentiate into various tissue forms. Cultivation of the ES cells on MEAs will enable us to combine long-term culture of differentiating cardiomyocytes with the characterization of their electrical activity.

Tissue culture

In culture, ES cells are maintained on feeder layers with the addition of LIF. For the preparation of EBs, a cell suspension is prepared and then cultivated in hanging drops as described previously. After 2 d, the drops are washed off the lid of the culture dish, and the 3-dimensional cell aggregates are maintained in suspension for another 5 d. After 7 d, the EBs are transferred into the culture dishes containing the MEA recording system. The culture dishes are filled with Iscove's medium, and, by careful manipulation with a pipet, the EBs are placed in the middle of the electrode array. After 1 d of incubation, the EBs attach to the bottom of the dish and start to grow out, they form a 3-dimensional tissue layer on top of the electrodes. At d 9, spontaneously beating areas of cardiomyocytes can be observed, and field potentials can be recorded.

Electrophysiological recording by using MEAs

We use the recording system of Multi Channel Systems, which consists of electrode arrays integrated in a culture dish, a heated recording system, a pre- and filter-amplifier, an A/D board (MC_Card), and a computer program (MCRack) for the acquisition and analysis of data.

The MEA consists of 60 substrate-integrated Ti/TiN electrodes with a diameter of 30 μm, arranged in an 8 × 8 array with an interelectrode spacing of 200 μm. The array is integrated in a reusable culture dish with an insulating cover of silicon nitrite. The system can be used for recordings from acute tissue preparations as well as for long-term culture. The data are sampled (up to 50 kHz) and can be analyzed on-line from all 60 channels with a PC-based acquisition system.

What do we learn from MEA recordings?

The MEA electrodes record a field potential that reflects the electrical activity of the cluster of cells growing on top of the electrode. Action potentials can be detected as negative spikes. Original voltage traces, representing a 2-s recording from cardiomyocytes, differentiated from ES cells can be seen. Each voltage trace represents the recording from one microelectrode. A large area of spontaneously active cells covered the electrode field. Corresponding to the location of this area, electrical activity could be detected on almost 3/4 of the electrodes. Two areas with distinct beating frequencies can be distinguished.

By the analysis of the delay in the occurrence of the spike at the different electrodes, the direction of excitation spread can be determined. The origin of excitation, as well as the direction of excitation spread, is stable in the differentiated clusters of cardiomyocytes. Moreover, Igelmund and coworkers could show that failures of propagation occur at narrow tissue pathways and by the Ca^{2+} channel blocker nimodipine.

The different components of the spike, representing the excitation or action potential of the cells covering the electrode can be seen. By the application of specific ion channel blockers, the different components of the spike can be characterized. Whereas the negative spike amplitude depends on the voltage-dependent Na^{+} current, the duration of the spike plateau can be modulated by Ca^{2+} and K^{+} channel blockers. The careful characterization of the signal will allow us to determine the effect of pharmacological agents on multicellular preparations of cardiomyocytes at the ion channel level. Besides the analysis of wild-type cells, the ES cell technique also allows us to examine cells that are genetically manipulated. In this way, we could also follow up the development of beating clusters of cardiomyocytes that are homozygous negative for the expression of Connexin 43, which is the major gap junction protein of ventricular muscle cells. These experiments are of special interest, since homozygous mice of this type are not viable.

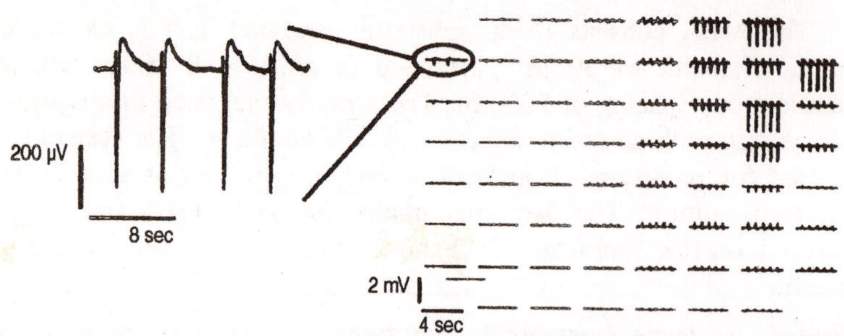

Fig. 14.5. Electrical activity recorded from cardiomyocytes differentiated from ES cells.

Since the MEA system is noninvasive, the cells in culture are not damaged by the experimental procedure. This enables us to follow up a single culture for several days and weeks and to monitor the excitability and excitation spread under developmental aspects or under the long-term influence of pharmacologic agents. Therefore, the ES cells in combination with the MEA provide an ideal system for the description of excitability and excitation spread in 3-dimensional preparations of cardiomyocytes.

Conclusion

There is an increasing number of techniques to evaluate the fundamental properties of ES cells differentiating within EBs. It should be mentioned that the functional analysis of differentiating ES cells is of increasing importance, since functionally intact cells are needed in approaches that use organotypic cells derived from ES cells as a source for cell or tissue transplantation. Scientific emphasis should, therefore, be drawn, not only on cell culture protocols that yield specific cell lineages with reproducible efficiency, but also on the basic physiological characterization of the ES cell-derived organotypic cells.

Notes

1. Since the functional characteristics of the isolated cardiomyocytes critically depend on their differentiation stage, the time after plating defines the stage of the development of the cells. For the preparation of very early stage (stage 0–1) cardiomyocytes, cells are plated for 2 d, for early stages 3 to 4 d and for the differentiated stage in which cells display action potentials typical of nodal-, atrial-, or ventricular-like cardiomyocytes, a plating time of 9–15 d is required. Although after longer plating times, cardiomyocytes can be still detected, enzymatic dispersion is increasingly complicated by the accumulation of interstitial tissue requiring longer digestion periods with collagenase. This results in a

significant deterioration of cell quality and poorer quality of electrophysiological data.

2. Under the above described conditions, pronounced run-down of I_{Ca} is observed. We have, therefore, used the perforated patch-clamp technique using amphotericin B. The pipet is backfilled with amphotericin B containing intracellular solution, and the pipet tip is shortly put into amphotericin-free intracellular solution. The final concentration of amphotericin B in the pipet solution is 1 mg/mL.

15

Osteoclast Cells Lineage

Osteoclasts are cells specialized to resorb bone. They originate from hematopoietic progenitors, which are widely distributed throughout the body. Osteoclast progenitors differentiate into cells that are positive for calcitonin receptor and *tartrate-resistant acid phosphatase* (TRAP), which are specific markers for this lineage, and then they fuse with each other and form fully functional multinucleated large cells. TRAP-positive osteoclasts are found tightly attached to bone matrix. Two cytokines, *macrophage colony-stimulating factor* (M-CSF) and receptor activator of NF-κB (RANK) ligand (RANKL; also known as OPGL, ODF, and TRANCE), are known to be responsible for this differentiation. In this chapter, we describe the procedure for the induction of differentiation into the osteoclast lineage from murine *embryonic stem* (ES) cells. We utilize co-culture systems with stromal cell lines, that is, bone marrow-derived ST2 cells or newborn calvaria-derived M-CSF-deficient OP9 cells. M-CSF is produced constitutively by ST2, and the expression of RANKL is induced on ST2 by 1,25-dihydroxyvitamin D3 [$1,25(OH)_2D_3$] and dexamethasone (Dex).

First, we describe the protocol for inducing osteoclasts within ES cell-derived growing colonies (single-step culture). After culturing

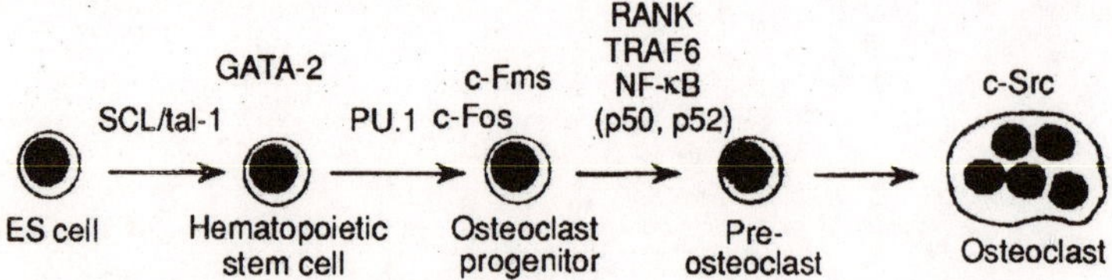

Fig. 15.1. Scheme describing the development of osteoclast lineage.

undifferentiated ES cells on the ST2 stromal cell line in the presence of $1,25(OH)_2D_3$ and Dex for 11 d, multinucleated osteoclasts are generated at the periphery of the colonies (single-step culture). Starting from single ES cells, the cells grow by piling up on each other thereby forming colonies. Hematopoietic cells are observed around the colonies after d 5. TRAP-positive mononuclear cells are first observed on d 8. Other cell lineages, for example, cardiac muscle cells and endothelial cells, are simultaneously observed with osteoclasts within the colonies. This is an appropriate system for examining the potential of ES cells to generate osteoclasts as these cultures are stable and enable us to assess the whole process of osteoclastogenesis without any manipulation except for regular medium changes.

Secondly, we describe a protocol to specifically induce osteoclasts using stepwise cultures. As described by Nakano et al., ES cells generate colonies containing hematopoietic progenitors after 5 d of culturing on OP9 cells. When these colonies are dissociated to a single-cell suspension, transferred onto ST2, and further cultivated for 6 d in the presence of $1,25(OH)_2D_3$ and Dex, most of the cells that are observed on the secondary plates are osteoclasts (2-step culture), suggesting that the colonies on d 5 also contain osteoclast progenitors. This system allows the separation of the maturation step by M-CSF and RANKL on ST2 from the step that lead to the generation of osteoclast progenitors during the initial 5 d culture on OP9. Alternatively, when the dissociated cells prepared on d 5 from ES/

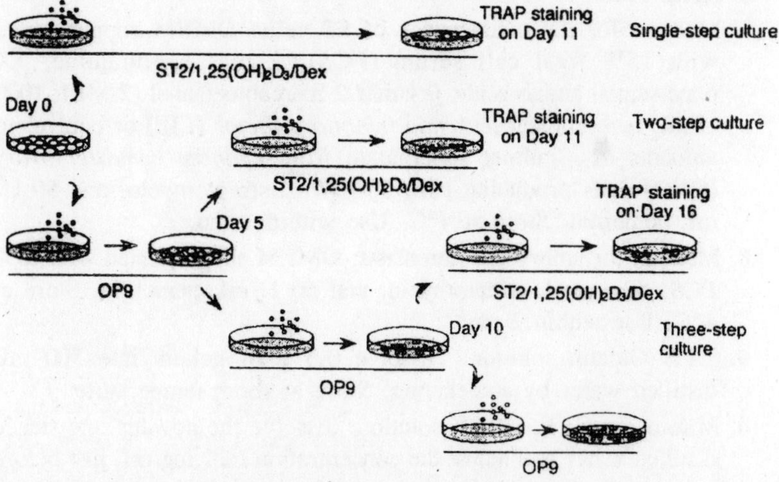

Fig. 15.2. Overview of the culture systems.

OP9 co-cultures are transferred onto fresh OP9 cells and cultured for an additional 5 d, immature hematopoietic cells including osteoclast progenitors selectively propagate and form hematopoietic clusters on OP9. Under this culture condition, very few cells have the tendency to pile up. When the hematopoietic cells are harvested by pipetting from the culture on d 10 and are cultured for an additional 6 d on ST2 in the presence of 1,25$(OH)_2D_3$ and Dex, we could efficiently and selectively induce osteoclasts (3-step culture). The number of osteoclast progenitors is increased 20-fold by the secondary step on OP9. Thus, the 3-step culture condition provides an excellent in vitro model system to study generation, expansion, and maturation of osteoclast progenitors.

Materials

1. 0.25% trypsin + 0.5 m*M* EDTA in phosphate-buffered saline (PBS).
2. 0.1% trypsin + 0.5 m*M* EDTA in PBS.
3. 0.05% trypsin + 0.5 m*M* EDTA in PBS.

   For trypsin-EDTA solutions, we purchased 2.5% trypsin solution from a commercial supplier, for example from Gibco BRL. Dispense into aliquots and store at –20°C. Dilute to appropriate concentrations with PBS and supplement with EDTA. Store at 4°C. Use within 1 mo.
4. Dulbecco's Modified Eagle Medium (DMEM).
5. α-Minimum Essential Medium (α-MEM).
6. RPMI-1640.
7. Medium for the maintenance of ES cells: DMEM supplemented with 15% fetal calf serum (FCS), 2 m*M* L-glutamine, 1X nonessential amino acids, 0.1 m*M* 2-mercaptoethanol (2-ME), 1000 U/mL recombinant *leukemia inhibitory factor* (LIF) or equivalent amounts of a culture supernatant from *Chinese hamaster ovary* (CHO) cells producing LIF, 50 μg/mL streptomycin, and 50 U/mL penicillin. Store at 4°C. Use within 1 mo.
8. Medium for embryonic fibroblasts: DMEM supplemented with 10% FCS, 50 μg/mL streptomycin, and 50 U/mL penicillin. Store at 4°C. Use within 2 mo.
9. 0.1% Gelatin solution: Dissolve 0.5 g of gelatin into 500 mL distilled water by autoclaving. Store at room temperature.
10. Mitomycin C: For 100X solution, dissolve the powder into sterile distilled water and adjust the concentration to 1 mg/mL just before use.

11. Medium for the maintenance of ST2: RPMI 1640 supplemented with 5% FCS, 50 μM 2-ME, 50 μg/mL streptomycin, and 50 U/mL penicillin. Store at 4°C. Use within 2 mo.
12. Medium for the maintenance of OP9: α-MEM supplemented with 20% FCS, 50 μg/mL streptomycin, and 50 U/mL penicillin. Store at 4°C. Use within 1 mo.
13. Medium for the differentiation of ES cells on ST2: α-MEM supplemented with 10% FCS, 50 μg/mL streptomycin, and 50 U/mL penicillin. Store at 4°C. Use within 2 mo.
14. $1,25(OH)_2D_3$. Store at –70°C at 10–4 *M* in ethanol under light-tight conditions. Stable at least for six mo. Dilute from this stock for each use.
15. Dex. Store original stocks at 10^{-2} *M* in ethanol at –70°C. Prepare the stock at 10–3 *M* in ethanol at 4°C. Dilute from this stock for each use. Stable at least for 1 yr (stocks at –70°C) or 3 mo (stocks at 4°C).
16. Medium for differentiation of ES cells on OP9: α-MEM supplemented with 20% FCS, 50 μg/mL streptomycin, and 50 U/mL penicillin. Store at 4°C. Use within 2 mo.
17. TRAP staining solution: for 500 mL of solution, dissolve 5.75 g of sodium tartrate dihydrate, 6.8 g of sodium acetate trihydrate, and 250 mg of naphthol AS-MX phosphate in an appropriate vol of dH_2O. Adjust the pH to 5.0 with acetic acid and adjust the vol to 500 mL with distilled water. Store the stock in a dark glass bottle at 4°C. Just before use, dissolve fast red violet LB salt in the volume you need at the final concentration of 0.5 mg/mL.

METHODS

For each manipulation, prewarm the medium to 37°C and do not leave cells out of the incubator for a long time. The cultures described here require a humidified incubator at 37°C in 5% CO_2. The volume of culture medium for the culture vessels that appear in the protocol.

Table 15.1. Volume of medium for culture vessels

Culture vessels	*Volume of medium (mL)*
25-cm^2 flask	6
60-mm dish	5
100-mm dish	10
24-well plate	1
6-well plate	2

Maintenance of ES Cells

We ordinarly maintain ES cells on mitomycin C-treated embryonic fibroblasts. ES cells are cultured in the medium described in the material section.

1. Prepare mitomycin C-treated confluent embryonic fibroblasts on gelatin-coated dishes.
2. Thaw ES cells on mitomycin C-treated embryonic fibroblasts.
3. Grow them up to a subconfluent state.
4. Replace the medium of the ES cell cultures with fresh medium 2–4 h before passage.
5. Wash three times with PBS and trypsinize the cultures with 0.25% trypsin/0.5 m*M* EDTA for 5 min at 37°C. Add medium and dissociate the cell clump by pipetting up and down. Centrifuge at 120*g* for 5 min at 4°C, aspirate the supernatant, and suspend the cell pellet in the medium. Count the number of ES cells.
6. Replate them onto the freshly prepared mitomycin C-treated embryonic fibroblasts. For 60 or 100-mm dishes, inoculate 10^6 or 3×10^6 ES cells, respectively.
7. Change the ES cell culture media the day after passaging.
8. Two days after the passage, the cultures must reach a subconfluent state. Maintain ES cells by the above-mentioned manner.

Maintenance and Preparation of ST2 Stromal Cells

ST2 cells are maintained in the RPMI 1640 media supplemented with 5% FCS and 50 μM 2-ME.

1. Trypsinize the confluent cells in a dish or flask with 0.05% trypsin/0.5 m*M* EDTA at 37°C for 1 to 2 min.
2. Add medium, dissociate the cells by pipetting, centrifuge them at 200*g* for 5 min, and split them 1:4.
3. Maintain by regularly passing them every 3 or 4 d.
4. To prepare feeder layers in 24-well plates, seed the cells of one confluent T25 flask or half of one confluent 100-mm dish into a 24-well plate. Two days later, the cells reach confluency and are ready to use for the differentiation of ES cells. Neither irradiation nor treatment with mitomycin C is needed.

Maintenance and Preparation of OP9 Stromal Cells

OP9 cells are maintained in α-MEM supplemented with 20% FCS.

1. Grow them to a subconfluent or confluent state.
2. Wash twice with PBS and trypsinize with 0.1% trypsin/0.5 m*M* EDTA at 37°C for 4 to 5 min.

3. Add medium, dissociate the cells well by pipetting up and down, centrifuge them at 200*g* for 5 min, and split them 1:4.
4. Change the medium 2 d later.
5. Passage them again the next day (subconfluent state) or 2 d after the medium change (confluent state).
6. To prepare a feeder layer in 6-well plates, seed half of the cells from a subconfluent or confluent 100-mm dish into a 6-well plate. Two days later, they reach confluency and are ready to use for the differentiation of ES cells. Neither irradiation nor treatment with mitomycin C is needed.

Single-Step Culture to Induce the Differentiation of Osteoclasts

1. Prepare ST2 feeder layer in 24-well plates.
2. Grow ES cells to a subconfluent state.
3. Replace the medium of the ES cell cultures with fresh medium 2–4 h before inducing the differentiation.
4. Wash three times with PBS and trypsinize the culture with 0.25% trypsin/0.5 m*M* EDTA for 5 min at 37°C. Add α-MEM/10% FCS and dissociate the cell clump by pipetting up and down. Centrifuge at 120*g* for 5 min at 4°C, aspirate the supernatant, and suspend the cell pellet in α-MEM/10% FCS. Count the number of ES cells.
5. Dilute to the appropriate cell density in α-MEM/10% FCS. Aspirate the medium from the plates of ST2 feeder cells. Dispense 1 mL of the cell suspension into each well. Supplement with 10–8 *M* $1,25(OH)_2D_3$ and 10^{-7} *M* Dex.
6. Cultivate in the incubator and change the culture medium every 2 or 3 d.
7. Stain the cultures for osteoclasts with a method detecting TRAP activity, on d 11, when mature osteoclasts have been generated.

Multistep Culture on OP9

1. Prepare an OP9 feeder layer in 6-well plates.
2. Grow ES cells to a subconfluent state.
3. Replace the medium of ES cell cultures with fresh medium 2–4 h before inducing differentiation.
4. Wash three times with PBS and trypsinize with 0.25% trypsin/0.5 m*M* EDTA for 5 min at 37°C. Add α-MEM/20% FCS and dissociate the cell clump by pipetting up and down. Centrifuge at 120*g* for 5 min at 4°C, aspirate the supernatant, and suspend

the cell pellet in α-MEM/20% FCS. Count the number of ES cells.

5. Dilute to 5 × 10^3 cells/mL with α-MEM/20% FCS. Aspirate the medium from the plates of OP9 feeder layers. Dispense 2 mL of cell suspension into each well (10^4 cells/well).
6. Maintain in the incubator at 37°C with 5% CO_2.
7. On d 2 or 3 of differentiation, replace half of the medium with fresh medium.
8. On d 5 of differentiation, colonies that have a differentiated appearance will be observed. Wash the cultures 3 times with PBS, and trypsinize them with 0.5 mL of 0.25% trypsin/0.5 m*M* EDTA for 5 min at 37°C. Add α-MEM/20% FCS and dissociate the cell clump by pipetting up and down vigorously. Transfer the dissociated cells into 15-mL centrifuge tube and centrifuge at 350*g* for 5 min at 4°C. Cell aggregates may be generated after trypsinization. These aggregates consist mostly of OP9 cell debris. To precipitate these persistent cell aggregates, let stand for 4 to 5 min, then transfer the upper cell suspension into fresh tubes, and centrifuge. Aspirate the supernatant, suspend the cells in fresh medium, and count the cell number with a hemocytometer. Do not count OP9 cells; they are large and easily distinguished from ES cell-derived cells. About 1–2 × 10^6 cells are obtained per well.
9. Dilute the cell suspension to 5 × 10^4 cells/mL with medium (α-MEM/20% FCS), and dispense 2 mL of the cell suspension into each well of a 6-well plate containing freshly prepared OP9 layers (1 × 10^5 cells/well).
10. Maintain in the incubator at 37°C with 5% CO_2.
11. On d 7 or 8 of differentiation, remove half of the medium gently from the culture plates and add 1 mL of fresh medium.
12. On d 10 of differentiation, hematopoietic clusters or colonies will have formed on the OP9 layers. Add 2 mL of fresh medium (α-MEM/20% FCS) to each well, and after pipetting vigorously, transfer the cells into centrifuge tubes, and let them stand for 4 to 5 min to precipitate debris of OP9 stromal cells. Transfer the supernatant into a fresh tube and centrifuge at 350*g* for 5 min at 4°C. Suspend the pellet in fresh medium and count the cell number. About 105 hematopoietic cells will be obtained per well. If you want to simultaneously analyze the other hematopoietic lineages, plate the cell suspension again onto fresh OP9 at 10^5 cells/well in 6-well plates.

Induction of Differentiation into Osteoclasts in Multistep Cultures

1. Prepare ST2 feeder layers in 24-well plates.
2. For 2-step cultures, dilute the cell suspension to 0.5–1 × 10^4 cells/mL with medium (α-MEM/10% FCS). For 3-step cultures, dilute to 10^3 cells/mL.
3. Aspirate the medium from the plates of ST2 feeder layers. Dispense 1 mL of the cell suspension into each well. Supplement with 10^{-8} *M* 1,25$(OH)_2D_3$ and 10^{-7} *M* Dex.
4. Cultivate in the incubator, and change the culture medium every other day.
5. Stain the cultures for osteoclasts with a method detecting TRAP activity 6 d after seeding when mature osteoclasts have been generated.

TRAP Staining

We ordinarily evaluate osteoclastogenesis by TRAP staining. It is also possible to check the functional activity of osteoclasts by the pit formation assay.

1. Aspirate the culture medium.
2. Add 1 mL of 10% formalin (3.7% formaldehyde) in PBS (v/v) to each well of the plates and fix them for 10 min at room temparature.
3. Wash once with PBS, aspirate, and cover with 0.5 mL of ethanol-acetone (50:50, v/v) for exactly 1 min at room temparature. After 1 min of treatment, immediately fill each well with PBS, aspirate the solution, and wash once more with PBS.
4. After aspiration of PBS, cover the fixed cells with 0.25 mL of TRAP staining solution and incubate for 10 min at room temparature. After the red color develops, wash the plates well with water. Insufficient washing will generate a high background staining.

Notes

1. Choose a lot of serum that well supports the growth of ES cells, does not generate differentiated cells, and has high plating efficiency. The culture at the clonal density without the addition of exogenous LIF would allow the selection of such lots. The batches of serum suitable for the maintenance of ES cells are also available from several commercial suppliers.
2. Batch-related variation of FCS affects the differentiation of ES cells. It is necessary to check the batches of the serum. In the single-step culture, most batches of serum support osteoclasto-

genesis. However, some lots are not good for the maintenance of ST2 feeder layer for 11 d, and the plating efficiency of ES cells or the number of osteoclasts generated on ST2 feeder layers varies according to the FCS lot. Carry out preliminary experiments to check the batches of the serum by seeding ES cells in the range of 10 to 1000 cells/well. Choose a lot of serum that is good for culturing ST2 and efficiently supports the formation of TRAP-positive osteoclasts. It should be noted that plating efficiency is not correlated with the ability to support osteoclasto-genesis. For multistep cultures on OP9, choose lots of FCS that support the growth and differentiation of ES cells on OP9. Some serum does not support the selective propagation of hematopoietic cells after the replating due to taking nonhematopoietic cells into the secondary plates. To induce osteoclasts in multistep cultures (the secondary or tertiary step in the 2- or 3-step culture, respectively), we recommend using a batch of serum that has been used for the differentiation of osteoclasts from an in vivo source such as bone marrow cells, spleen cells, or peritoneal cells.

3. At least 3 ES cell lines D3, J1, and CCE, gave similar results using our protocol. It is also possible to use ES cells that are adapted to feeder-independence.
4. Treat embryonic fibroblasts with 10 μg/mL mitomycin C for 2.5 h. After washing the cells, seed 10^6 or 3×10^6 cells into a 60- or 100-mm dish, respectively. It is also possible to freeze mitomycin C-treated cells. In this case, thaw the 6×10^6 cells from a stock vial into a 100-mm dish or three 60-mm dishes. Dishes should be coated with a 0.1% gelatin solution at 37°C for 30 min or at room temperature for 2 h. Remove the gelatin solution just before the plating of embryonic fibroblasts.
5. The passage numbers of ES cells should be kept as low as possible. Prepare sufficient amounts of young stocks and recover when needed.
6. ST2 cultures grown long periods sometimes change in appearance (i.e., changing to a more dendritic shape or senescent appearance). Discard such cultures and use freshly thawed young ST2 cell stocks.
7. ST2 (RCB0224) and OP9 (RCB1124) cells have been registered in the RIKEN Cell Bank.
8. For the purpose of preparing feeder layers, it is also possible to seed at higher or lower density, and culture for a shorter or

longer times until the stromal cells are in a confluent state when ES cells are inoculated.

9. OP9 are highly sensitive to old media. Use fresh medium and change the medium 2 d after the passage. OP9 must not to be left in their confluent state for prolonged periods, because this makes it difficult to prepare the single-cell suspension. Pass them when they are subconfluent or have just reached confluency. Discard them if their growth is bad or if they have a senescent appearance.
10. Do not use ES cells that have just been thawed from frozen stocks. Use ES cells that have been cultured for at least 1 d after being thawed.
11. If you want to remove the embryonic fibroblasts, replate the ES cells on gelatin-coated dishes and incubate for 30 min in the culture medium, then collect the nonadherent cells by pipetting and count them. However, it does not cause any problems to directly seed the mixture of ES cells and embryonic fibroblasts onto ST2 or OP9 layers.
12. The efficiency with which ES cells form colonies on the ST2 feeder layer changes according to the lot of serum from only 1% to about 40%. Seed ES cells so that ≤20 colonies are generated per well in the 24-well plates. Seeding at higher density prevents osteoclastogenesis because spaces for osteoclasts on ST2 are occupied by the growing colonies. Plating efficiency also varies depending on the ES cell line used. Even the same cell line sometimes gives different plating efficiencies if it is obtained from different laboratories. Carry out preliminary experiments to check the plating efficiency using your lines and serum by seeding ES cells in the range from 10 to 1000 cells/well.
13. It is difficult to find ES cells on the first 1 or 2 d of differentiation. After this period, small ES cell colonies derived from single ES cells are discernible under the microscope. They will grow up to <3 mm (including the area of osteoclasts) in diameter. It has been observed that under these conditions, ST2 cells may generate adipocytes. This is not problematic and should not influence experimental outcomes.
14. Under these conditions, OP9 cells give rise to adipocyte cells, but this does not influence experimental outcomes.
15. When you replate the cultures in which the growth of ES cells on OP9 is not good on d 5, nonhematopoietic cells might be

predominantly replated into the secondary plate. In such cases, delay the day of replating until the colonies grow up to sufficient level.

16. Colonies that have an undifferentiated appearance are also observed on d 5. Such colonies may comprise, at most, one-third of the total colonies.
17. As described by Nakano et al. (1994), cultivation on OP9 for an additional 3 to 4 d generates various hematopoietic lineages, including the erythroid lineage, granulocytes, the mono/macrophage lineage, megakaryocytes, mast cells, and even lymphoid lineages.
18. Pit formation assay on dentine slices should be carried out using cells from multistep cultures. Although it is possible to check the osteoclast function by directly culturing undifferentiated ES cells and ST2 cells on the dentine slices in the presence of $1,25(OH)_2D_3$ and Dex, this method is not recommended as the results fluctuate greatly and are not suitable for quantitative analyses.

16

LINEAGE TRACERS

Methods to study the establishment and distribution of embryonic cell lineages have increased our understanding of the diverse events that comprise normal development. The resultant fate maps have also provided an important framework for systematically analyzing genotype–phenotype relationships uncovered by mutagenesis. Until recently, vertebrate fate maps have been plotted principally in avian systems because of the ease of manipulating tissue in ovo. These studies, using methods comprised of injecting retroviral or fluorescent lineage tracers or of grafting quail cells into chick embryos, have provided the core of what we know about vertebrate development. In contrast to chicken embryos developing in eggs, mouse embryos developing in utero are much less accessible, making the established tracing methods significantly more difficult. To circumvent these difficulties, noninvasive methods have recently been developed to genetically activate lineage tracers in mice using site-specific recombination.

Two recombinase systems from the λ integrase family have been established in mice for catalyzing gene (in)activation: the Cre-*loxP* system from the bacteriophage P1 and the FLP-*FRT* system from the budding yeast *Saccharomyces cerevisiae*. Cre and FLP each cleave DNA at distinct 34 bp target sequences (designated *loxP* or *FRT* sites, respectively) and then ligate it to the cleaved DNA of a second identical site to generate a contiguous strand. The orientation of *loxP* (or *FRT*) sites relative to each other on a segment of DNA determines the type of modification catalyzed by the recombinase, with directly oriented identical sites leading to excision of the intervening DNA. This excision reaction can be exploited to activate lineage tracers in vivo.

Current recombinase-based fate mapping systems are comprised of two basic elements: (i) the recombinase mouse, expressing the site-specific recombinase Cre or FLP in a gene-specific fashion; and (ii) the indicator mouse, harboring a transgene that "*indicates*" a recombination event in a given cell and "*remembers*" or "provides a permanent record" of that recombination event in a heritable manner far after the time of recombinase expression.

In embryos harboring both the recombinase and indicator transgenes, the recombinase will catalyze excisional recombination between two target sites in the indicator transgene, thereby activating the indicator (e.g., switching-on expression of the β-galactosidase (β-gal)-encoding *lacZ* to irreversibly mark a progenitor population and its descendant cells. Cell fates are then determined by colocalization of the indicator molecule (e.g., β-gal) with specific cell-identity markers (molecular and morphological). In this way, a fate map is generated by noninvasive means that not only links a progenitor population to its final array of fates, but also links progenitor- or subtype-specific gene expression to those fates.

Applied to the developing nervous system, this approach has led to a significant breakthrough in being able to determine how embryonic cell identity (embryonic gene expression) links to the final location and properties of those neurons in the adult. This link is made regardless of the transient nature of the expression profiles characterizing many genes, and holds irrespective of intervening cell migration or morphogenetic movement. In other words, transient gene expression patterns can be transformed by site-specific recombination to become neutral tracers of cell fate.

Recombinase-Expressing Mice

In recombinase-based fate mapping experiments, the progenitor population to be tracked is set by the pattern of recombinase expression; thus, it is critical to be able to restrict recombinase activity to a specific progenitor population during development. Such progenitor-specific recombinase transgenics can be generated in one of two ways: (i) zygote injection of a transgene that can express the recombinase; or (ii) targeted insertion of the recombinase transgene into an endogenous gene (a knock-in targeting experiment). Each method has both unique advantages as well as drawbacks.

The approach of zygote injection has at least two advantages. First, specific enhancer elements can be selected to drive recombinase expression in a very restricted and transient fashion, enabling both a spatially and temporally defined subgroup of progenitor cells to be traced. Second, the method allows for the relatively rapid generation of recombinase transgenics. While this is an essential and widely exploited approach, there are two drawbacks worth mentioning. Constructing a recombinase transgene requires previous isolation and characterization of promoter and enhancer elements; these elements may not always be readily available (although the number is increasing significantly as genome analyses expand).

Additionally, recombinase expression from an integrated transgene (which is typically comprised of a head-to-tail array of multiple transgene copies) can be mosaic, with only a subset of the expected cell population expressing the transgene. This variation is usually stochastic and can be counter-balanced in fate mapping studies by analyzing a larger number of recombinase–indicator double transgenics in order to achieve full coverage of the lineage(s) under investigation.

If the goal is to fate map every cell that at some point in its history expressed a given gene (versus just a subset of expressing-cells in which a discrete enhancer element is active), the recombinase knock-in approach is preferable. Moreover, by placing a single copy of the recombinase transgene under the transcriptional control of an endogenous locus, recombinase expression should be optimally regulated with minimal mosaicism. However, when interpreting fate maps generated through the use of recombinase knock-in strains, it is important to remember that recombinase knock-ins can have continuous and changing expression patterns, reflecting the dynamic nature of the wild-type locus. Thus, at a given developmental stage, indicator-positive ("*mapped*") cells may reflect either recombinase-expressing cells or recombinase-negative lineal descendants of earlier recombinase-expressing cells.

Constructing a recombinase transgene for zygote injection involves the following steps: (i) the appropriate promoter–enhancer sequences are coupled to Cre- or FLP-encoding sequences; (ii) the translation start codon is bracketed with Kozak consensus nucleotides for efficient translation of the recombinase; (iii) splice donor–splice acceptor sequences and a polyadenylation signal sequence (pA) are incorporated to maximize transgene expression; and (iv) unique restriction sites are engineered into the construct so that the transgene can be isolated away from the bacterial plasmid backbone. Consideration should also be given to flanking the transgene with insulator sequences to minimize the effects of integration site on proper transgene expression.

Essential components of a knock-in targeting vector have been well described and include: (i) two arms of isogenic DNA homologous to the genomic target locus; (ii) the Cre- or FLP-encoding transgene positioned between the homology arms in such a way that it will be under the transcriptional control of the endogenous locus following homologous recombination; (iii) a positive selection gene placed between the homology arms; (iv) a negative selection gene positioned at the end of a homology arm; (v) a unique restriction site to linearize the vector outside the homologous sequences; and (vi) directly oriented recombinase target sites flanking the positive selection marker. Later recombinase-mediated removal of the selection marker ensures that the strong regulatory elements driving expression of the selectable marker do not interfere with normal regulation of either the knock-in allele or neighboring endogenous genes.

Whether recombinase strains are generated by zygote injection of a transgene or by gene replacement techniques, it is critical that the recombinase activity profile by adequately determined for each recombinase mouse strain. Expression analyses should be performed at cellular resolution and should include: (i) detection of recombinase mRNA by in situ hybridization or by the detection of a histochemical or fluorescent marker encoded on a bicistronic transcript that also encodes the recombinase; (ii) visualization of recombinase protein by immunodetection or by fluorescence using a version of the recombinase tagged with either *green fluorescent protein* (GFP) or one of its spectral variants (red, yellow, and cyan fluorescent proteins); and (iii) detection of recombinase activity by crossing to an indicator mouse strain and analyzing recombinase–indicator double transgenic progeny for sites of reporter expression. As a means to assess recombinase expression, it is important to emphasize that analyses of recombinase–indicator double transgenics, alone, are insufficient. This is because recombinase and indicator profiles will initially be coincident, but later will diverge, with indicator-positive cells reflecting either new sites of recombinase expression or recombinase-negative cells that are lineal descendants of the earlier recombinase-expressing cells. Distinguishing between these two possibilities requires analyzing recombinase expression by an independent method, such as in situ mRNA hybridization. Overall, expression analyses should be performed on a developmental series of staged embryos as well as on adult tissues. Especially critical is to rule out ectopic recombinase expression. Any unexpected recombinase expression would confound subsequent fate mapping studies by switching on the indicator (e.g., β-gal) in unrelated cells that would be erroneously interpreted as part of (or lumped into) a given lineage.

Indicator Mice

In contrast to the transient expression of most recombinase transgenes, the critical feature for indicator transgenes is permanent reporter expression in all arms of a given lineage. Success, therefore, hinges on using a promoter with constitutive and ubiquitous activity to drive the indicator transgene such that following a recombination event in a given cell, that cell and all progeny cells will be marked regardless of subsequent fate. In constructing an indicator transgene, such constitutive promoter sequences are placed upstream of the reporter gene, but separated from the reporter sequence by a "stop" segment of DNA flanked by either directly oriented *FRT* or *loxP* sites. One efficient and strong stop DNA segment is a head-to-tail array of four *simian virus* 40 (SV40) pA sequences coupled with translational stop codons in all reading frames. Using such a *loxP*- or *FRT*-disrupted indicator transgene, the reporter molecule is not expressed unless the disrupting stop DNA is excised by the appropriate recombinase. Following such an excision event, the reporter is switched-on and cell marking becomes heritable and dependent on constitutive reporter expression only (reporter expression is now independent of the recombinase).Thus, in designing an indicator transgene, the first key decision involves choosing the most suitable promoter.

The second critical step concerns choosing a reporter molecule that will best highlight the cell population under study, including specialized features of that cell type (e.g., dendrite or axon), and that will enable colocalization of the reporter molecule with informative cell identity markers (nuclear or cytoplasmic). As for any transgene, the indicator transgene should also contain Kozak consensus nucleotides surrounding the initiator ATG for efficient translation of the reporter, and splice donor–splice acceptor sites and pA sequences to maximize reporter expression following recombinase-mediated activation.

Ideally, the most suitable promoter is one capable of achieving robust and ubiquitous indicator expression following site-specific recombination, as this would enable mapping all cell types. The identification of such a *promoter::indicator* combination has not been a trivial task, having only been approached recently through two different strategies: either targeting indicator transgenes to the constitutive *ROSA26* locus or by using the *cytomegalovirus* (CMV) enhancer/chicken β-*actin* (*CAG*) promoter to drive single-copy indicator transgenes. The

ROSA26 locus was identified through the gene-trap strain *ROSA βgeo 26*, in which a βgeo reporter was constitutively expressed during embryonic development. Multiple genes have now been targeted to the *ROSA26* locus, with *ROSA26* regulatory sequences driving generalized transgene expression from preimplantation onward.

The second approach, using the *CAG* promoter, involved an extensive screen for *embryonic stem* (ES) cell clones that both harbored a single integrated copy of a *CAG::indicator* transgene and robustly expressed the indicator following transient recombinase expression. Such permissive ES clones were then used to generate transgenic mice. In addition to these two approaches, other promoters have been used to drive indicator trangenes. While of less universal use, these include promoter–enhancer sequences from the mouse *Hmgcr* gene (encoding hydroxy-methylglutaryl-coenzyme A reductase), the chicken β-actin gene, and the mouse *BT5* gene-trap locus.

To date, the most widely used reporter molecule in indicator transgenes is β-gal encoded by the bacterial *lacZ* gene. In situ β-gal activity can be visualized either by 5-bromo-4-chloro-3-indolyl-β-D-galactopyranoside (X-gal) histochemistry or by immunodetection using anti-β-gal antibodies, the latter method enabling immunolocalization of β-gal with other cell-specific antigens.

As a cytoplasmic protein, β-gal typically gives good staining of cell bodies. In some cases, however, such staining can make it difficult to distinguish individual cells within a cluster of β-gal-positive cells. This potential limitation can be circumvented by using a form of β-gal that localizes to the cell nucleus. In addition to providing clear visualization of individual cells, nuclear β-gal both increases the sensitivity of the indicator system by concentrating the β-gal signal and is distinguished from any confounding endogenous β-gal activity that would be cytoplasmic.

In many fate mapping experiments, such as when studying neurons where the dendritic arborization and axon morphology reflect very specialized functions, additional insights could be obtained if entire cells were outlined by the reporter molecule. While β-gal serves as an excellent cell tracer, it does not completely fill cells, particularly long processes. Membrane-localized reporters, such as heat-resistant human *placental alkaline phosphatase* (PLAP) and the farnesylated version (EGFPF) of the *enhanced green fluorescent protein* (EGFP) are capable of outlining long cellular processes without seeming to compromise viability.

PLAP is anchored to the outer (noncytoplasmic) surface of the plasma membrane by a *glycan-phosphatidyl inositol* (GPI) moiety. Histochemical detection of PLAP activity is so sensitive that single axons can be traced to their terminals and individual growth cones and dendrites visualized quite readily. In addition, monoclonal and polyclonal antibodies specific to PLAP are available and can be used to immunolocalize PLAP expression with cell-identity markers.

EGFPF is a fusion between EGFP and the carboxy-terminal 20 amino acids of c-Ha-Ras containing membrane-targeting farnesylation and palmitoylation sequences. In *Drosophila* and mammalian cells, EGFPF provides a robust fluorescent signal high-lighting long cellular processes such as dendrites and axons. To date, no defects in neural connections or cells viability have been found associated with either EGFPF or PLAP, as have been associated with some other axon-targeted reporter molecules. Membrane-localized EGFPF should allow for unequivocal marking of long cellular processes (such as axons) and should enable visualization of migrating cells in live tissue.

Before using an indicator strain for lineage studies, it is important to determine the expression profile of the activated reporter molecule. This profile determines the range of cell types that can be mapped reliably by the indicator strain, establishing its overall utility. If a given cell is incapable of expressing the indicator transgene, then it would not be possible to determine by reporter detection whether the recombinase was ever expressed in that particular lineage. To unambiguously determine the scope of the potential indicator profile, a derivative mouse strain should be generated in which every cell harbors a recombined copy of the indicator transgene while maintaining the chromosomal environment of the starting (unrecombined) transgene. Generating such a derivative ("*activated*") indicator strain is involves crossing the indicator strain to a recombinase-expressing deleter strain, which will catalyze indicator recombination in the germ line and transmit this recombined indicator transgene to progeny. Since any chromosomal position effects on indicator expression will be identical between the derivative activated strain and the parental strain, the reporter profile for the activated strain will precisely reflect the expression potential of the parental indicator strain.

Expanding the Capabilities of Recombinase-Based Tools

Through the production of new FLP variants, unique capabilities have been added to recombinase-based fate mapping systems, allowing for an important degree of flexibility in experimental design. The three variants include: FLP-wild-type (FLP-wt), FLPe, and FLP-L. "*Enhanced FLP*" (FLPe) harbors four amino acid changes that collectively render FLPe more thermostable, thereby improving recombinase activity in vitro 4-fold at 37°C and 10-fold at 40°C as compared to FLP-wt. Importantly, when expressed in mice, FLPe has been able to achieve maximal target gene recombination. In contrast, FLP-L, containing an alternative point mutation, exhibits at least a 5-fold decrease in activity as compared to FLP-wt. These activity differences can be exploited in vivo. For example, maximal target gene recombination, such as approached with FLPe, is desirable for mapping cell populations en masse when examining either gross morphogenetic movements or the development of an entire neural system.

The more modest recombination frequencies achieved with FLP-L are advantageous when visualization of fewer cells improves the resolution of a developmental process, such as following discrete axon trajectories of lineally related neurons. In addition to the expanded flexibility afforded by the different variants, use of the FLP system for fate mapping in general has the advantage that mouse strains with FLP-marked lineages can be combined with Cre-induced mutations and deficiencies without concern for potentially confounding cross-reactions between the fate mapping (FLP) and mutagenesis (Cre) systems (which could occur should the same recombinase system be used for both purposes). It then becomes possible to analyze mutant gene activities directly for their effect on the FLP-marked lineage and its fate. Due to these advantages, the number of available FLP-expressing mouse strains is on the rise.

Alternatively, if Cre is used as the fate mapping tool, the large number of already-generated Cre transgenics can be exploited. Regardless of the recombinase used, it is essential that the recombinase expression patterns be detailed both spatially and temporally for each strain prior to their use in order to make proper sense of the resultant fate maps.

In addition to using a single-recombinase system for fate mapping (e.g., FLP or Cre alone), it is also possible to exploit both FLP and Cre together for fate mapping purposes, enabling the generation of higher resolution maps. For example, indicator transgenes have been generated in which cell lineages are marked only if they have expressed both FLPe and Cre at some point in their history. Such "*intersectional*" fate mapping can be used to determine the fate of more restricted progenitor populations, subsets demarcated by specific pair-wise combinations of expressed genes. Additional recombinase systems are being developed that will likely increase capabilities even further. These include the integrase from the Streptomyces phage ϕC31 and the Cre-FLP hybrid enzymes, Clp and Fre, which recognize novel DNA target sites. Moreover, all of these systems can gain finer temporal control of recombination events either by using inducible promoters (broadly active or lineage-specific) to express the recombinase or by using a recombinase–steroid receptor fusion protein that can be activated at will by systemic administration of hormone.

While the described Cre and FLP fate mapping techniques are limited by our current knowledge of early gene expression patterns, the number of well-characterized progenitor-specific genes is growing rapidly due to transcript profiling at the genome level through the use of such methods as DNA microarrays and the SAGE method for serial analysis of gene expression. Thus, the possibilities for highly sophisticated spatial and temporal genetic manipulations appear without limit.

Recombinase-based fate mapping approaches have been used principally in vivo; however, it should be feasible to exploit this basic strategy, or at least certain aspects of the method, in the in vitro study of lineages differentiated from cultured stem cells. The underlying principles and transgenes are the same. These methods, however, should be applied to stem cell cultures with some degree of caution because the efficiency of each recombinase variant has not been unequivocally established in ES cells. In this chapter, we describe methods to assess recombinase expression patterns, to determine the overall utility of a given indicator strain, and to histochemically detect either β-gal or PLAP activity in tissue sections in order to track cell populations during development. These approaches will be presented within the context of in vivo analyses.

Materials

Profiling Recombinase Expression by In Situ mRNA Hybridization

Pretreatment and cryosection of tissues

1. Phosphate-buffered saline (PBS): 137 m*M* NaCl, 2.6 m*M* KCl, 10 m*M* Na_2HPO_4, and 1.8 m*M* KH_2PO_4. Adjust pH to 7.4 with HCl, autoclave, and store at room temperature.
2. 4% Fix: 4% paraformaldehyde in PBS. Prepare fresh from frozen 20% paraformaldehyde-PBS stock. Store at 4°C for up to 24 h.
3. 30% Sucrose solution: 30% sucrose (w/v) in PBS. Filter-sterilize (0.2 μm) and store at 4°C.

4. OCT: embedding resin.
5. Equipment: dissecting instruments (forceps and scissors), embedding molds, Superfrost/Plus glass slides, freezer-safe elastic sealing tape, freezer-safe slide boxes, and a cryostat.

Synthesis of digoxigenin-labeled riboprobes for in situ detection of recombinase mRNA

Precautions should be taken to avoid any contamination of reagents or equipment with ribonucleases. Sterile, disposable plasticware is essentially free of RNases and should be used whenever possible.

1. Linearized plasmid (1μg/μL) containing T3, T7, or SP6 promoter sequence and the cDNA of interest as template for riboprobe synthesis.
2. Diethyl pyrocarbonate (DEPC)-water: to inhibit any contaminating ribonucleases, add 1 mL of DEPC per liter of distilled water. Note that DEPC is suspected to be a carcinogen and should be handled with care. Stir the 0.1% DEPC in water overnight in a fume hood at room temperature. Autoclave to inactivate the DEPC. Store at room temperature.
3. Stock solutions (prepare in DEPC-water): 10 mg/mL proteinase K (store in aliquots at –20°C); 0.5 *M* EDTA, pH 8.0; 4 *M* LiCl; 10 mg/mL ethidium bromide.
4. 10X Transcription buffer: 400 m*M* Tris-HCl, pH 8.0, 60 m*M* $MgCl_2$, 100 m*M* dithioerythritol, 20 m*M* spermidine, 100 m*M* NaCl, and 1 U/mL RNase inhibitor.
5. 10X NTP labeling mixture. 10 m*M* ATP, 10 m*M* CTP, 10 m*M* GTP, 6.5 m*M* UTP, and 3.5 m*M* digoxigenin-UTP in Tris-HCl, pH 7.5.
6. RNase inhibitor, 40 U/μL.
7. T3, T7, or SP6 RNA Polymerase, 20 U/μL.
8. DNase I (RNase-free), 10 U/μL.
9. TE_{50}: 50 m*M* Tris-HCl, pH 8.0, 1 m*M* EDTA, pH 8.0. Autoclave and store at room temperature.
10. TE: 10 m*M* Tris-HCl, pH 8.0, 1 m*M* EDTA, pH 8.0. Autoclave and store at room temperature.
11. 100 and 70% ethanol. Store at –20°C.
12. Electrophoresis-grade agarose.
13. Digoxigenin-labeled RNA standards, 0.1 μg/μL.
14. RNase-free microfuge tubes and pipet tips.

***In situ* mRNA detection**

1. Stock solutions (prepare in DEPC-water): 1 *M* triethanolamine; 5 *M* NaCl; 1 *M* KCl; 1 *M* Na_2HPO_4; 1 *M* KH_2PO_4; 1 *M* levamisole (store in aliquots at –20°C); 1 *M* Tris-HCl, pH 7.5 and pH 9.5; 20% paraformaldehyde in PBS (store in aliquots at –20°C); heat-inactivated sheep serum; 20% *sodium dodecyl sulfate* (SDS); polyoxyethylene-sorbitan monolaurate; acetic anhydride; 10 mg/mL proteinase K (store in aliquots at –20°C); 10 mg/mL yeast tRNA (store in aliquots at –20°C); 10 mg/mL heparin (store in aliquots at –20°C); DEPC-water.
2. PBS: 137 m*M* NaCl, 2.6 m*M* KCl, 10 m*M* Na_2HPO_4, and 1.8 m*M* KH_2PO_4. Adjust pH to 7.4 with HCl, autoclave, and store at room temperature.
3. PBS_{DEPC}; 137 m*M* NaCl, 2.6 m*M* KCl, 10 m*M* Na_2HPO_4, and 1.8 m*M* KH_2PO_4. Adjust pH to 7.4 with HCl. Add 1 mL DEPC/L PBS, stir overnight in fume hood, autoclave, and store at room temperature.
4. 4% Fix: 4% paraformaldehyde in PBS_{DEPC}. Prepare fresh from frozen 20% paraformaldehyde/ PBS stock. Store at 4°C for up to 24 h.
5. Glycine-PBS_{DEPC}: 2 mg/mL glycine in PBS_{DEPC}, prepared just before use.
6. 20X standard saline citrate (SSC): 3 *M* NaCl, 0.3 *M* Na-citrate in DEPC-water. Adjust to pH 4.5 with 0.6 *M* citric acid.
7. Hybridization buffer: 50% formamide, 5X SSC, 50 µg/mL yeast tRNA, 1% SDS, and 50 µg/mL heparin in DEPC-water. Store at –20°C.
8. Riboprobe: 0.1 µg/µL in TE_{50}.
9. Wash I: 50% formamide, 5X SSC, 1% SDS in DEPC-water.
10. Wash II: 0.5 *M* NaCl, 10 m*M* Tris, pH7.5, 0.1%. Tween-20 in DEPC-water.
11. Wash III: 50% formamide, 2X SSC in DEPC-water.
12. Rnase A Wash: 25 µg/mL Rnase A in Wash II.
13. Anti-digoxigenin Fab fragments conjugated to alkaline phosphatase (αdigAP), 0.75 U/µL. Aliquot and store at 4°C.
14. 10X TBST: 1.4 *M* NaCl, 27 m*M* KCl, 250 m*M* Tris-HCl, pH 7.5, and 10% polyoxyethylene-sorbitan monolaurate (Tween-20). Store at room temperature.

15. TBST-L: 1X TBST, 2 m*M* levamisole. Prepare fresh prior to use.
16. Antibody solution: αdigAP at 1:5000 in TBST-L with 1% heat-inactivated sheep serum.
17. NTMT-L: 100 m*M* NaCl, 100 m*M* Tris-HCl, pH 9.5, 50 m*M* $MgCl_2$, 1% polyoxyethylene-sorbitan monolaurate (Tween 20), and 2 m*M* levamisole. Prepare immediately prior to use.
18. NBT: 4-nitro blue tetrazolium chloride, 100 mg/mL.
19. BCIP: X-phoshate-5-bromo-4-chloro-3-indolyl-phosphate, 50 mg/mL.
20. Alkaline phosphatase staining solution: 4.5 μL NBT and 3.5 μL BCIP/mL NTMT-L. Prepare immediately prior to use.
21. Gelvatol or other aqueous mounting media.
22. Equipment: plastic staining dishes (capacity for 24 slides in 250 mL of solution), slide holders with handles (each holds 24 slides), slide mailers (capacity for 5 slides in 15 mL of solution), 24 × 60 mm glass coverslips.

Preparation of gelvatol aqueous mounting media

1. Stock solutions: 10 m*M* KH_2PO_4; 10 m*M* Na_2HPO_4. Store each at room temperature.
2. Polyvinyl alcohol resin (grade 205).
3. NaCl.
4. Glycerol.
5. Equipment: a 2-L beaker, a 6-inch stir bar, 250-mL polypropylene centrifuge bottles, and 50-mL polypropylene conical tubes.

Profiling Recombinase Expression by Crossing Recombinase Mice to a lacZ Indicator Strain

Detection of β-gal activity on tissue cryosections by X-gal histochemistry

1. Stock solutions: 10% Nonidet P-40 (NP40); 1% Na deoxycholate; 1 *M* Na-phosphate (combine 1 *M* NaH_2PO_4 and 1 *M* Na_2HPO_4 to pH 7.4); 0.25 *M* EGTA, pH 7.4; 0.5 *M* piperazine-N,N'-bis(2-ethanesulfonic acid) (PIPES) pH 6.9; 1 *M* $MgCl_2$; 20% paraformaldehyde in PBS (store in aliquots at –20°C); 0.5 *M* K_3 $(Fe[CN]_6)$; 0.5 *M* $K_4(Fe[CN]_6)$; 25 mg/mL X-gal in N,N-dimethyl-formamide (DMF) (store at –20°C protected from light).
2. 0.2% Fix solution: 0.2% paraformaldehyde (from frozen 20% paraformaldehyde-PBS stock), 0.1 *M* PIPES, pH 6.9, 2 m*M* $MgCl_2$, and 5 m*M* EGTA, pH 7.4. Prepare immediately prior to use. Store at 4°C for up to 24 h.

3. 2.0% Fix solution: 2.0% paraformaldehyde (from frozen 20% paraformaldehyde-PBS stock), 0.1 *M* PIPES, pH 6.9, 2 m*M* $MgCl_2$, and 5 m*M* EGTA, pH 7.4. Prepare immediately prior to use. Store at 4°C for up to 24 h.
4. Reaction buffer: 100 m*M* Na-phosphate, pH 7.4, 2 m*M* $MgCl_2$, 0.1% Na-deoxycholate, 0.2% NP40.
5. X-gal staining solution: 5 m*M* $K_3(Fe[CN]_6)$, 5 m*M* $K_4(Fe[CN]_6)$, and 1mg/mL X-gal in rinse solution. X-gal staining solution can be reused several times if stored at 4°C protected from light.
6. PBS: 137 m*M* NaCl, 2.6 m*M* KCl, 10 m*M* Na_2HPO_4, and 1.8 m*M* KH_2PO_4. Adjust pH to 7.4 with HCl, autoclave, and store at room temperature.
7. PBS-M: 2 m*M* $MgCl_2$ in PBS.
8. Gelvatol or other aqueous mounting media.
9. Equipment: glass staining dishes, slide-racks, and 24 × 60 mm glass coverslips.

Determining the Utility of an Indicator Strain

Germ line transmission of an activated indicator transgene

1. Deleter mice: a recombinase (Cre or FLP)-expressing mouse strain that is capable of mediating site-specific excisional recombination in the germ line.
2. Indicator mice: a mouse strain harboring a *loxP*- or *FRT*-disrupted reporter transgene that is capable of indicating a recombination event and providing a permanent record of that recombination event in a heritable manner far after the time of recombinase expression.
3. Nontransgenic mice of the same genetic background as the indicator strain.

DNA extraction from tail biopsies

To minimize contamination of samples with unwanted plasmid or genomic DNA, all solutions should be prepared using polymerase chain reaction (PCR)-grade reagents, dispensed using aerosol-resistant tips, and stored and/or incubated in autoclaved microfuge tubes or containers.

1. Stock solutions: proteinase K (store in 10 mg/mL aliquots at –20°C); phenol:chloroform (1:1); chloroform:isoamyl alcohol (24:1); 3 *M* NaOAc.
2. Phenol: phenol equilibrated with TE, pH 8.0.
3. 100% Ethanol. Store at –20°C.

4. STES buffer: 100 m*M* NaCl, 50 m*M* Tris-HCl, pH 7.4., 10 m*M* EDTA, pH 7.5, and 1% SDS.
5. TE: 10 m*M* Tris-HCl, pH 8.0, 1 m*M* EDTA, pH 8.0. Autoclave and store at room temperature.
6. Equipment: polypropylene microfuge tubes (autoclaved), aerosol resistant tips, and glass capillary tubes.

Identifying recombinase and indicator transgenics by PCR-genotyping DNA isolated from tail biopsies

1. 10X PCR buffer: 100 m*M* Tris-HCl, pH 9.0, 500 m*M* KCl, 1% Triton X-100.
2. 25 m*M* $MgCl_2$.
3. *Taq* DNA polymerase, 5 U/μL.
4. 50X dNTP mixture: 10 m*M* dATP, dTTP, dGTP, and dCTP. Aliquot and store at –20°C.
5. PCR primers stored at –20°C as concentrated stocks ≥ 3 μg/μL) and as working stocks (20 μ*M*). For example, *FRT7* specific indicator primers:
   219 (5'-CTAGAGGATCCCCGGGTACCG-3'),
   218 (5'-GCATCGTAACCGTGCATCTGCC-3'),
   90 (5'-CAGTTCATTCAGGGCACCGGACAGG-3'),
   204 (5'-CACGAGCATCATCCTCTGCATG-3'),
   205 (5'-CAGCGACTGATCCACCCAGTCC-3'),
   FLP-specific primers:
   222 (5'-CCCATTCCATGCGGGGTATCG-3'),
   223 (5'-GCATCTGGGAGATCACTGAG-3').

Generation of a Recombinase-Based Fate Map

Identifying recombinase–indicator double transgenic embryos by PCR-genotyping DNA from yolk sac membranes

1. PBS: 137 m*M* NaCl, 2.6 m*M* KCl, 10 m*M* Na_2HPO_4, and 1.8 m*M* KH_2PO_4. Adjust pH to 7.4 with HCl, autoclave, and store at room temperature.
2. Yolk sac lysis buffer: 50 m*M* KCl, 1.5 m*M* $MgCl_2$, 10 m*M* Tris-HCl, pH 8.5, 0.01% gelatin, 0.45% NP40, 0.45% polyoxyethylene-sorbitan monolaurate. Store at room temperature.
3. 10 mg/mL Proteinase K (store in aliquots at –20°C).
4. Equipment: autoclaved microfuge tubes, aerosol-resistant tips, and microfuge tube covers (or screw cap tubes) to prevent splash contamination.

Detection of progenitor cells and their descendants by staining for heat-resistant alkaline phosphatase activity

1. Stock solutions: 1 *M* Tris-HCl, pH 9.5; 5 *M* NaCl; 1 *M* $MgCl_2$; 0.5 *M* EDTA, pH 8.0; 20% paraformaldehyde in PBS (store in aliquots at –20°C); 1 *M* levamisole (store in aliquots at –20°C).
2. PBS: 137 m*M* NaCl, 2.6 m*M* KCl, 10 m*M* Na_2HPO_4, and 1.8 m*M* KH_2PO_4. Adjust pH to 7.4 with HCl, autoclave, and store at room temperature.
3. PBS-M: 2 m*M* $MgCl_2$ in PBS.
4. 4% Fix: 4% paraformaldehyde in PBS. Prepare fresh from frozen 20% paraformaldehyde-PBS stock. Store at 4°C for up to 24 h.
5. Detection buffer: 100 m*M* Tris-HCl, pH 9.5, 100 m*M* NaCl, and 50 m*M* $MgCl_2$.
6. NBT: 100 mg/mL.
7. BCIP: 50 mg/mL.
8. EDTA wash solution: 20 m*M* EDTA (from 0.5 *M* EDTA, pH 8.0 stock) in PBS.
9. Gelvatol or other aqueous mounting media.
10. Equipment: glass staining dishes, slide-racks, and 24 × 60 mm glass coverslips.

METHODS

Profiling Recombinase Expression by In Situ mRNA Hybridization

Pretreatment and cryosection of tissues

1. Dissect embryos or adult tissue in PBS, removing extra-embryonic membranes and opening any cavities to avoid trapping of reagents. Place tissue in cold PBS on ice until all dissections are complete. Wash tissue with cold PBS to remove any residual blood.
2. Replace PBS with 4% fix solution and gently rock at 4°C for 2 h to overnight.
3. Wash three times with PBS for 5 min each to eliminate residual fixative.
4. To preserve tissue architecture on later sectioning, soak tissues in 30% sucrose solution, rocking gently at 4°C for 6 h to overnight.
5. Carefully dip or roll tissue in an aliquot of OCT to remove residual sucrose solution prior to embedding.
6. Submerge and position tissue in an OCT-containing embedding mold. Gently "*push out*" air bubbles with forceps, and immediately freeze

by floating mold in a dry ice/ethanol bath until OCT has turned from clear to white.

7. Carefully remove mold from ice bath, wiping off excess ethanol. Wrap mold tightly in aluminum foil and place in an airtight plastic bag to avoid specimen dehydration. Store indefinitely at –80°C.
8. Cryosection tissue (10–12 μm) and collect on glass slides. Air-dry 2 h to overnight or dry under vacuum with desiccant. Store sections in sealed box at –80°C.

Synthesis of digoxigenin-labeled riboprobes for in situ detection of recombinase mRNA

Digoxigenin-labeled complementary single-stranded (antisense) RNA probes (riboprobes) are transcribed from recombinant plasmids containing the cDNA sequence of interest downsteam from a promoter for either T3, T7, or SP6 RNA polymerase. Synthesis of such probes has been described previously in detail. First, plasmid DNA is cut downstream of the cDNA sequence of interest by digestion with an appropriate restriction endonuclease. This ensures that the generated riboprobe is free of contaminating vector sequence. Second, riboprobe is transcribed from the truncated plasmid template using either T3, T7, or SP6 RNA polymerase, as determined by the upstream promoter sequence. Riboprobes ranging in size from 0.3 to 1.5 kb have been found to give good signal-to-noise results—short enough to enable thorough penetration to target mRNAs and long enough to give a strong hybridization signal. Probe fragments under 100 bp should be avoided since the melting temperature for hybridization decreases exponentially with fragment length, making it difficult to control the stringency of hybridization. Sense strand probes provide useful controls on nonspecific background and are synthesized by transcribing in the opposite direction on the template DNA.

Use T3 RNA polymerase and *Eco*RV-digested pBS-FLP-L to generate a FLP antisense riboprobe; use T7 RNA polymerase and *Eco*RV-digested pBS-FLP-L to generate control sense probes. Precautions should be taken to avoid any contamination of reagents or equipment with RNases. Sterile disposable plasticware is essentially free of RNases and should be used whenever possible.

1. To generate approx 10 μg of a given riboprobe, mix these reagents in the following order at room temperature: 13 μL DEPC-water, 2 μL 10X transcription buffer, 2 μL 10X NTP labeling mixture, 1 μL linearized plasmid (1μg/μL), 1 μL RNase inhibitor (40 U/μL),

and 1 μL T3, T7, or SP6 RNA polymerase (20 U/μL). Incubate at 37°C for 2 h.

2. Remove a 1-μL aliquot and electrophorese on a 1% agarose gel containing 0.5 μg/mL ethidium bromide. An RNA band approximately 10-fold more intense than the linearized plasmid band should be seen, indicating that about 10 μg of riboprobe has been synthesized.
3. To digest plasmid from the riboprobe synthesis reaction, add 1 μL DNase I (10 U/μL) and incubate 25 min at 37°C.
4. Add 2 μL 0.5 *M* EDTA.
5. Add 100 μL TE50, 10 μL 4 *M* LiCl, and 300 μL 100% ethanol. Mix well and incubate at –20°C for 30 min.
6. Spin out precipitate at approx 10,000*g* at 4°C for 15 min in a microcentrifuge. Discard supernatant while ensuring that the riboprobe pellet remains in the microfuge tube. Wash pellet with 70% ethanol and air-dry.
7. Resuspend pellet in 120 μL TE_{50}. Precipitate again by adding 10 μL 4 *M* LiCl, and 300 μL 100% ethanol. Mix well and incubate at –20°C for 30 min.
8. Spin out precipitate at approx 10,000*g* at 4°C for 15 min in a microcentrifuge. Discard supernatant while ensuring that the riboprobe pellet remains in the microfuge tube. Wash pellet with 70% ethanol and air-dry.
9. Resuspend pellet in 100 μL Te_{50} to achieve a final concentration of ~0.1 μg/μL and store at –80°C. Use approx 0.5–1 μg of iboprobe per mL of hybridization mixture.
10. To more precisely estimate riboprobe concentration, remove 1- and 5-μL aliquots and electrophorese on a 1% agarose gel containing 0.5 μg/mL ethidium bromide. Next to the riboprobe, load digoxigenin-labeled RNA standards (100, 200, 400, and 800 ng). Compare band intensities to estimate the amount of riboprobe synthesized.

In situ mRNA detection

The following protocol has been derived from previously published methods and has been optimized for detection of *FLP* mRNA. Some steps may need to be altered to best detect other mRNAs of differing abundance and composition. The protocol is written for processing 24 slides (one rack) and can readily be expanded by processing additional slide-filled racks, each starting at approx 30-min intervals. Incubations

Table 16.1. In situ hybridization quick-reference summary of incubations

In situ hybridization	
PBSDEPC	5 min at RT
4% fix	15 min at RT
PBSDEPC	5 min at RT
Proteinase K	5 min at RT
Glycine/PBSDEPC	5 min at RT
PBSDEPC	5 min at RT
4% fix	15 min at RT
Acetic anhydride	10 min at RT
PBSDEPC	5 min at RT
Prehybridization	15 min at 65°C
Hybridization	O/N at 70°C
Prewash (Wash I)	dip at 65°C
Wash I-1	15 min at 65°C
Wash I-2	15 min at 65°C
Wash I-3	15 min at 65°C
Wash II: Wash I (1:1)	10 min at 65°C
Wash II	3 × 5 min at 65°C
RnaseA	30 min at 37°C
Wash II	5 min at 65°C
Wash III-1	15 min at 65°C
Wash III-2	15 min at 65°C
Wash III-3	15 min at 65°C
TBST-L	3 × 10 min at RT
Block	30 min at RT
Antibody solution	2 h at RT or O/N
TBST-L	Prewash dip
TBST-L	5 min at RT
TBST-L	4 × 15 min at RT
NTMT-L	3 × 5 min at RT
Alkaline phosphatase staining solution	30 min–4 d at RT
To mount	
PBS	10 min at RT
4% fix	15 min at RT
PBS	10 min at RT

are done in plastic staining dishes containing 250 mL of solution; and performed in slide mailers containing 15 mL of solution. To expedite this multistep protocol, label each plastic staining dish (total of 14) as either: PBS, 4% fix, proteinase K, glycine, acetic anhydride, wash I-1, wash I-2, wash I-3, wash II, wash II: wash I (1:1), Rnase A wash III-1, wash III-2, wash III-3, TBST-L, block, or NTMT-L.

Precautions should be taken to avoid any contamination of reagents or equipment with RNase. Sterile disposable plasticware is essentially free of RNase and should be used whenever possible; glassware should be baked. Posthybridization washes do not require these precautions, because hybrids formed in situ are resistant to RNase activity in the presence of high salt.

1. Bring the sealed box containing tissue sections from –80°C to room temperature prior to opening; this prevents unwanted water condensation on the tissue sections.
2. Load slides into a rack and place into PBS_{DEPC}, in the "PBS"-labeled staining dish. Incubate 5 min at room temperature.
3. Transfer slides into 4% fix solution in the "4% fix"-labeled staining dish. Incubate 15 min at room temperature, after which discard used fix solution as hazardous waste.
4. Transfer slides to fresh PBS_{DEPC} in the "PBS"-labeled staining dish. Incubate 5 min at room temperature.
5. Transfer slides into 1 μg/mL proteinase K in PBS_{DEPC} (prepared from freshly thawed 10 mg/mL proteinase K stock) in the "*proteinase K*"-labeled staining dish. Incubate 5 min at room temperature.
6. To quench proteinase K activity, transfer slides into glycine/PBS_{DEPC} solution in the "*glycine*"-labeled staining dish. Incubate 5 min at room temperature.
7. Transfer slides to fresh PBS_{DEPC} in the "PBS"-labeled staining dish. Incubate 5 min at room temperature.
8. Refix tissue by transferring slides to fresh 4% fix solution. Incubate 15 min at room temperature.
9. Transfer slides into 0.25% acetic anhydride in 0.1 *M* triethanolamine in the "*acetic anhydride*"-labeled staining dish. The acetic anhydride should be added to the triethanolamine solution just prior to use. Incubate 10 min at room temperature. Discard acetic anhydride as hazardous waste.
10. Transfer slides to fresh PBS_{DEPC} in the "PBS"-labeled staining dish. Incubate 5 min at room temperature.

11. Prehybridization of tissue sections in slide mailers: transfer slides into slide mailers containing prewarmed (70°C) hybridization buffer (slide mailers hold 5 slides and 15 mL of solution). Incubate for 15 min at 70°C.
12. Hybridization: transfer slides into slide mailers containing prewarmed (70°C) hybridization solution with 0.5–1 μg/mL digoxigenin-labeled RNA probe. Seal slide mailers tightly and incubate overnight at 70°C.
13. Transfer slides from mailers to a rack and dip 3 times in prewarmed (65°C) wash buffer I in "wash-I-1" staining dish. Discard used wash buffer as hazardous waste and replace with fresh prewarmed (65°C) wash I-1. Incubate 15 min at 65°C. Discard used wash buffer as hazardous waste.
14. Transfer slides to fresh prewarmed (65°C) wash buffer I in the "wash I-2"-labeled staining dish. Incubate 15 min at 65°C.
15. Transfer slides to fresh prewarmed (65°C) wash I in the "wash I-3"-labeled staining dish. Incubate 15 min at 65°C.
16. Transfer slides to fresh prewarmed (65°) Wash II: Wash I (1:1 ratio) in the "Wash II-Wash I"-labeled staining dish. Incubate 10 min at 65°C.
17. Transfer slides to fresh prewarmed (65°C) Wash II in the "Wash II"-labeled staining dish. Incubate 5 min at 65°C. Repeate twice using fresh Wash II.
18. Transfer slides to fresh Rnase A wash in the "Rnase"-labeled staining dish. Incubate 30 min at 37°C.
19. Transfer slides to fresh prewarmed (65°C) Wash II in the "Wash II"-labeled staining dish. Incubate 5 min at 65°C.
20. Transfer slides to fresh prewarmed (65°C) wash III in the "wash III-1"-labeled staining dish. Incubate 15 min at 65°C.
21. Transfer slides to fresh prewarmed (65°C) wash buffer III in the "wash III-2"-labeled staining dish. Incubate 15 min at 65°C.
22. Transfer slides to fresh prewarmed (65°C) wash buffer III in the "wash III-3"-labeled staining dish. Incubate 15 min at 65°C.
23. Transfer slides to TBST-L in the "TBST-L"-labeled staining dish. Incubate 10 min at room temperature. Repeat two additional times using fresh TBST-L.
24. Transfer slides to TBST-L containing 10% heat-inactivated lamb serum in the "block"-labeled staining dish. Incubate 30 min at room temperature.

25. Add 15 mL of antibody (αdigAP) solution into each of an appropriate number of slide mailers. Transfer slides from step 20 into each mailer. Incubate at room temperature for 2 h or overnight at 4°C.
26. Transfer slides into rack and dip in fresh TBST-L in the "TBST-L"-labeled staining dish.
27. Replace with fresh TBST-L and incubate 5 min at room temperature.
28. Replace with fresh TBST-L and incubate 15 min at room temperature. Repeat 3 additional times using fresh TBST-L.
29. Transfer slides into fresh NTMT-L in the "NTMT-L"-labeled staining dish. Incubate 5 min at room temperature. Repeat two additional times.
30. Transfer slides into slide mailers containing freshly prepared alkaline phosphatase staining solution. Keep in the dark as much as possible by wrapping containers in aluminum foil. Incubate at room temperature until desired degree of color development has occurred (a few hours to a few days depending on the abundance of the mRNA under detection). Change alkaline phosphatase staining solution after the first 8 h and then every 24 h thereafter to avoid unwanted precipitate.
31. When the color has developed to the desired extent, wash twice in NTMT for 5 min each.
32. Transfer slides to rack and place in PBS for 10 min at room temperature.
33. Transfer slides to 4% fix solution and incubate 15 min at room temperature. Discard fix solution as hazardous waste.
34. Transfer slides to fresh PBS and incubate 10 min at room temperature.
35. Remove slides one at a time, dabbing sides with a Kimwipe to remove excess PBS. Mount slides in aqueous media such as prewarmed (37°C) gelvatol or Vectashield. Apply approximately three drops of the prewarmed mounting media per slide. Gently place glass coverslip on top, being careful not to introduce bubbles. Press down on the coverslip lightly to remove excess gelvatol and small air bubbles.
36. Air-dry slides until gelvatol is firm. Wipe slides clean of excess mounting media and examine by light microscopy. Slides can be stored at –80°C to prevent any unwanted additional color development.

Preparation of gelvatol aqueous mounting media

The following protocol, modified from Rodriguez and Deinhardt makes approximately 1.25 L of gelvatol.

1. Adjust a 10 m*M* KH_2PO_4 solution to pH 7.2 by adding 10 m*M* Na_2HPO_4. Prepare 500 mL.
2. Place a 6-inch stir bar into a 2-L beaker. Add 500 mL of 10 m*M* KH_2PO_4/Na_2HPO_4 solution and 4.1 g NaCl (0.14 *M* NaCl final concentration). Stir at room temperature.
3. Slowly add 125 g of polyvinyl alcohol resin. Stir at room temperature for 3 h.
4. Remove stir bar and microwave mixture until it reaches a boil (about 4 min).
5. Replace stir bar and mix at room temperature overnight.
6. While stirring, gradually add 525 mL glycerol. The solution will become very viscous. Continue stirring mixture at room temperature for 3 d.
7. To remove undissolved particles, aliquot into 250-mL polypropylene centrifuge bottles and spin in Sorvoll GSA rotor at 16,000*g* at room temperature for 30 min.
8. Pool gelvatol supernatants and readjust the pH to between 6.0 and 7.0 using either 10 m*M* KH_2PO_4 or 10 m*M* Na_2HPO_4 as needed
9. Aliquot gelvatol into polypropylene tubes (50 mL size) and store at 4°C.

Profiling Recombinase Expression by Crossing Recombinase Mice to a lacZ Indicator Strain

Detection of β-gal activity on tissue cryosections by X-gal histochemistry

Process staged embryos and adult tissues isolated from recombinase-indicator double transgenics with the following modifications:

In step 2, gently rock embryos or adult tissues in 0.2% fix solution (rather then 4% fix solution) at 4°C for 1 h to overnight. Fixation times will vary with the size and nature of the tissue. Early- to mid-gestation mouse embryos should be exposed to the fixative for shorter time periods (e.g., 1 h), while late-stage embryos and adult tissues should be exposed for longer time periods (e.g., 4 h to overnight). In general, shorter fixation times are preferable, as X-gal staining may be decreased by lengthy fixation.

Collecting thicker cryosections (60 μm as opposed to 10 μm) allows more tissue to be analyzed per given section. All histochemistry is performed in glass staining dishes: one for the fixative, one for the reaction buffer, and one for the X-gal staining solution. The following protocol is for 30 slides (one rack) and can readily be expanded by processing additional racks at approximately 10-min intervals.

1. Bring a sealed box containing tissue sections from –80°C to room temperature prior to opening; this prevents unwanted water condensation on the tissue sections.
2. Load slides into rack and place into cold 2% fix solution for 10 min on ice. Do not discard used fix solution; store at 4°C until step 6.
3. Transfer slides into cold rinse solution for 10 min on ice.
4. Transfer slides to X-gal staining solution. Protect from light by wrapping staining dish with aluminum foil and incubate at 37°C until color has developed to the desired extent (typically 2–48 h). Optional: place slides in X-gal staining solution at 4°C for an additional 24 h to maximize precipitation of the blue X-gal product.
5. Place slides in PBS-M for 5 min at room temperature to remove any residual X-gal staining solution.
6. Refix slides by placing in the 2% fix solution from step 2 for 5 min at room temperature. Discard fix solution as hazardous waste.
7. Transfer slides to PBS-M to remove any residual fix solution. Coverslip tissue using aqueous mounting media such as prewarmed gelvatol (37°C) or Vectashield.
8. Air-dry until gelvatol is firm. Wipe slides clean of excess mounting media and examine by light microscopy.

As an alternative approach to X-gal detection of β-gal activity, antibody detection of β-gal can also be employed. We have had success using a rabbit polyclonal antibody to β-gal at a 1:500 dilution. As a secondary antibody, we use either *lissamine rhodamine B sulfonyl chloride* (LRSC)-conjugated goat anti-rabbit IgG at 1:200 dilution or *fluorescein isothiocyanate* (FITC)-conjugated goat anti-rabbit IgG at 1:200.

Determining the Utility of an Indicator Strain

Before using an indicator strain for lineage studies, it is important to determine the expression profile of the activated reporter molecule. This profile determines the range of cell types that can be mapped reliably by the indicator strain, establishing its overall utility. To unambiguously determine the scope of the potential indicator profile, it is necessary to generate a derivative mouse strain, in which every cell harbors a recombined copy of the indicator transgene, while maintaining the chromosomal environment of the starting (unrecombined) transgene. As described below, this derivative strain is generated by germ line transmission of the recombined indicator transgene.

The following protocols were worked out for assessing our *FRT*-disrupted *lacZ* indicator strain, *Hmgcr*::FRTZ, and involve using the deleter strain *hACTB*::FLPe.9205 and standard X-gal detection assays. The same general strategy applies to either *FRT*- or *loxP*-disrupted indicator strains harboring other reporter genes such as PLAP or EGFPF.

Germ line transmission of an activated indicator transgene

1. Cross deleter (e.g., *hACTB*::FLPe.9205) and indicator (e.g., *Hmgcr*::FRTZ) mice.
2. Identify double transgenic (*hACTB*::*FLPe.9205*/ *Hmgcr*::*FRTZ*) F1 progeny by PCR-genotyping DNA isolated from tail biopsies. These mice will harbor recombined activated indicator transgenes (*Hmgcr*::*FRTZ-A*) in their germ cells.
3. To generate the control derivative mouse strain (*Hmgcr*::*FRTZ-A*), in which every cell harbors a recombined copy of the indicator transgene, outcross double transgenic F1 mice to nontransgenic wild-type mice. By PCR-genotyping DNA isolated from tail biopsies, identify progeny harboring the recombined indicator transgene, *Hmgcr*::*FRTZ-A*, in the absence of the recombinase transgene. These *Hmgcr*::*FRTZ-A* mice constitute the derivative control strain.

DNA extraction from tail biopsies

This protocol yields high-quality genomic DNA, suitable for both PCR and Southern hybridization analyses. All solutions should be prepared using PCR-grade reagents and dispensed using aerosol-resistant pipet tips to minimize unwanted contamination of samples with other plasmid or genomic DNA.

1. To a 0.5-cm biopsy add 500 μL STES buffer and 10 μL proteinase K (from a freshly thawed 10 mg/mL stock). Mix well and incubate at 55°C overnight or until tissue is digested.
2. Add 500 μL phenol:chloroform solution to each sample. Mix well (do not vortex samples from this step onward, as this may sheer the genomic DNA) and spin for 5 min at 10,000*g* at room temperature in a microcentrifuge. Transfer the top aqueous layer

containing the genomic DNA to a fresh microfuge tube. Discard the bottom layer as organic waste.

3. Add 500 μL chloroform:isoamyl alcohol (24:1). Mix well and spin for 5 min at 10,000*g* at room temperature in a microcentrifuge. Transfer the top aqueous layer containing the genomic DNA to a fresh microfuge tube. Discard the bottom layer as organic waste.
4. Add 40 μL 3 *M* NaOAc and 1 mL ice cold 100% ethanol. Mix well. Strands of genomic DNA will immediately precipitate. Scoop-out the DNA on the sealed and hooked-end of a glass capillary tube. Briefly air-dry 1 to 2 min.
5. Resuspend genomic DNA in 70 μL of TE (the final DNA concentration will be roughly 0.5 μg/μL). If the DNA does not readily release from the glass capillary, break off DNA-containing glass tip directly into the TE. Store genomic DNA at 4°C.
6. Use approx 0.5–1 μL of tail DNA per 25 μL PCR.

Identifying recombinase and indicator transgenics by PCR-genotyping DNA isolated from tail biopsies

To distinguish the indicator transgene, *Hmgcr::FRTZ*, from the recombined derivative, *Hmgcr::FRTZ-A*, we typically analyze tail DNA using three distinct PCR amplifications:

Primer set 219/218 amplifies a 1.6-kb product for the *Hmgcr::FRTZ* transgene and a 0.45-kb product for the recombined *Hmgcr::FRTZ-A* transgene when used under the following cycle conditions.

1. Combine 12.5 μL sterile distilled water, 2.5 μL PCR buffer (10X), 2 μL $MgCl_2$ (25 m*M*), 3 μL primer 218 (20 μ*M* working stock), 3 μL primer 219 (20 μ*M* working stock), 0.5 μL dNTP stock (50×), 0.5 μL *Taq* DNA polymerase (5 U/μL), and 1 μL of tail DNA (0.5 μg/μL).
2. Amplify by cycling at 94°C for 5 min, 58°C for 1 min, and 72°C for 3 min for 1 cycle; 94°C for 1 min, 58°C for 1 min, and 72°C for 3 min. for 39 cycles; 72°C for 10 min (final extension).
3. Load entire reaction on a 1% agarose gel with 0.5 μm/mL ethidium bromide.

Primer set 90/218 amplifies a 0.85-kb product for the *Hmgcr::FRTZ* transgene when used under the following cycle conditions. These primers do not amplify the recombined *Hmgcr::FRTZ-A* transgene, as the 90 primer site has been removed by FLP-mediated excisional recombination.

1. Combine 12.5 μL sterile distilled water, 2.5 μL PCR buffer (10X), 2 μL $MgCl_2$ (25 m*M*), 3 μL primer 218 (20 μ*M* working stock), 3

μL primer 90 (20 μ*M* working stock), 0.5 μL dNTP stock (50X) 0.5 μL *Taq* DNA polymerase (5 U/μL), and 1 μL of tail DNA (0.5 μg/μL).

2. Amplify by cycling at 94°C for 5 min, 60°C for 1 min, and 72°C for 1 min. for 1 cycle; 90°C for 1 min, 60°C for 1 min, and 72°C for 1 min, for 39 cycles; 72°C for 10 min (final extension).
3. Load entire reaction on a 1 to 2% agarose gel with 0.5 μg/mL ethidium bromide.

Primer set 204/205 amplifies a 0.6-kb fragment from the *lacZ* sequence shared by both the *Hmgcr*::FRTZ and *Hmgcr*::FRTZ-A transgenes and serves to simply detect the presence of the indicator transgene regardless of configuration. This amplification provides a useful control for interpreting the above PCR results.

1. Combine 12.8 μL sterile distilled water, 2.5 μL PCR buffer (10X), 1.7 μL $MgCl_2$ (25 m*M*), 3 μL primer 204 (20 μ*M* working stock), 3 μL primer 205 (20 μ*M* working stock), 0.5 μL dNTP stock (50X), 0.5 μL *Taq* DNA polymerase (5 U/μL), and 1 μL of tail DNA (0.5 μg/μL).
2. Amplify by cycling at 94°C for 5 min, 60°C for 1 min, and 72°C for 1 min, for 1 cycle; 90°C for 1 min, 60°C for 1 min, and 72°C for 1 min, for 39 cycles; 72°C for 10 min (final extension).
3. Load entire reaction on a 1 to 2% agarose gel with 0.5 μg/mL ethidium bromide.

Primer set 222/223 amplifies a 0.7-kb product for the FLP recombinase transgene when used under the following cycle conditions.

1. Combine 12.5 μL sterile distilled water, 2.5 μL PCR buffer (10X), 2.0 μL $MgCl_2$ (25 m*M*), 3 μL primer 222 (20 μ*M* working stock), 3 μL primer 223 (20 μ*M* working stock), 0.5 μL dNTP stock (50X), 0.5 μL *Taq* DNA polymerase (5 U/μL), and 1 μL of tail DNA (0.5 μg/μL).
2. Amplify by cycling at 94°C for 5 min, 65°C for 1 min, and 72°C for 1 min. for 1 cycle; 90°C for 1 min, 65° for 1 min, and 72°C for 1 min, for 39 cycles; 72°C for 10 min (final extension).
3. Load entire reaction on a 1 to 2% agarose gel with 0.5 μg/mL ethidium bromide.

Determining the β-gal profile of an activated indicator strain by X-gal histochemistry on tissue cryosections

Process staged embryos and adult tissues isolated from the activated (derivative) indicator strain as described above. The resultant β-gal profile determines the range of cell types that can be mapped reliably using the parental indicator strain, establishing its overall utility.

Generation of a Recombinase-Based Fate Map

Identifying recombinase–indicator double transgenic embryos by PCR-genotyping DNA from yolk sac membranes

Embryos can be genotyped by PCR amplification of transgene DNA directly from a yolk sac lysate. This is especially useful for early stage embyos (<9.5 dpc), in which the yolk sac is still quite small and the yield of DNA by conventional extraction methods is poor. To minimize contamination of samples with unwanted plasmid or genomic DNA, all solutions should be prepared using PCR-grade reagents, dispensed using aerosol-resistant tips, and stored or incubated in autoclaved microfuge tubes.

1. Isolate yolk sac membranes from each embryo individually, rinse in PBS, and place in microfuge tube on dry ice until all dissections are complete. Store at –20°C.
2. Without thawing the yolk sac tissue, add 100 μL of yolk sac lysis buffer and 1 μL proteinase K to each sample. Mix well and incubate at 55°C for 4 h to overnight or until completely digested.
3. Heat-inactivate the proteinase K by placing samples at 95°C for 15 min (use microfuge tube covers to prevent splash contamination). Store at –20°C
4. Use 1 to 2 μL/25 μL PCR genotyping reaction.

Detection of progenitor cells and their descendants by staining for heat-resistant alkaline phosphatase activity

Many different indicator mouse strains exist for use with Cre and FLP transgenics, each harboring different reporter molecules (e.g., transgenes encoding either β-gal, PLAP, or EGFP. We present a protocol previously for β-gal detection, and here we present a method for PLAP detection.

Prepare tissue sections as described above with the following two modifications and then proceed with step 1 below:

A. Replace PBS with PBS-M.
B. Fixation (in 4% fix solution) is typically reduced to 2–4 h at 4°C, although extended incubations may help reduce unwanted endogenous alkaline phosphatase activity.

1. Bring sealed box containing tissue sections from –80°C to room temperature prior to opening; this prevents unwanted water condensation on the tissue section.

2. Place slides in rack and submerge in 4% fix solution for 10 min on ice, after which discard fix solution as hazardous waste.
3. Transfer slides to ice-cold PBS-M for 10 min on ice to eliminate residual fixative. Repeat twice.
4. Transfer slides to prewarmed (65°C) PBS-M and incubate at 65°C for 30–90 min. This step will inactivate endogenous (heat-sensitive) alkaline phosphatase activity that would otherwise result in confounding background activity. The duration of this step will vary for different tissues depending on the amount of endogenous alkaline phosphatase activity.
5. Transfer slides to detection buffer for 30 min at room temperature.
6. Transfer slides to freshly prepared detection buffer containing 0.1 mg/mL BCIP, 1 mg/mL NBT, and 0.5 m*M* levamisole. Protect samples from light by wrapping the staining dish with aluminum foil. Incubate at room temperature until color has developed to the desired extent (typically 1–48 h). Replace staining solution after the first 8 h and then every 24 h thereafter to prevent any unwanted precipitate from forming.
7. Transfer slides to EDTA wash solution for 10 min at room temperature. Repeat this step two additional times.
8. Coverslip slides as described above, using aqueous mounting media such as prewarmed (37°C) gelvatol or Vectashield. EDTA (20 m*M*) may be included in the mounting media to inhibit further development of the alkaline phosphatase reaction.
9. Air-dry slides in a dark place until gelvatol is firm. Wipe slides clean of excess mounting media and examine by light microscopy. Slides may be stored at –80°C to prevent further development of the alkaline phosphatase reaction.

Notes

1. To inactivate any phosphatases present in the sheep serum that could cause nonspecific background, incubate at 70°C for 30 min, swirling every few minutes to prevent hardening. Aliquot and store at –80°C.
2. While most X-gal staining reactions use a reaction buffer comprised of 100 m*M* Na-phosphate, pH 7.4, 2 m*M* $MgCl_2$, 0.01% Na-deoxycholate, 0.02% NP40, we have found that increasing the concentration of Na-deoxycholate and NP40 10-fold (to 0.1 and 0.2%, respectively) can improve the β-gal detection without compromising tissue architecture.

3. To ensure that the aliquot of riboprobe is not degraded during the electrophoresis, precautions should be taken to avoid any contamination of the electrophoresis buffer and apparatus with RNases.

   Electrophoresis tanks suspected of contamination should be cleaned with a detergent solution, rinsed well in water, dried with ethanol, and then filled with a solution of 3% H_2O_2. After 10 min at room temperature, the electrophoresis tank should be rinsed thoroughly with DEPC-water.
4. We have observed a significant reduction in nonspecific background staining when riboprobes are precipitated twice with LiCl and 100% ethanol.
5. Riboprobe penetration is increased by treating tissue sections with low concentrations of proteinase K (1–3 μg/mL). It is useful to vary both the concentration and duration of this treatment in order to determine the conditions that yield the best hybridization signal.
6. Prehybridization and hybridization solutions (containing probe) can be stored in slide mailers at –80°C and reused several times. With some probes, the degree of nonspecific background staining will decrease as the hybridization solution is reused.
7. While recommended, levamisole can be eliminated here and in all subsequent steps, without deletions increases in nonspecific alkaline phosphatase activity.
8. To prevent crystal deposits on tissue sections, change the alkaline phosphatase staining solution after the first 8 h and then every 24 h thereafter.

   To ensure that even small crystals are removed, dip slides into fresh NTMT-L and wash the slide mailer with NTMT-L between changes of the alkaline phosphatase staining solution. These steps will minimize nonspecific background staining.
9. Reducing the viscosity of gelvatol by prewarming to 37°C will facilitate mounting applications.
10. Some sediment will collect on tube bottoms after gelvatol centrifugation. Carefully aliquot the gelvatol into 50-mL polypropylene tubes, leaving sediment behind. Store indefinitely at 4°C.
11. While β-gal encoded by the bacterial *lacZ* gene is usually quite stable, enzyme activity can be diminished by over-fixation.

Protocols show a range of fixatives including 0.2, 2.0, and 4.0% paraformaldehyde in PBS, as well as 0.2 and 2.0% glutaraldehyde. We have seen consistently strong X-gal precipitate with minimal to no endogenous background β-gal activity when tissue is fixed in 0.2% paraformaldehyde-PIPES buffer.

12. Background staining resulting from endogenous alkaline phosphatase activity can be reduced effectively by heat (the exogenous PLAP encoded by the transgene is heat-resistant). In addition to heating tissue sections, we recommend heating whole tissues to 65°C prior to incubating in the sucrose solution.

17

CARCINOMAL STEM CELLS

Embryonal carcinoma (EC) cells are the stem cells of teratocarcinomas. These tumors, which present a caricature of embryogenesis, have fascinated pathologists for many hundreds of years. Indeed, Wheeler (1983), in his excellent review of the history of these tumors, mentions that the earliest reference to what is evidently a teratoma, a benign form, is found on clay tablets from the Chaldean Royal Library of Nineveh dating from 600 to 900 B.C. Among these tablets, devoted to methods of predicting the future, one sign is described, *"When a woman gives birth to an infant that has three feet, two in their normal position (attached to the body), and the third between them, there will be great prosperity in the land"*. Such an optimistic forecast perhaps foretells the value that modern biologists have found in EC cells as tools for the study of cell differentiation in embryonic development and cancer.

Teratomas and teratocarcinomas occur in a range of manifestations. The most common are ovarian dermoid cysts. These form from oocytes that are parthogenetically activated and begin development but eventually become disorganized, giving rise to a teratoma containing a haphazard array of embryonic tissues. Such tumors are generally benign, but they can grow to very large sizes. Teratomas also occur, although more rarely, in other sites, including the base of the spine in newborn infants, and it is a tumor of this type that is evidently the subject of the Chaldean writer.

Of greater clinical significance to cancer biologists are the teratocarcinomas of the testis. These tumors are a subgroup of cancers that, because of their histological complexity, site of origin, and apparent caricature of embryogenesis, have generally been thought to

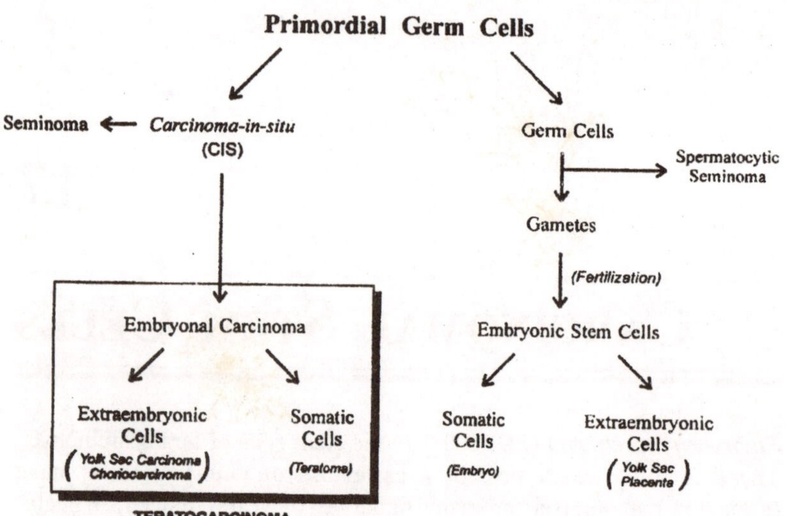

Fig. 17.1. Schematic representation of a common view of germ cell tumor histogenesis and its parallel with embryogenesis.

arise from germ cells. *Germ cell tumors* (GCT) account for almost all testicular cancers and are always malignant. They occur typically in young post-pubertal men, with a peak incidence in the third decade of life. Although they are rare, their incidence has increased markedly over the past 50 years, and their medical significance also reflects the young age of the patients. On the other hand, GCT are among the most treatable cancers since the advent of *cis*-platinum-based therapy in the 1970s.

Testicular GCT are usually divided into seminomas and non-seminomas. Seminomas are, histologically, relatively homogeneous tumors consisting of cells that resemble primordial germ cells, in contrast to the non-seminomas, which are histologically heterogeneous and may, confusingly, also contain elements of seminoma. More striking, however, is the frequent occurrence of somatic tissues such as nerve, bone, or muscle in non-seminomatous GCT as well as a histologically undifferentiated cell type known as *embryonal carcinoma* (EC). EC cells are now known to be the key malignant stem cells of these tumors, capable of differentiating into the wide range of somatic cells that comprise the teratomatous elements. The term teratocarcinoma is used to refer to GCT containing both embryonal carcinoma and teratoma components. Other elements of non-seminomatous GCT include highly malignant cells corresponding to the yolk sac (yolk sac carcinoma) and trophoblast (choriocarcinoma). Embryonal carcinoma, yolk sac carcinoma, and, rarely, choriocarcinoma may all occur alone, or combined with one another, as well as with teratoma elements.

TERATOCARCINOMAS IN THE LABORATORY MOUSE

Pathologists have long held the notion that the formation of teratocarcinomas in some way reflects the processes of embryonic development, and that EC cells perhaps resemble undifferentiated stem cells from the early embryo. However, the rarity of the human tumors and the sporadic nature of similar tumors in animals limited their study until Stevens and Little (1954) discovered that about 1% of male mice of strain 129 develop testicular teratomas. This provided the starting point for a detailed study that led through the characterization of murine EC cells to the eventual development of *embryonic stem* (ES) cell lines from mouse embryos in 1981.

The teratomas of 129 male mice can be observed by 15 days of embryonic development, as structures described as embryoid bodies within the seminiferous tubules of the fetal gonad. Stevens (1964) estimated that such teratomas originated between 11 and 12 days of development. He also showed that the incidence of these tumors can be increased by transplanting the genital ridges of early embryos to the testis capsule of adult mice. Furthermore, although no other strain of mouse regularly produces testicular teratomas spontaneously, transplantation of the genital ridges from embryos of the A/He strain also resulted in the formation of teratomas. There is a very narrow window in development when the genital ridges are susceptible to such manipulation. Primordial germ cells, which are first identifiable in the extraembryonic yolk sac, migrate through the hindgut to the genital ridges, arriving at about 11 days of development. The greatest incidence of induced teratomas occurred after transplanting genital ridges from 12 to 12.5 day embryos, and fell dramatically when genital ridges from older embryos were transplanted. An upper limit of about 13.5 days implies some changes in the germ cells after arriving in the genital ridge rendering them resistant to transformation. This could be associated with their entering mitotic arrest soon after their arrival in the genital ridge.

Confirmation of the germ cell origin of the spontaneous and experimental testicular teratomas came from studies of mice homozygous for mutations at the *Steel (Sl)* locus. The *Sl* locus encodes *stem cell factor* (SCF), a growth factor required for survival of the various stem cells including melanocytes, hematopoietic stem cells, and primordial germ cells. Viable *Sl/Sl* homozygotes are infertile since the primordial germ cells do not survive migration. After crossing *Sl* on to the 129 background, Stevens found that the genital ridges from *Sl/Sl* mice did not yield teratomas, confirming the origin of those tumors from primordial germ cells. Curiously, however, males heterozygous for *Steel* (*Sl*/+) exhibit an increased incidence of teratomas. The reason for this has never been elucidated. However, the receptor for SCF is encoded by the *W* locus, which encodes the c-*kit* oncogene. One wonders whether, in the presence of suboptimal SCF levels that might occur in *Sl*/+ mice, expression of *W*/c-*kit* might be up-regulated, with subsequent effects on germ cell proliferation.

Subsequently, Stevens identified another strain of laboratory mouse, the LT strain, that exhibits ovarian teratomas at high frequency. In this case, it is evident that oocytes within the ovary frequently undergo parthogenetic activation and initiate embryonic development that becomes progressively disorganized, forming tumors that appear to be the counterparts of benign ovarian cysts in humans.

Despite its long history, the genetic basis for the susceptibility of 129 mice to teratoma development remains unclear. A mutation, *Ter,* occurring in a subline 129/terSv, causes teratomas to develop with a much higher frequency than in the original 129/Sv line. Backcross analysis of 129/ter mice has shown that the *Ter* gene is located on chromosome 18 within 0.6 cM of *Grl1*, but its molecular identity is currently unknown. Strangely, *Ter/Ter* mice exhibit a reduced number of germ cells. Thus, counterintuitively, the mutations at two loci (*Sl* and *Ter*) known to increase susceptibility to germ cell tumors also appear to reduce germ cell viability and/or proliferation. To date, no mutations causing hyperproliferation of mouse germ cells are known. *Ter* is a modifying gene that is neither sufficient nor necessary for GCT formation and does not cause tumors on backgrounds other than 129. Evidently, susceptibility to teratomas is polygenic and is presumably influenced by environmental or stochastic factors, since penetrance is low.

The testicular tumors formed spontaneously in 129 mice or following genital ridge transplantation can be divided into those that can be retransplanted into other syngeneic mice, and those that cannot. The retransplantable tumors, termed teratocarcinomas, contain groups of cells identified as EC cells by comparison with the EC cells recognized in human teratocarcinomas. Kleinsmith and Pierce (1964) provided the first evidence that these EC cells are the malignant stem cells of teratocarcinomas: Transplantation of single EC cells from one tumor to a new host proved sufficient to result in the formation of a new teratocarcinoma containing the full range of differentiated elements seen in the parental tumor. Thus, not only are the EC cells the malignant stem cells of the tumor, they are also the repository of the pluripotent nature of these tumors. Non-retransplantable teratomas did not contain EC cells.

Although testicular teratomas can only be induced in a limited number of strains, teratomas can also be formed from many other strains of mice if rather earlier embryos, notably at about 7 days of development at the egg cylinder stage, are transplanted to ectopic sites. As in the ovarian parthogenetic tumors, these embryos continue to grow, become disorganized, and form teratomas or, in some cases, retransplantable teratocarcinomas containing EC cells. Whether retransplantable teratocarcinomas are formed depends on the host strain into which the embryo is transplanted, and not on the genotype of the embryo itself. Solter and Damjanov (1979) also showed that the outcome of these experiments is dependent on the immune status of the host. Thus, whereas C3H embryos transplanted to C3H hosts typically form teratocarcinomas, teratomas were mostly formed if the C3H host had been rendered immunodeficient by neonatal thymectomy and sublethal irradiation. This phenomenon remains unexplained. One possibility is that the persistence of EC cells is fostered by a factor produced by the immune system; alternatively, it was suggested that higher levels of natural killer cells in mice deficient for T-lymphocytes might be detrimental to EC cell survival.

Murine Embryonal Carcinoma Cell Lines

Cell cultures of EC cells derived from murine teratocarcinomas were first reported by Kahn and Ephrussi in 1970. Subsequently, several groups established murine EC cell lines, although with some difference in emphasis and techniques. Many of these lines can be maintained as undifferentiated EC cells in vitro but are able to form teratocarcinomas when transplanted back to an appropriate mouse host. They are said to be "*pluripotent*"; however, some EC cell lines evidently lose this ability to differentiate and are termed "*nullipotent*".

Some of the pluripotent lines also differentiate in culture, but the circumstances vary between lines and between laboratories. For example, among EC lines derived by Jakob and her colleagues, PCC3 EC cells remain undifferentiated if kept proliferating in subconfluent cultures. However, if allowed to reach confluence and maintained in that condition for several days, these cells differentiate spontaneously, yielding a variety of cell types, including nerve and muscle. Maintenance of other pluripotent EC lines in an undifferentiated state, however, was found to depend on culture on feeder layers of transformed mouse fibroblasts. The STO line of transformed mouse fibroblasts rapidly became the standard feeder widely used for such cells. In this case, differentiation can be induced by removing the EC cells from the feeder cells. It was found particularly advantageous to force the cells to grow in suspension without attachment to a substrate by culture in bacteriological petri dishes. Under these conditions, the cells aggregate and form structures known as embryoid bodies, in which an inner core of EC cells is surrounded by a layer of cells resembling the visceral endoderm of an early mouse conceptus. Gradually these embryoid bodies become histologically complex and cystic, and a wide variety of differentiated cells grow out when the embryoid bodies are plated on a substrate that permits attachment.

Relationship to the Early Embryo

It had long been hypothesized from studies of human teratocarcinomas that EC cells might resemble stem cells from the early embryo. During the 1960s and early 1970s, research in a different sphere, in the biology of the immune system and the genetics of transplantation, focused attention on the role of the cell surface in regulating cell behavior. The area of immunogenetics that led to the development of techniques for producing antisera to specific cell-surface antigens had also led to the notion that these antigens might play a role in the regulation of cell differentiation. The coincidence of these ideas with the availability of cultured EC cells quickly led to experiments to identify specific *embryonic* cell-surface antigens. One approach, adopted by Artzt et al. (1973), was to immunize adult 129 mice with EC cells that were also of 129 origin. Prevailing concepts of immunology suggested that, because of tolerance, the only antigens to which the adult 129 mice would form antibodies would be those expressed only on the "*embryonic*" EC cells and not on any adult cells. Anti-F9 sera produced in 129 mice did indeed detect, in cytotoxicity and immunofluorescence assays, an antigen expressed by EC cells but not by a range of other more differentiated cells. This same antigen was also expressed by cells of the *inner cell mass* (ICM) of the embryonic blastocyst. This relationship, as well as similar expression of other markers like alkaline phosphatase, taken together with a comparable capacity for differentiation, led to the notion that EC cells are a malignant counterpart of ICM embryonic cells.

The nature of the F9 antigen became controversial because of suggestions that it was related to key cell-surface molecules encoded by the *T-locus*, the complex genetics of which was then poorly understood. The difficulties of working with polyclonal antisera meant that the precise nature of this anti-F9 activity was never resolved, although a link to the *T-locus* became progressively unlikely. Nevertheless, the hypothesis provided a strong stimulus to mouse developmental biology both in ideas and technology, just as approaches based on molecular genetics and monoclonal antibodies became available. The advent of monoclonal antibodies led to the production of reagents that identify antigens with similar characteristics to the F9 antigen. Perhaps the most notable of these monoclonal antibody-defined EC cell antigens is *stage-specific embryonic antigen*-1 (SSEA1), which was subsequently shown to involve an oligosaccharide epitope known as the Lewis-X (Le^x) antigen. SSEA1 is commonly expressed by murine EC cells and embryonic ICM cells, and it seems likely that the polyclonal 129 anti-F9 serum contained significant levels of antibodies recognizing this epitope.

The proposition that EC cells are indeed the counterpart of ICM cells was soon tested directly by transferring small numbers of EC cells to blastocysts, which were subsequently re-implanted into pseudopregnant females. In the first experiments, reported by Brinster (1974), EC cells derived from an agouti mouse were injected into a blastocyst from an albino strain: A mouse was born that had patches of agouti fur as well as albino fur, indicating that some of its fur derived from the implanted EC cells. Others subsequently made more detailed experiments and demonstrated that the implanted EC cells in some cases contributed to almost all tissues of the host embryo, and in rare cases it was reported that the germ line of the resultant chimeras was derived from the EC component. Not only did the experiments serve to demonstrate the close relationship of EC cells to the ICM, but they also suggested that their malignant character, seen in their ability to form retransplantable teratocarcinomas, was suppressed in the embryo. Indeed, as long suggested from the human studies, their differentiated derivatives were themselves generally not malignant, supporting the ideas of Pierce that the formation of cancers, and not only of teratocarcinomas, is associated with defects in the normal mechanisms of stem cell differentiation. In fact, it became evident that suppression of malignancy is not always complete, perhaps reflecting subtle genetic abnormalities in some of the EC cell lines. Nevertheless, the concept of suppression of malignancy survives even though there are many exceptions, and oncologists treating human GCT are generally wary of persistent teratoma lesions in successfully treated patients, because of their potential for regaining malignancy.

Although murine teratocarcinomas were initially described as being euploid, it gradually became evident that small karyotypic changes frequently occurred. It seems likely that continued growth as a tumor, or extended growth in culture, leads to the accumulation of mutations that promote a transformed phenotype. That some EC cells lose their ability to differentiate could well reflect selection for mutations that interfere with differentiation. Indeed, cell hybrids formed between EC cells and somatic cells, notably thymocytes, often continue to exhibit an EC phenotype with a greater capacity for differentiation than the parental EC cells. This suggests that wild-type alleles derived from the somatic cell parent complement "*anti-differentiation*" mutations derived from the parental EC cells.

Differentiation of EC Cells in Culture

Initial studies of EC cell differentiation focused on their ability to differentiate spontaneously under a variety of circumstances, most notably after producing embryoid bodies when cultured in suspension. However, although this differentiation to a wide range of cell types is intriguing, the range of cell types produced and the uncontrolled nature of the differentiation make study of the underlying processes difficult. A significant advance was the discovery that an apparently nullipotent EC cell line, F9, can be induced to differentiate by exposure to retinoic acid. After exposure to both retinoic acid and cAMP, F9 generated cells that closely resemble parietal endoderm. Although retinoic acid had been known for many years to play a key role in epidermal cell differentiation, the results with F9 cells focused the attention of developmental biologists on the role of this important derivative of vitamin A in regulating embryonic development. Subsequently, if F9 cells are cultured in suspension in the presence of retinoic acid, they form embryoid bodies in which the outer layer of cells resembles visceral endoderm, while the inner cells retain the EC phenotype. Thus, apparently, F9 cells can be switched between differentiating into visceral or parietal endoderm by altering the conditions of induction. It was later shown that retinoic acid is able to induce the differentiation of a number of mouse EC cell lines such as PCC7. Perhaps the most well known of such EC lines is P19, which differentiates in a predominantly neural direction when exposed to retinoic acid, but in a mesodermal direction, with the fonnation of muscle cells, when exposed to another agent, *dimethylsulfoxide* (DMSO). EC cell differentiation can also be induced by other agents such as *hexamethylene bisacetamide* (HMBA).

There is no doubt that studies of mouse EC cells in culture provided insights into molecules that play a role in embryonic development and regulate differentiation of embryonic cells. Moreover, the experience with EC cells provided the foundations for the development of ES cell lines. However, with the rapid advances in molecular genetics that have allowed access to the early mouse embryo, the availability of ES cells has reduced the necessity for studying EC cells. On the other hand, although their capacity for differentiation is substantially less than that of ES cells, EC cell lines may be more robust and simpler for some experimental purposes.

Human EC Cells

While embryogenesis in the laboratory mouse became progressively more accessible to experimental study, analysis of human development remained, and indeed still remains, severely restricted, not only by ethical considerations but also by the logistical problems of working with human embryos. Nevertheless, although recent developments in biology indicate a strong conservation of regulatory mechanisms throughout phylogeny, stretching from the nematode worm all the way to mammalian development, there is no doubt that each species presents unique features and that human development differs in significant ways from that of other mammals. Human EC cell lines provide an opportunity to investigate mechanisms that regulate embryonic cell differentiation in a way that is pertinent to early human development, while also shedding light on a medically significant form of cancer.

Cell lines were first derived from human germ cell tumors and maintained as xenografts in the 1950s. Later, several lines were established in culture, notably TERA1, TERA2, and SuSa, described during the 1970s. Initial studies of these cell lines highlighted similarities with the mouse EC cells. In particular, some of the human GCT-derived cell lines were reported to express the F9 antigen and, later, SSEA1. It was first assumed that this was consistent with human EC cells expressing the F9 antigen, like murine EC cells. However, a comparative study by Andrews et al. (1980) of a range of cell lines derived from GCT, and a more detailed analysis of one of these,

2102Ep, led to the conclusion that human EC cells differ in a number of respects from their murine counterparts. In particular, SSEA1, which had become a hallmark of murine EC cells, appeared *not* to be expressed by human EC cells, in contradiction to the earlier studies, although it is expressed by some derivative cells following differentiation. On the other hand, two new antigens, SSEA3 and SSEA4, which are expressed by cleavage-stage murine embryos but not ICM or EC cells, are present on human EC cells.

SSEA3 and -4 are epitopes associated with globoseries glycolipids expressed on the cell surface. In contrast, SSEA1 is an epitope associated with a lactoseries glycolipid that contains a different core structure, although it is synthesized from the same precursor, lactosylceramide. Murine EC cells and early embryos express the Forssman antigen, which also possesses a globoseries core structure. However, the terminal disaccharide, galactosaminyl galactosamine, which forms the Forssman epitope, occurs in the mouse but not in humans. One possibility is that the terminal structures forming the SSEA3 and -4 epitopes in humans might not occur in mouse EC cells because of competition by the enzyme forming the terminal Forssman epitope. Perhaps the more significant issue is the similarity between human and murine EC cells in their expression of globoseries glycolipid core structures, rather than differences in the terminal modifications of these oligosaccharides.

The SSEA3 and -4 epitopes are members of the P blood group system, and the red blood cells of most people express both epitopes. However, a very small number of individuals lack the ability to synthesize the P blood group substance, which has been identified with globoside, and these individuals are also unable to produce the SSEA3 or -4 antigens. Interestingly, women lacking the P blood group antigens have a high rate of spontaneous abortions, perhaps because of their ability to mount an immune reaction to antigens such as SSEA3 and –4, which studies of EC cells suggested are expressed on the very early embryo. It also transpires that red blood cells of about 1% of Caucasians do not express SSEA4, which has been equated to a previously identified blood group antigen called Luke. It is not known whether Luke (–)/SSEA4(–) individuals also have high rates of spontaneous abortions, or what other consequences might flow from this polymorphism.

Another set of antigens that have been identified in human EC cells are epitopes associated with keratan sulfate, notably TRA-1-60 and TRA-1-81, as well as GCTM2, K21, and K4. It appears that these epitopes are commonly expressed by human EC cells, and indeed some, notably TRA-1-60, have been shown to be useful serum markers in germ cell tumor patients as they are shed by EC cells. A workshop to compare expression of a variety of antigens by a large panel of human GCT cells confirmed that SSEA3, SSEA4, TRA-1-60, TRA-1-81, and GCTM2 are all characteristic markers of human EC cells.

Differentiation of Human EC Cells

A striking feature of many established human EC cell lines is their lack of ability to differentiate into well-recognizable cell types. This might, in part, reflect their evolution in tumors, since an ability to differentiate would tend to limit tumor growth and so provide a selective disadvantage for stem cells, whereas acquisition of an inability to differentiate would provide a strong selective advantage. Unlike murine EC cells, human EC cells are highly aneuploid, and it is easy to envisage that genetic changes which inhibit their differentiation might occur readily during their development. This facet of human GCT biology makes difficult the definition of EC cells, since an ability to differentiate is generally taken as a key diagnostic feature of an EC phenotype, and of course, the particular interest of EC cells to developmental biologists lies in their ability to differentiate. Recently, we observed that hybrids formed between 2102Ep, a relatively nullipotent EC line, and NTERA2 pluripotent EC cells appeared to retain an ability to differentiate. Therefore, we concluded that 2102Ep cells fail to differentiate because of a loss of some function rather than acquisition of an active inhibitor of differentiation.

Nevertheless, a number of human EC cell lines do show morphological changes, accompanied by changes in expression of various markers, when cultured under different conditions. In particular, many undergo a transition to a "*large flat*" phenotype when cultured at low cell densities. This has been examined in closest detail in the 2102Ep EC cell line. Low-density culture of these cells results in down-regulation of SSEA3, and the appearance of SSEA1. The low-density, SSEA1 (+) cells also activate expression of fibronectin, and some produce *human chorionic gonadotropin* (HCG) and resemble trophoblastic giant cells.

The mechanism of low-density-induced differentiation is unclear. It is evidently not due to low levels of an autocrine factor produced by the EC cells themselves, as conditioned medium from high-density cultures does not inhibit the phenomenon. Additionally, inhibition of cadherin-mediated cell:cell adhesion by culture in medium with low levels of Ca^{++} does not induce this type of differentiation. Nevertheless, some short-range signal between cells must be involved.

The apparent trophoblast differentiation seen in low-density cultures of 2102Ep, and some other human EC cell lines, reflects a notable difference between murine and human GCT; namely, the frequent occurrence of trophoblastic elements in human but not mouse teratocarcinomas. The observation that murine EC cells appear to make trophoblastic elements only rarely, if at all, correlates with the notion that murine EC cells are equivalent to late ICM, or primitive ectoderm cells which have lost the capacity for trophoblastic differentiation. A corollary of these observations is that human EC cells correspond to an earlier stage of embryonic development than mouse EC cells (e.g., cleavage-stage embryos), or that the embryonic cells to which the human cells are related possess a wider range of potency than the corresponding mouse cells at the same stage of embryonic development.

Although many human EC cells do not appear to differentiate in response to retinoic acid, a number of lines that do differentiate extensively have been described. For example, GCT27 is an EC cell line that requires maintenance on feeder layers to prevent differentiation. When removed from feeders, the cells undergo differentiation into a variety of cell types that include extraembryonic endodermal cells and cells with neural properties. They respond to retinoic acid, yielding cells resembling extraembryonic endoderm, and also differentiate in response to BMP2. Other EC lines that differentiate extensively include NCR-G3, NCC IT, and NEC14. However, perhaps the most extensively studied is the TERA2 line.

TERA2 Pluripotent EC Line

The TERA2 teratocarcinoma cell line is one of the oldest extant human GCT lines, but it was several years before it was recognized as a pluripotent EC cell because cultures of these cells frequently contain multiple cell types and may, depending on culture conditions, contain very few EC cells. A more robust subline, NTERA2, was derived by Andrews et al. (1984b) after passage of TERA2 through an athymic (*nu/nu*) (nude) mouse in which it formed a xenograft tumor with marked teratoma features. TERA2 and NTERA2 xenografts contain multiple cell types, most notably glandular structures and neural elements. Single-cell clones of NTERA2 were isolated, and NTERA2 clone D1 (often abbreviated NT2/D1) became the standard line that is now widely used. NTERA2 cells express characteristics in common with other human EC cells such as 2102Ep, for example, SSEA3 and SSEA4, as well as TRA-1-60 and high levels of the liver isozyme of alkaline phosphatase. Interestingly, unlike many other human EC lines, the TERA2-derived lines do not express any placental-like ALP activity and show no evidence of trophoblastic differentiation when cultured at low cell density, although induction of both SSEA1 and fibronectin occurs.

Not only do NTERA2 EC cells form well-differentiated teratomas when grown as xenografts in nude mice, they also respond to retinoic acid and other agents in culture. After exposure to 10^{-5} or 10^{-6}M retinoic acid, NTERA2 cells rapidly lose their EC phenotype, acquiring a substantially different growth pattern and cellular morphology. Cultures exposed to retinoic acid typically lose expression of EC markers such as SSEA3, SSEA4, or TRA-1-60 over a 1- to 2-week period. At the same time, a variety of other antigens, notably ganglioseries glycolipids, appear on the surface of the cells. Generally, a 2- to 3-day exposure to retinoic acid is sufficient to commit almost all the cells to differentiate, and within 2–3 weeks, EC cells are not detectable in the cultures.

Differentiation of NTERA2 EC cells is characterized not only by changes in surface antigen expression, but also by changes in susceptibility to infection with certain viruses, notably *human cytomegalovirus* (HCMV) and *human immunodeficiency virus* (HIV). For example, NTERA2 stem cells are resistant to infection with HCMV and HIV, whereas the differentiated cells are permissive for the replication of both viruses. In the case of HCMV, resistance results from inactivity of the major immediate early promoter of the virus in the EC cells.

Many genes also show a marked regulation during NTERA2 differentiation. For example, *Oct4*, which is characteristically expressed by EC cells and ES cells, is down-regulated following retinoic acid induction of NTERA2 cells. At the same time, a number of other genes are induced. Among these is a member of the *Wnt* family, *Wnt13*, that is not expressed by NTERA2 or other human EC cells, but is induced strongly upon retinoic acid induction. Curiously, we have not detected expression of other members of the *Wnt* family during NTERA2 differentiation whereas, for example, *Wnt1* has been noted to be induced during differentiation of the mouse EC line P19. Members of the *Frizzled* family of genes that encode putative receptors for *Wnt* are also expressed in various patterns during NTERA2 differentiation, and we have speculated that this may indicate a possible role for *Wnt* signaling in controlling the types of cells that are generated during differentiation. Lithium, an inhibitor of GSK3β, which is a component of the Wnt signaling pathway, is also able to induce NTERA2 differentiation, and we have speculated that this might indicate a potential for EC cell differentiation to be modulated by Wnt signaling.

One gene family that is subject to marked up-regulation following retinoic acid treatment is the *Hox* family. Mammalian *Hox* genes are encoded by four separate clusters located throughout the mammalian genome. These clusters are related in organization to those that occur in lower vertebrates and invertebrates such as *Drosophila*, and the temporal and spatial pattern of expression of the genes along the anterior–posterior axis of the developing embryo is related to their positions in the clusters. *Hox* genes located at the 3' ends of the clusters have a more anterior pattern of expression than those found in the 5' ends of the clusters.

Mavilio and his colleagues observed that many *Hox* genes are induced during differentiation of NTERA2 cells and, moreover, their expression is induced in a retinoic acid-dosage-dependent manner that relates to the position of *Hox* genes within the gene clusters. Thus, *Hox* genes located at the 3' ends of the *Hox* clusters are inducible to maximum level by low concentrations of retinoic acid (less than 10^{-7}M) whereas genes located at the 5' ends of the clusters require much higher concentrations of 10^{-5} and 10^{-6} M retinoic acid for maximum induction. It was suggested that this feature reflects a possible role of retinoic acid in patterning the anterior–posterior axis of the developing embryo. The temporal sequence of *Hox* gene activation is also related to their position in the gene clusters, 3'-end genes appearing substantially before 5'-end genes.

As retinoic-acid-induced differentiation of NTERA2 EC cells progresses, neural markers become evident and neurons expressing neurofilament proteins and a typically neuronal morphology appear, most usually during the second week after first exposure to retinoic acid. Neurons derived from NTERA2 after retinoic acid treatment probably comprise only 2–5% of all the differentiated cells, but they are the most obvious and prominent ones. These neurons express tetrodotoxin-sensitive sodium channels and exhibit regenerative membrane potentials, as well as a wide variety of other neural characteristics, such as glutamate receptors and voltage-gated calcium channels. Neurons may be purified from th cultures by techniques involving differential trypsinization and treatment with mitotic inhibitors. Recently, it has been suggested that these NTERA2-derived neurons could be implanted into the central nervous system to correct neural deficits resulting from various diseases. Thus, in rats, such neurons will apparently survive and integrate functionally to correct partial defects resulting from experimentally induced stroke.

The differentiation of NTERA2 cells into neurons appears in many ways to recapitulate the steps that occur during embryonic development of the nervous system. For example, *nestin*, a gene that encodes an intermediate neurofilament protein characteristic of proliferating neuroprogenitors, is rapidly up-regulated soon after NTERA2 EC cells are exposed to retinoic acid. This transient rise of *nestin* expression is immediately followed by increases in expression of *neuroD1*, a bHLH transcription factor localized in postmitotic neuroblasts of the developing nervous system. Subsequently, genes typical of terminally differentiated neurons become expressed; for example, neuron-specific enolase and synaptophysin.

NTERA2 EC cells are susceptible to induction not only with retinoic acid but also with a number of other agents, most notably hexamethylene bisacetamide (HMBA) and members of the BMP family. NTERA2 cells induced with 3 mM HMBA are distinct from those induced with retinoic acid, and neural elements are mostly not evident, although one good marker of differentiation induced by HMBA is NCAM. BMPs, notably BMP7, also induce NTERA2 differentiation, yielding again a distinct set of cells.

On the other hand, there are some overlaps in the nature of the cells induced by these agents. For example, we have found the induction of smooth muscle actin following exposure to BMP7, but also, at rather lower levels, following retinoic acid induction. Although many of the markers typical of retinoic acid induction are not expressed soon after treatment with HMBA, the HMBA-induced cultures eventually often express some of the markers that are typical of retinoic-acid-treated cultures, and a few neurons are occasionally observed. Thus, although the pathways of differentiation induced by these three agents, retinoic acid, HMBA, and BMP7, are predominantly different, they nevertheless seem to overlap with one another and the pathways are not mutually exclusive.

The nature of the nonneural cells seen in differentiating cultures of NTERA2, whether induced by retinoic acid, HMBA, or BMP7, are not fully characterized. Although the appearance of smooth muscle actin suggests the appearance of smooth muscle cells, we have never noted expression of *MyoD*, or the induction of cells corresponding to skeletal or cardiac muscle, which would indicate mesodermal differentiation; nor has endodermal differentiation been observed. It appears that NTERA2 EC cells are committed to ectoderm but that the nature of the ectodermal pathways induced depends on the culture conditions applied.

Under some conditions, for example after retinoic acid induction, the cells adopt a fate more akin to dorsal ectoderm and give rise to neural derivatives, whereas in other circumstances they may give rise to more ventral types of ectoderm. Smooth muscle induction may be indicative of differentiation corresponding to neural crest derivatives. However, not only the NTERA2 EC cells, but also a number of other human EC lines express *Brachyury* even though mesodermal differentiation has not been observed. Whether this expression of *Brachyury* indicates a competence for mesodermal differentiation that is not realized under the available culture conditions, or whether *Brachyury* is not an indicator of competence for mesodermal differentiation in humans, remains to be ascertained. On the other hand, the gross aneuploidy of NTERA2 cells, in common with other human EC cells, may be indicative of genetic changes that interfere with their developmental potential.

Primate ES Cells Lines

Although human EC cells resemble mouse EC cells in some respects, they differ from them in a number of important areas. Thus, the relationship of human EC cells to human *embryonic stem* (ES) cells was, to some extent, in question not least because of the extensive chromosomal changes seen in EC cells. However, recently, primate ES cell lines have been isolated from rhesus monkeys, common marmosets, and humans, and they have proved to share many of the characteristics previously defined for human EC cells.

The primate ES cells were derived by plating an isolated ICM on fibroblasts in the presence of serum, but in the absence of other exogenous growth factors. These conditions are similar to those used for the derivation of some human EC cell lines, and for the derivation of mouse ES cells prior to the identification of *leukemia inhibitory factor* (LIF) as a critical mediator of mouse ES cell self-renewal. Human *embryonic germ* (EG) cell lines have also been isolated from fetal germ cells plated on fibroblasts in the presence of serum, but their derivation required supplementation of the medium with LIF, basic fibroblast growth factor, and forskolin.

The factors produced by fibroblasts that are required for human ES and EG cell self-renewal are unknown. Mouse ES cells remain undifferentiated and proliferate in the absence of fibroblasts if LIF, or other LIF cytokine family members such as *ciliary neurotropic factor* (CNF) or oncostatin M, are present. Each of these factors acts through the LIF receptor complex, a heterodimer of the LIF receptor and the IL-6 signal transducer, gp130. However, primate ES cells cultured in the presence of LIF and in the absence of fibroblasts uniformly differentiate or die within about 7–10 days.

Rhesus ES cells grown on primary fibroblasts derived from homozygous LIF-knockout mice continue undifferentiated proliferation for at least three passages. Some human EC cell lines are feeder-dependent, and LIF, oncostatin M, and CNF also fail to prevent their differentiation. Given the reported importance of LIF in the culture of human EG cells, further work is warranted to clarify whether the gp130 signaling pathway is active at all in primate ES cells, or whether an entirely different signaling pathway mediates undifferentiated proliferation.

Human, rhesus monkey, and common marmoset ES cells all express a shared repertoire of cell-surface antigens that is similar to that of undifferentiated human EC cells, including SSEA3, SSEA4, TRA-1-60, TRA-1-81, and alkaline phosphatase. Human EG cells similarly express these markers, but they also express the lactoseries glycolipids, SSEA1, which is not expressed by human EC and primate ES cells. The pattern of markers expressed by both primate ES cells and human EG cells differs from those expressed by mouse ES cells, underscoring basic differences in early mouse and human embryology.

The morphologies of primate ES cells and human EC cells are very similar, but are distinct from the reported morphology of human EG cells. Primate ES cells form relatively flat, compact colonies that dissociate into single cells easily in trypsin or in Ca^{++}/Mg^{++}-free medium, whereas human EG cells form tight, more spherical colonies that are refractory to routine dissociation methods.

In general, undifferentiated growth of human EC cells is promoted at high cell densities. In contrast, primate ES cells differentiate rapidly if they are allowed to pile up after the cultures grow to confluence. It is not yet clear whether the cell-surface marker and morphological differences between human ES cells and human EG cells reflect fundamental biological differences, or merely differences in culture conditions.

Primate ES cells share with human EC cells the ability to differentiate to extraembryonic lineages, including yolk sac and trophoblast. Mouse ES cells rarely contribute to trophoblast in chimeras, and it is unclear whether the rare ES-cell-derived cells that integrate into the trophoblast are functional trophoblast cells. Primate ES cell lines will spontaneously differentiate in vitro to endoderm (probable extraembryonic endoderm) and trophoblast, evidenced by the secretion of α-fetoprotein and chorionic gonadotropin into the culture medium. In the mouse embryo, the last cells capable of contributing to derivatives of both trophectoderm and all three embryonic germ layers are early ICM cells of the expanding blastocyst.

The timing of commitment to ICM or trophectoderm has not been established for any primate species, but because human EC/ES cells can differentiate to trophoblast and to derivatives of all three germ layers, and because they express SSEA3 and SSEA4, human EC/ES cells may resemble an earlier stage of embryogenesis than mouse EC/ES cells. However, the fact that mouse ES cells will contribute to the trophoblast even rarely in chimeras suggests that they may have the developmental potential to form trophoblast, but that the environmental conditions in chimeras simply do not allow efficient differentiation to trophoblast by ES cells. Indeed, it has recently been demonstrated that repression of Oct4 in mouse ES cells induced trophectoderm differentiation.

Like their mouse ES/EC counterparts, primate ES cell lines have both more advanced and more consistent developmental potentials than human EC cell lines, possibly reflecting the malignant origin and the severe karyotypic abnormalities of the EC cells. Human and rhesus ES cells injected into SCID mice consistently form teratomas with advanced differentiated structures of all three embryonic germ layers, including smooth and striated muscle, bone, cartilage, gut and respiratory epithelium, keratinizing squamous epithelium, neurons, and ganglia.

What is remarkable about the primate ES cell teratomas is not only the range of individual cell types, but also the formation of complex structures requiring coordinated interactions between different cell types. For example, the development of neural tube-like structures, gut-like structures, and hair follicles requires highly coordinated interactions between different cell types. Finally, perhaps the most impressive example of organized epithelial–mesenchymal interactions in these primate ES cell tumors is the formation of well-organized tooth buds, complete with ameloblasts, odontoblasts, and intervening dentin. The formation of teratomas by human EG cells has not been described, but differentiation of derivatives of all three embryonic germ layers in vitro has been reported.

Concluding Remark

The past 20 years have been marked by a dramatic increase in our understanding of mammalian development, most notably in the laboratory mouse. In part, this has been a consequence of the revolution in molecular genetics, and the finding that many regulatory mechanisms have been highly conserved throughout phylogeny. It has also had its roots in the study of teratocarcinomas and their EC stem cells. Those studies provided a stimulus to thought and approaches to embryonic development at a time when other tools were limited. They provided access to key molecules and genes that play a role in development, and they continue to provide a convenient experimental system for addressing some questions about the mechanisms that regulate embryonic cell differentiation. The culmination of that work was the development of techniques for culturing ES cells, which now provide the key route to genetic manipulation of the laboratory mouse.

Although human embryogenesis undoubtedly resembles that of other mammalian species, there are certainly differences at the morphological as well as at the cellular and molecular levels. Given the logistical as well as ethical and legal difficulties surrounding work with human embryos, teratocarcinomas and their EC stem cells derived from human germ cell tumors provide a useful tool for helping to translate findings in species more tractable to experimental study to the human situation. For example, we have investigated the pattern of expression of various glycolipid antigens, and there are certainly aspects of their expression that are distinct between human and mouse EC cells and, evidently, embryos.

The recent characterization of human and monkey ES cells has validated the presumption that human EC cells do relate to pluripotent cells of the very early embryo, as in the mouse model, so that results from EC cell lines will continue to provide important pointers for future study. In some ways, the capacity of EC cell differentiation is limited, and the availability of human ES cell lines will extend the range of questions that can be addressed in vitro. On the other hand, the limitations of EC cells can sometimes be put to advantage, as EC cells provide a simpler and more robust experimental system. Human EC and ES lines are likely to remain complementary tools.

Apart from their value as experimental tools, pluripotent stem cells are likely sources of specific differentiated cell types for tissue replacement therapies for a whole host of diseases. Some have already made a start in this direction using EC-cell-derived neurons. Eventually, ES cells may provide the source of choice for such treatments because of their evident "*normality*" and the lesser likelihood of reversion to a malignant phenotype. Nevertheless, our current ability to culture human ES cells, and our understanding of their biology, rests on many decades of work with their tumor-derived counterparts, the EC cells of murine and human teratocarcinomas.

INDEX